Tank and Well Irrigation Crisis: Spatial, Environmental and Social Issues

About the Editor

Olivia Aubriot, at C.N.R.S. (National Centre for Scientific Research, France), with a joint qualification in agronomy and anthropology has been working on social water management in Nepal and South India for the last 20 years. Her centres of interest are peasant water management and its underlying logic, water rights, dynamics of spatial organisation, technical anthropology and nowadays, the impact of climate change on changes in water management practices.

From mid-2005 to mid-2008 she was at the French Institute of Puducherry, a period during which she developed a joint research programme on competition between collective traditional irrigation and individual modern irrigation in the hinterland of Puducherry, South India.

Contributors

Pere Ariza-Montobbio - Olivia Aubriot - Judith D'Souza - Elena Galan-del-Castillo - Martine Gestin - Pierre Jorcin - Olivier Lefebvre - P. Ignatius Prabhakar - Victoria Reyes-García - Marc Roesch - Loes Schenk-Sandbergen - Tarik Serrano Tovar

Tank and Well Irrigation Crisis
Spatial, Environmental and Social Issues

Cases in Puducherry and Villupuram Districts (South India)

Edited by

Olivia Aubriot

CONCEPT PUBLISHING COMPANY PVT. LTD.
NEW DELHI-110059

ISBN-13 : 978-81-8069-786-9

First Published 2013

Published and Printed by

Concept Since 1974

Concept Publishing Company Pvt. Ltd.
A/15-16, Commercial Block, Mohan Garden
New Delhi-110059 (India)
T: 25351460, 25351794, *F*: 091-11-25357109
E: publishing@conceptpub.com, W: www.conceptpub.com
Editorial Office: H-13, Bali Nagar, New Delhi-110 015, India.

Cataloging in Publication Data--Courtesy: D.K. Agencies (P) Ltd. <docinfo@dkagencies.com>

Tank and well irrigation crisis : spatial, environmental and social issues : cases in Puducherry and Villupuram districts (South India) / edited by Olivia Aubriot.

p. cm.
Contributed articles.
Includes bibliographical references and index.
ISBN 9788180697869

1. Irrigation—India—Puducherry (District)—Regional disparities. 2.rigation India—Viḻuppuram Irāmacāmip Pataiyātciyār Māvaṭṭam (District)—Regional disparities.3. Tank irrigation—Environment aspects—India Puducherry (District) 4.Tank irrigation—Environmental aspects—India—Viḻuppuram Irāmacāmip Paṭaiyāṭciyār Māvaṭṭam (District) 5 Tank irrigation—Social aspects—India Puducherry (District) 6. Tank irrigation—Social aspects—Inoia—Viḻuppuram Irāmacāmip Paṭaiyātciyār Māvaṭṭam (District) 7. Well water—India—Puducherry (District) 8. Well water—India—Viḻuppuram Irāmacāmip Paṭaiyāṭciyār Māvaṭṭam (District) 9. coastal zone management—India—Puducherry (District)10. Coastal zone management India—Viḻuppuram Irāmacāmip Paṭaiyāṭciyār Māvaṭṭam (District) I. Aubriot, Olivia.

DDC 333.913095486 23

Foreword

Despite important differences according to the scale considered and notable divergences on the analysis of their root causes, water crises are manifest and multifarious in many parts of the world. Southern India, although often conjuring up images of a land blessed by abundant monsoon rains, is no exception, as water has long been a central resource of both agriculture and livelihoods. Alongside with large-scale public irrigation schemes, largely inherited from the British and occupying large plains and deltas, the region is also peppered with a constellation of small reservoirs – *the tanks* – and, more recently, dotted with hundreds of thousands of individual wells. This diversity of sources is enough to suggest that water uses and circulation are likely to exhibit mind-boggling patterns and to pose serious challenges to managers and policy-makers. But if one adds to this state of affairs the region's fragmented and segregated social structure, the issue takes on a more political and societal dimension, intermingling the circulation of power and the control of water.

This books - *Tank and well irrigation crisis: spatial, environmental and social issues* - stands at the confluence of two lasting debates in South-India: the dynamics and the place of small tanks in regional history and institutions; and the bonanza-cum-curse associated with the rapid and massive development of groundwater resources: as stated in the introduction of this volume, this takes us to a journey between the past and present, communities and the states, collective and individual management or ownership of resources, surface water and groundwater, low and high casts, men and women.

Based on statistical data, maps, surveys and seven case studies carried out in Puducherry hinterland and in Villupuram district, Tamil Nadu, the book first offers the reader useful chapters on a typology of wells and tanks that associates physical and climatic factors, technical features, and social dimensions, and explains why there is no obvious relationship between tank or well density and water availability or irrigation intensity. The following chapters address economic, environmental, social and policy issues and these different analyses and points of view combine to reveal a wealth of observations and information on current social and environmental dynamics.

These successive zooms unpack important issues, such as the social inequities associated with access to water, the logics underlying the shift to groundwater and the weakening of tank management and collective action, and the economic and social dimension of irrigated paddy production. The book brings many important points to on-going policy debates regarding, for example, the function of the tanks, the means of controlling over-abstraction, or the need to establish Water user Associations; it shows the gap between some intended regulations, state action and local dynamics. The recent emphasis on the groundwater recharge function of the tanks is eloquently shown to parallel the vested interests of well drillers and owners, or villagers willing to use tanks to raise fish. As in most situations where wells have displaced or destroyed communal systems (e.g. qanats), or replaced the collective (albeit often dysfunctional) management of canals in large schemes (e.g. the Nile delta), it is hard to imagine a way back to former management patterns: farmers have gained security and flexibility, often diversifying cropping patterns and calendars; here they are also weary to be denied water from well owners, in the case they don't own their own well. Everything looks as if the transaction costs of collective action had risen to the point of non retour, making individuation – and the partial shift from a collective resource to one appropriated, and sold, by the elites – appear as a "cheaper" solution, provided that environmental costs and resulting inequities are not factored in.

Dropping water tables in all environments (rocky, alluvial, or sedimentary areas), and attendant ever deeper (or deepened) wells pose the crucial question of sustainability. Here like, sadly, in most parts of the world, physical constraints (depth/costs, salinity, availability) are the ultimate regulatory device, with little observed contribution from legal provisions, market mechanisms, technology or institutional innovations. Farmers tend to perceive groundwater as a stock that is replenished, just like tanks are each year, and which can or even should be fully used. Even if they share a more qualified understanding of groundwater dynamics, short-term constraints and hand-to-mouth living conditions leave little scope for averting the tragedy of open access.

Wells, indeed, as the book shows, improve the social status of its owner, reducing his vulnerability to rainfall vagaries and allowing diversification, and tend to be synonymous to dignity. But investing in wells or in rice cultivation can, in many cases, hardly be understood in pure economic terms; and social gains are also paralleled with a more mundane concern for food security at the household level, a primary objective that leaves many farmers with little choice but to continue farming, investing non-agricultural revenue or borrowing money to keep the activity going.

The relationships between institutional change, power and resource distribution, and equity are also explored. As is generally the case elsewhere, interventions by NGOs or state agencies promoting community management or Water User Associations provide some room for manoeuvre to a few individuals who succeed in challenging the prevailing social order. In more cases, though, resources or institutional reforms brought from outside are diverted by local elites which are able to sustain or even increase their power. On balance, very few associations established for tank rehabilitation projects or "management transfer" can claim to have improved the situation of, or empowered, the poorest, with the poor, landless farmers, and women often losing out in a process that favours deep tubewell owners and large farmers in canal irrigation areas.

In all cases, the focus on WUAs tend to obscure the fact that beneficiaries extend much beyond the group of farmers drawing water from the tank; while the tanks have a clear economic value to some villagers, involved or not in irrigation, they also provide environmental services, and have important social and cultural functions. While the knowledge and perceptions of the tank by different social or occupation groups vary, all value and use the tank differently but with a similar intensity.

This well-documented series of chapters makes up a commendable journey through well and tank systems in Tamil Nadu, bringing light on the diversity of both the physical and human landscapes, and the way they intermingle to produce patterns and dynamics that often resist generalization. It will, no doubt, be a significant addition to the rich existing literature on irrigation and tanks in India that will appeal to both researchers and decision-makers.

Alep, the 8th of December 2010 **François Molle**

Preface

This collective book is deeply rooted in the description of case studies and their analysis through the lens of various disciplines (geography, anthropology, sociology, agro-economics) as well as through the eyes of individuals with different personal experience (researchers or NGO workers). Notwithstanding these various approaches, some shared ideas emerge from the different texts in the book: water is a common resource and should therefore be managed on a collective basis, not as an economic asset –which is far from being the case; there are social inequalities in accessing water and in decision-making that need to be described and their dynamics analysed; there may be a dual scale of analysis to encompass both village and regional scales given the diversity of situations which results either from surface water or groundwater characteristics and from the human environment in a restricted area: here the hinterland of Puducherry, thus explaining the complexity of the issues at stake.

Moreover, the book may be characterised by the themes that are given prominence here, since they are not often addressed in Indian literature on irrigation:

- spatial and social inequalities of water access and use;
- interdependency of groundwater and surface water in the decision-making process with regard to local water management;
- perceptions of water users;
- knowledge about water (especially groundwater) and water use;
- political and factional dynamics in the running of water institutions.

Undeniably, the book does not offer any overall ready-made solution or suggestion, since irrigation is shown to be the result of the interaction between technical, environmental, social and political spheres, and is thus specific to local constraints in a given socio-economic and political context. Moreover, local dynamics may differ from one place to another depending on the weight of the past.

Paris, 11th July, 2011

Olivia Aubriot

Acknowledgements

Most of the research presented here was funded by two institutions to which I owe my thanks: the French Institute of Pondicherry (IFP) and through it the French Ministry of Foreign Affairs that financed my work and made a contribution to logistically supporting the individual training of students; and secondly, the AUF (Agence Universitaire pour la Francophonie) which funded a 18-month project for inter-university cooperation, including most of the research carried out by I. Prabhakar and by a group of French students who attended a joint training course.

The research project could not have been carried out without the help of a number of persons. I am deeply appreciative of the support given by Dr. Muller, director of IFP during the project's duration. I would also like to acknowledge the contributions of those who assisted me, colleagues and students in the field without whom collecting information would have been virtually impossible: G. Venkatasubramanian, T. Sudhakar and David T. Chakravarthy, who interacted at various times during the project. Next, I would personally like to thank the IFP staff, whom I have been working alongside, or whose services I have called upon, from those in the laboratory of geomatics to those in the secretariat, service employees or library staff, persons who sorely contribute to ensuring that this institution plays its role as a meeting platform for researchers and students. Another major stage in the research work is the drafting of papers, and here two persons in particular have to be thanked. Mary P. Boseman translated parts of my chapters, and Bernadette Sellers, entrusted with the hard work of correcting our rather shaky English, was involved in proofreading some parts of this book.

Lastly, as project coordinator and editor, I wish to express my gratitude to colleagues and students who have actively contributed to the dynamics of exchange, to training students in research, and to all the valuable teamwork. Indeed, many people deserve to be acknowledged, since collaborative work took place both within IFP programmes —within the Social Science Department itself or with the Ecology Department—, but also with other institutions, such as Puducherry University, the Autonomous University of Barcelona (Spain), the Université de Provence (France), Montpellier SupAgro (France) and Universite de Paris 7. The project also significantly benefited from collaborative programmes run by C.Z. Guilmoto (demographer at I.R.D. on secondment at

IFP during the first programme) from the late 1990s to the mid-2000s and which resulted in mapping the 1991 census data of South India (especially the EMIS programme). Finally, special thanks go to my colleague, I. Prabhakar, with whom I shared many weeks in the field, many discussions and moments of interaction, both as a colleague and friend, and whose help in this project has been invaluable.

Ultimately, the work could not have been carried out without the help of villagers who welcomed each of us, and always accepted to spend time explaining the situations or answering our questions.

CONTENTS

LIST OF TABLES

List of Figures

List of Photographs

List of Boxes

About the Contributors

Pere Ariza-Montobbio

Pere Ariza-Montobbio is a PhD student in Ecological Economics and Political Ecology. His research focuses on Societal Metabolism and Ecological Distribution Conflicts and Environmental Justice. He is currently working on interdisciplinary research on the relations between energy and land use governance. His contribution here is part of a common field research experience with E. Galan-del-Castillo and T. Serrano Tovar conducted under the supervision of V. Reyes-Garcia, on the social perception and the multi-functionality of rain water harvesting systems as ecosystems for decentralized water management in Tamil Nadu (India).

Ariza-Montobbio, P., S. Lele, G. Kallis and J. Martinez-Alier. 2010. The political ecology of Jatropha plantations for biodiesel in Tamil Nadu, India. *Journal of Peasant Studies* 37 (4): 875-897.

Ariza-Montobbio, P. and S. Lele. (2010). Jatropha plantations for biodiesel in Tamil Nadu, India: Viability, livelihood trade-offs, and latent conflict. *Ecological Economics* 70: 189-195.

Olivia Aubriot

Olivia Aubriot is an agro-anthropologist at the C.N.R.S. (National Centre for Scientific Research), and member of the Centre for Himalayan Studies. She is conducting research on social water management, in Nepal and South India. She was on secondment at the French Institute of Pondicherry for a period of three years (2005-2007), during which she developed the research programme "Local Practices and Public Policies, Sustainability of the Resource Availability for irrigation in Tamil Nadu and Pondicherry: needs and management", the main results of which are presented in this book.

Prabhakar, I. and O. Aubriot. 2009. "Unrestrained quest for groundwater: what about the more vulnerable? Cases from Tamil Nadu and Pondicherry, South India". In *Eaux, pauvreté et crises sociales. Water poverty and social crisis.* Edited by H. Ayeb and T. Ruf. (from the International Symposium of Agadir, Morocco, *Water Poverty and Social Crisis* December 11-15, 2005). CD-Rom. Paris: IRD Editions (Collection Colloques et séminaires).

Aubriot, O. 2004. Irrigation history in central Nepal: the interface between agriculture and technology. *Water Nepal* 11 (2): 19-43.

Mathieu, Paul, Ahmed Benali and Olivia Aubriot. 2001. Water-rights, investments and meanings: conflict and change in a traditional irrigation system in northern Morocco. *International Journal of Water* 1 (3/4): 270-284.

Judith D'Souza

Judith D'Souza has been working on gender and rural sociology issues in India. She has a Masters Degree in Social Work. Her work has focused on building systems and processes in organisations to address gender justice concerns in operations and projects. She joined Auroville Water Harvest as Project Manager for "Salinity Moderating and Proofing of Coastal Aquifers". Judith is currently with the UN, working for the specialised agency IFAD at the India country office.

D'Souza, J. 2007. "Water management Awareness for villagers", Conference on *Water: spatial dynamics, competitive claims and governance,* November 30-December 1, 2007, Puducherry.

Boulicot, G. and J. D'Souza. N.d. "*Water – Dynamics in the Auroville Bioregion*", Harvest website, http://www.auroville.org/environment/harvest/harvest.htm

Elena Galan-del-Castillo

Elena Galan-del-Castillo (MSc, Environmental Studies, UAB, Spain) is a PhD student in Historical Economics at Universitat de Barcelona (UB), Spain. She has been working as a researcher at the Centre d'Investigació i Informació en Consum (CRIC), Spain, an independent research centre on responsible consumption. She has also been working on the "water footprint" concept applied to hidden flows of water in Spanish biofuel policies.

Galan-del-Castillo, E. and E. Velazquez. 2010. From water to energy: The virtual water content and water footprint of biofuel consumption in Spain. *Energy policy* 38 (3), March 2010: 1345-1352. *http://dx.doi.org/10.1016/j.enpol.2009.11.015*

Galán, E. 2009. "La huella hídrica de los agrocombustibles". In *Agrocombustibles. ¿Otro negocio es posible?* Edited by Mónica Vargas. Barcelona: Icaria editorial.

Martine Gestin

Martine Gestin has a PhD in Social Anthropology and is a freelance researcher affiliated to the Social Anthropology Laboratory of the *College de France*. She has written several papers on gender, kinship, and sexuality in South India. South Indian oral traditions are also part of her research.

Gestin, M., 2009. "A brillant gioss for Tamil social history: pre-marital courtship and marriage at the time of Nakkirar" pp. 145-167, in E. Wilden, *Between Preservation and Recreation. Tamil Traditions of Commentary*. Ed. IFP-EFEO.

Pierre Jorcin

Pierre Jorcin is a geographer specialized in GIS and Remote Sensing, with a Masters in GIS (University of Caen, France). After taking part in research with the French Institute of Pondicherry, he joined Harvest, a non-profit organization that comes under the Centre for Scientific Research (CSR) of the Auroville Foundation, an organization acting as an integrated development agency for water conservation in Puducherry and the Tamil Nadu region. He worked there for 6 years on hydro-geological studies and rural development programmes in South India. He is now working on geospatial applications in Malaysia.

Vincent, A., S. Violette, R.S. Krishnamurthy, K. Sivasubramaniam, G. Boulicot and P. Jorcin. 2004. "Preliminary report on the hydro geological study of the Kaluvelly-Pondicherry coastal sedimentary basin" (31p), In progress Report (December 2004), program HELP UNESCO.

Vincent, A., S. Violette, G. de Marsilly, P. Jorcin, K. Sivasubramaniam and R.S. Krishnamurthy, 2006. "Infiltration quantification at the watershed scale using hydrological models, in a semi-arid context (Kaluvelly-Pondicherry coastal sedimentary basin, Pondicherry territories and Tamil Nadu, India)", European Geosciences Union Annual Meeting, Vienna (April 4-7 2006).

Vincent, A., S. Violette, G. de Marsilly, G. Boulicot, P. Jorcin, R.S. Krishnamurthy and K. Sivasubramaniam, 2005. "Hydro geological study of the Kaluvelly-Pondicherry coastal sedimentary basin". International Association of Hydrological Sciences, VIIth Scientific Assembly, Iguacu, Brazil (April 3-9, 2005).

Olivier Lefebvre

Olivier Lefebvre (Master of Science in Social Water Management, engineering degree in Water Sciences and Technologies) is currently project manager in a water access and renewable energy project for a French NGO "Initiative Development" in China, Yunnan. He previously worked on the Reunion Island, coordinating an irrigation project on the west coast. He spent his six-month training period for his degree course at the French Institute of Pondicherry, in the Social Water Management Programme, then worked as a research assistant on the preparation of an irrigation atlas for the Puducherry hinterland.

P. Ignatius Prabhakar

P. Ignatius Prabhakar has a PhD in Anthropology from the University of Madras. He is currently working with SEEDS, an Indian NGO, undertaking interventional activities on disaster mitigation among fishermen in South India. He has research experience in the fields of tribal studies, medical anthropology, development studies, people's participation and social capital among Self Help Groups. His contribution to this book follows his postdoctoral fellowship at the French Institute of Pondicherry, about the socio-political impact and outcome of irrigation institutions.

Tamizoli, P. and P. Ignatius Prabhakar. 2010. "Traditional Governance System of Fishing Communities: Internal Mandate, Interfacing and Integrating Development", pp.139-154. In *Communities and Coastal Zone Management*. Edited by Rajib Shaw & R.R. Krishnamurthy. Singapore: Research Publishing Services.

Tamizoli, P. and P. Ignatius Prabhakar. 2007. "Legal Identity and Natural Resource Management: The Foraging Irula Women of Killai, Cuddalore District, Tamil Nadu", pp. 82-97. In *Women's Livelihood Rights: Recasting Citizenship for Development*. Edited by Sumi Krishna, New Delhi: Sage Publication.

Tamizoli, P. and P. Ignatius Prabhakar. 2006. "Social Capital and Rural Development in India: Role of Self Help Groups in Development", pp. 193-215. In *Potential of Social Capital for Community Development*. Edited by Mr. Shigeki Yokoyama and Dr. Takeshi Sakurai, Tokyo: Asian Productivity Organisation.

Victoria Reyes-García

Victoria Reyés-García (PhD in Anthropology, 2001) is ICREA Professor at the Institut de Ciència i Tecnologia Ambientals (ICTA), Universitat Autònoma de Barcelona (Spain). Her research addresses the benefits generated by local ecological knowledge and the effects of the integration to the market economy on this knowledge. She coordinates the Ethnoecology Laboratory at ICTA: *http://icta.uab.es/Etnoecologia/mission.htm*. Two of her major publications (from a list of 80) are:

Reyes-García, V. *et al.* 2003. Indigenous people share botanical knowledge. Estimates from an Amerindian society in Bolivia. *Science* 299: 1707.

McDade, TW, V. Reyes-García *et al.* 2007. Ethnobotanical knowledge is associated with indices of child health in the Bolivian Amazon. *Proceedings of the National Academy of Science*. 104 (15): 6134-6139.

Marc Roesch

Marc Roesch is an agronomist at CIRAD Montpellier (Centre International de Recherche Agronomique pour le Développement) with a PhD in Agro-Economics. After working for more than 20 years in Africa and Asia, especially in Agronomic Research, he took charge of CIRAD's Training Service for 9 years. Since 1999 he has been a research fellow specialised in socio-economic aspects of rural economy at farm level and more particularly in microfinance in Africa and Asia. Since 2005, his research focuses on the impact analysis of formal and informal financial markets on rural households in South India and Morocco, and particularly on indebtedness.

Roesch, M. 2007. "Chapitre 18: Financement et trésorerie dans les exploitations familiales africaines", pp. 279-288. In *Exploitations agricoles familiales en Afrique de l'Ouest et du Centre.* Edited by M. Gafsi, P. Dugué, J.-Y. Jamin and J. Brossier. Éditions Quae (Collection Synthèses).

Roesch, Marc, Venkatasubramanian and Isabelle Guérin. 2008. "Bonded Labour in the Rice Mills: Fate or Opportunity?" pp. 284-311. In *Bonded Labour in India.* Edited by J. Breman. Oxford University Press.

Roesch, M., and O. Hélies. 2008, "Microfinance: a tool of risk management or of risk through over-indebtedness? The case of South India" pp. 419-440. In *Micro Finance and Poverty Eradication: Indian and Global Experiences.* Edited by Daniel Lazar. New Delhi, India: New Century Publications.

Loes Schenk-Sandbergen

Loes Schenk-Sandbergen is a retired Associate Professor at the Department of Sociology and Anthropology of the University of Amsterdam. She has published extensively on gender issues, linking them with poverty and the process of empowerment. Besides, she is a gender consultant and has worked on many development cooperation projects in India, Laos, Vietnam and Bangladesh

3 water-related references:

Schenk-Sandbergen, Loes & Outhaki Choulamany-Khamphoui. 1995. *Women in Rice Fields and Offices: irrigation in Laos, gender specific case studies in villages.* Heilbo, The Netherlands: Empowerment.

Schenk-Sandbergen, Loes & Niren Choudhury. 2003. *From Heroines to Beneficiaries; from beneficiaries to heroines? The impact of a small-scale irrigations project on gender in the West Bengal Terai.* New Delhi: Manohar (IDPAD Series).

Schenk-Sandbergen, L. 2004. "Some Gender Dimensions in Managing Water Scarcity", pp. 371-400. In *Managing Water Scarcity, experiences and prospects.* Edited by A. Vaidyanathan & H.M. Oudshoorn. New Delhi: Manohar (IDPAD Series).

Tarik Serrano-Tovar

Tarik Serrano-Tovar (MSc Environmental Studies, UAB, Spain) is working on the development of integrated analytical methods to be applied to rural systems, energy systems and institutional networks (power relations issue). He has experience working on rural systems analysis in developing countries (India, Laos and Guatemala). His research interests include: integrated assessment, societal metabolism, rural systems, energy analysis, institutional analysis.

Reyes-García V., O. Aubriot, P. Ariza-Montobbio, E. Galán-del-Castillo, T. Serrano-Tovar and J. Martinez-Alier. (2011). Local perception and use of the multifunctionality of water tanks in two villages of Tamil Nadu, South India, *Society and Natural Resources* 24:485-499.

Galán del Castillo, E., T. Serrano Tovar and P. Ariza Montobbio. 2007. Local ecosystemic perception for integral management of water tanks in Tamil Nadu, South India, *Periferia* 7 (December 2007). *www.periferia.name*

INTRODUCTION

Olivia Aubriot

India is facing a water crisis, with situations of greater scarcity and increasing competition for this resource. This crisis is not only brought about by growing water needs due to industrialisation, urbanisation and demographic growth – which rose between 1951 and 2008 from 361 to an estimated figure of 1,130 million—, but also by how the available water is used. Thus for R.R. Iyer (2007), the crisis is not due to a water shortage but to serious mismanagement in various spheres: drinking water for urban and rural areas, irrigation in large-scale and medium-scale projects, water use in agriculture, inter-State disputes, groundwater use, floods and water pollution. Given the diversity of the problems, many authors talk of different water crises. Moreover, whatever one's point of view, the crisis can be analysed in various ways: for engineers, it is mainly a water availability crisis; for neo-liberal economists, the crisis is caused by a failure to recognise water as an economic asset and the absence of a clear definition of property rights; while for social activists, it is due to the loss of control over natural resources by both civil society and the State (*ibid.*).

Using the term "crisis" is not devoid of meaning, as it not only announces important changes with potentially disastrous consequences for the environment, but it also prompts people to react and, in the case of an environmental crisis, to integrate ecological imbalances at the very heart of the political, economic and social process[1] (Villalba 2006: 384); "the ways of explaining the crisis, of presenting the facts and of exposing the consequences are crucial in understanding its ability to penetrate institutional systems" (*ibid*: 386). However, this term is not used systematically, as R.R. Iyer (2007) remarks at the beginning of his book: some official reports talk of the "difficult situation", others of the "turbulent future"; whatever the term used, "water is going to be a major area of concern for us".

The agricultural sector, which on an Indian scale uses 83 per cent of the collected and diverted resource (GoI 2007), is particularly affected. As for other sectors, water needs will increase if more farm products are to be produced to cope with the growing demand as a result of demographic growth, a phenomenon likely to heighten competition for the resource. The rise in the demand for food could be met by improving water productivity in irrigated and rainfed agriculture, rather than by expanding the area of production, since water productivity in irrigated farming is relatively low in South Asia (Molden 2007: 15-16). To achieve this goal, changes in water management would be necessary, especially a reduction in the quantity of water used per surface unit. This latter concept of quantity reduction must be applied with caution, since it has to be analysed on a water cycle scale in order to be effective. For this, it is important to differentiate between water use and water consumption, the difference being the amount of water that is lost by direct evaporation plus the water that returns to the soil or the river—the latter can be reused (see Molden 2007; Venot 2008). Thus, if surface water irrigation is reduced, then the quantity of water returning to the upper aquifer will be reduced as well as the amount available for wells or for springs downstream, which is not necessarily the aim in reducing water use[2]. However, we will see that a reduction in the quantum of water tapped in deep aquifers is crucial. Nevertheless, if a reduction in the amount of water tapped is combined with an expansion of the area irrigated by groundwater, total water extraction may not be reduced (Kemper, 2003).

Reducing water use and changing water management are not easy tasks. Yet, since Independence, this sector has only undergone political incentives to develop irrigation, to intensify agricultural cycles and to increase irrigated surfaces. The gross irrigated area was indeed multiplied by 3.5 from 1950 to 2000 (cf. Fig. 1). Groundwater has played a major part in this increase (88 per cent of net area increase in India between 1960 and 2000 according to NSSO data), and now represents the main irrigation source in India. Tamil Nadu, a State already well equipped with hydraulic infrastructures, shows a less spectacular increase in its total irrigated area (multiplied by 1.6 over the same period), yet the increase in the area irrigated by groundwater is nonetheless impressive (net area multiplied by 3.4) and still rising, with the result that groundwater is also the major source for irrigation there nowadays (Fig 1.7 in Chap .1). Groundwater irrigation concerns millions of individuals who have their own well or borewell (20 million such infrastructures in India, 1.8 million in Tamil Nadu, according to NSSO data, 2007). Besides, this rate of growth has been achieved in this State at the expense of traditional collective irrigation through "tanks", semi-embanked seasonal lakes which shape the southern Indian landscape and are found in every village within the study area, located in South-East India in the Puducherry hinterland.[3]

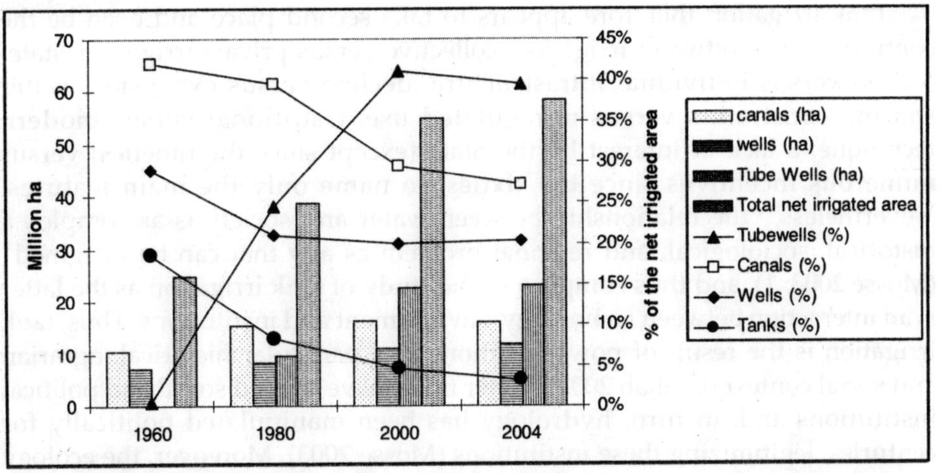

Fig. 1: Change in the irrigated area in India: net area and relative proportion of the four main types of irrigation (canals, tubewells, wells and tanks) (from NSSO data, 2007)

The frantic quest for the underground resource is ongoing, leading to the overexploitation of groundwater in many places, with an alarming environmental, social and socio-economic impact. Irrigation using groundwater (GW), through individual bores, offers numerous advantages, which explains the keen interest shown by farmers in this technique. The user is thus free of the many constraints of collective irrigation, and to some extent is no longer reliant on rainfall. In addition, the right to extract groundwater is linked to land ownership, with no limitation regarding the amount of water or the bore's depth, which gives the well owner total liberty to sell the water he extracts. The technique also offers farmers the possibility of accessing irrigation on land that was previously rainfed. Lastly, heavily subsidised electricity for agricultural purposes and even free electricity in Tamil Nadu (this State being the first to introduce the measure in 1991) make this technique economically attractive and justify the political support behind this practice.

On the contrary, tank irrigation presents various constraints: it is mostly dependent on rainfall for tank filling, so that there is constant uncertainty about the available amount of water and its period of availability; tank storage capacity may be limited, and further reduced by silt deposits and encroachment; infrastructures are seldom in good working order, which complicates their management and reduces their storage capacity; since this involves a collective irrigation technique, access to water is dependent on field location, informal or formal distribution rules, and even on social power relations. To access water, a tank irrigator is thus dependent on others, especially if his fields are located in the downstream part of the irrigated area.

Tank irrigation therefore appears to take second place and even be the contrary of groundwater irrigation: collective versus private irrigation, state-owned versus individual infrastructure, decline versus expansion, water sharing constraints versus unregulated use, traditional versus modern technique, a lack of interest by the State (except since the nineties) versus numerous incentives since the sixties, to name only the main features. Nevertheless, "the relationship between water and society is as complex a historical, sociological, and regional problem as any that can be imagined" (Mosse 2003: 1), and thus complicates the study of tank irrigation as the latter is an interaction between technology, environment and institutions. Thus, tank irrigation is the result of power relations in a particular historical, agrarian and social context (E. Shah 2003). Water flows have shaped social and political institutions and, in turn, hydrology has been manipulated politically for centuries, legitimizing these institutions (Mosse 2003). Moreover, the ecology of tank irrigation systems is far from uniform (*ibid*: 6), with tanks themselves showing diverse features (size, storage capacity, source of supply, management, etc. with some regional disparities, see Vaidyanathan 2001), while groundwater development impacts this irrigation system in a very indiscriminate manner. A good number of in-depth studies on tanks and on the impact GW development has on them already exist, especially those undertaken since the nineties by MIDS, TNAU[4] and Annamalai University (Chennai), and we will refer to their findings throughout the book. Let us point out here that we will focus on tank irrigation, groundwater irrigation and how they interact, since each village in the study area (Villupuram district in Tamil Nadu and Puducherry district) is equipped with both types of infrastructure, tank(s) and wells—open wells and/or tubewells. We will not discuss the other main type of surface irrigation in India, canal irrigation.[5]

The aim of the book is to analyse the way farmers use water, as well as their perception of the situation in a specific political, developmental and environmental context and to show how these elements influence irrigators' practices. The idea is to favour both a geographical approach to irrigation management dynamics and an ethnology-based approach that provides us with the perception water users have of their milieu, of the water crisis, as well as of changes in water management and local institutions. We will not adopt a quantitative approach to water usage (except in some parts of Chap. 5). We will document the diversity of the situations and expose the difficulties of analysis that this entails. However, no clear-cut answers will be offered to the questions raised—which may render the reader somewhat frustrated: first of all, because we are conscious that our approach is limited given that the weight of each element in the system has not been specifically analysed; and secondly, because we believe that solutions must be implemented locally according to the given environmental and social constraints. Furthermore, in some parts of the book, we have adopted an experimental research approach, and this requires more in-depth work before it can provide solutions for global use.

Nevertheless, we argue that the crises facing these two types of irrigation (tank and groundwater) are heightened by the absence of an integrated and global approach to water resource management. Indeed, due to the split management of surface water and underground water, the latter come across as separate resources, since water is not considered from a larger perspective, which would include the various hydraulic transfers within and between watershed, basin and aquifers. We will in fact see (Part 1) that tank use cannot be analysed without taking into account access to groundwater. Though we support the idea that it is necessary to manage the entire water resource in an integrated way since the two types of water are interrelated, the management of groundwater and tank irrigation are nevertheless presented here in separate parts (namely Part 2 and 3), notably because they are managed differently. We will highlight the factors that exacerbate the issue. Due to their many differences, these two types of irrigation are not undergoing the same sort of crisis.

Environmental crisis regarding groundwater

Groundwater irrigation is clearly facing an environmental crisis, which is marked by an overexploitation of groundwater and a drop in the water tables. In fact, nearly 45 per cent of the *blocks*—administrative units for development implementation—are overexploited or in a critical state in Tamil Nadu, compared to 15 per cent in India (Ministry of Water Resources, GoI 2007). One may talk of an environmental crisis according to Van der Leeuw's definition (2006), in the sense that society has yet to find solutions to curb the rapid degradation of the environment which undermines the sustainability of practices. Legal, technical and institutional solutions may exist. Some government steps have been taken (national water reform in accordance with the National Water Policy 2002, Tamil Nadu Groundwater Act, 2003, subsidies for drip irrigation, groundwater recharge programmes), but they prove to be inadequate, and even incomplete in the case of the underground water law for which the decree has yet to be published—as of June 2009, date of writing this introduction and book.

It is crucial here to stress the major difference between upper aquifers and deep aquifers, since the hydrogeological dynamics are different, as are the water crises. The water table of the upper aquifer naturally varies, dropping during summer and rising during the rainy season, when recharged by surface water (rain, river, ponds, tanks, canals and irrigation, especially from irrigated paddy-fields). For the upper aquifer, we may speak of a crisis when the increase in the number of pumps implies a rapid and lasting fall in the water table so that current practices using well irrigation are not sustainable as such, because of the high frequency of water shortages at the end of the cropping season. Another consequence of such groundwater deficiency is the drying out of shallow wells and springs that supply canals. Regarding other aquifers, their recharge process is more complex and may involve a long-term process. The overexploitation of such aquifers (when the quantity of water extracted is greater than the recharge) is due to tubewell irrigation, as described below.

This rapid water depletion is caused by several factors: technical and agronomic progress, political incentives (Vaidyanathan 1996) to which, according to our findings, one may add a key factor: a lack of measures to regulate exploitation of the resource, even though water extraction techniques have rapidly evolved. The conditions for exploiting the resource are indeed defined by a law dated 1882 (*Easement Act*), a time when water was drawn using animal or human traction and only at a depth of about ten metres. This law simply associates the right to exploit water with land rights. And yet a radical change in practices has been observed since the sixties in terms of the quantity of water extracted, and since the eighties in terms of depth and of the aquifer accessed, with no amendment to the law regarding the conditions in which the resource has to be exploited. These combined conditions have created genuine "Tubewell Capitalism" (to quote the title of Dubash's book, 2002), in the sense that tubewell owners take advantage of the lack of limitations to drill ever deeper to access large amounts of water, sell the resource, and thus consolidate the inequitable power relation within local communities (Janakarajan 1994, Janakarajan and Moench 2002). They therefore wield some power locally because they enjoy a monopoly over accessing groundwater (Palanisami *et al.* 1998). Yet, as Dubash (2002) shows through case studies carried out in Gujarat, water market systems do not all function in the same way. They cannot therefore be reduced to a single 'market rationality' or explained by a single school of thought as the neoclassical economist theory claims.

Indeed, studies carried out in Tamil Nadu[6] in the nineties show that the overdevelopment of groundwater irrigation led to greater environmental risks (a number of contributions in Agarwal 1997), particularly for the poor, and to the latter's dependency on the wealthiest well owners (Janakarajan 1997, 1999). Since then, the situation has continued to deteriorate in alluvial or sedimentary areas, as well as in rocky areas (cf. Dubash 2002, Kumar 2007, Janakarajan and Moench 2002, Palanisami *et al.* 2000), also affecting some villages with regard to their access to drinking water (Coste *et al.* 2007).

Though terms such as "overexploitation" and "overuse" are regularly used in reports, so far no effective measures have been taken, and rare are the remarks made by government authorities, such as those clearly expressed in the report made by the expert group on groundwater management for the 10th Plan Committee: "Clearly, the groundwater situation in the State [of Tamil Nadu] is among the worst in the country and is a matter of concern" (2007: 25). How can one explain the government's delay in reacting? Several reasons may be evoked. Firstly, limiting water use is a politically sensitive issue. Secondly, by building dams and checkdams, as well as by implementing groundwater recharge programmes, the State plays the role of a provider of water resources rather than that of its protector (Rajagopal *et al.*: 15), even if recent water reforms (National Water Policy 2002, Groundwater Model Bill 2005) tend to reinforce the State's role as water regulator, and to limit its responsibility regarding water management and maintenance through a transfer to user groups (Cullet 2006). Thirdly, ten to fifteen years ago, the overexploitation problem was seen as a local issue, concentrated round places such as North Gujarat, West Tamil Nadu, etc., even though it was assumed that it would take on national proportions (T. Shah in Agarwal 1997). Lastly, it is not easy to find a solution, and very diverse and complex hydrogeological situations prevent any implementation of a simple means of regulation. Everybody agrees that the answer to the problem is not to be found through a single solution but through a combination of measures that may be technical, legal or institutional, and whose impact is still being debated by experts (Chap. 1.2). Changes must be made to the conditions for groundwater exploitation. Some[7] recommend that, to start with, groundwater property rights, which are far too imprecise and even ambiguous in terms of ownership rights and user rights, be changed, while others—including groundwater experts for the 10th Plan Commission (2007)—suggest not altering these property rights, but simply taking action to regulate water usage. Given the ineffectiveness of purely legal regulations in India, several solutions have been recommended: acting on factors influencing water production, especially the cost and/or supply of electricity, or advocating community management or promoting dialogue and cooperation between state agencies, NGOs and communities, etc. We will describe some of the current stumbling blocks.

The crisis regarding groundwater irrigation (whether water is from upper or deep aquifers) is indeed related to a management problem, notably the absence of any regulation of its use, which encourages farmers on their relentless quest for free access to this resource. We intend to show that for the time being the natural limitation of water availability is the only factor that regulates pumping.

How is the degradation of the environment perceived by farmers who are the main water users and the first to be affected? The "frantic pumping" that is done in the study area is rather bewildering when compared to the state of the resource. Indeed, how can one explain that the damage caused to the environment does not prompt farmers to curb their water consumption? What social, political, economic or cultural factors explain this phenomenon? The decline in the water table leads to constantly deepening borewells, overdigging wells, purchasing ever more powerful pumps (Janakarajan 1999, Ramasamy *et al.* 1999) and therefore to ruinous costs, while the agricultural sector is described as being non-profitable (Narayamoorthy 2006). Farmers are perfectly aware of these aspects, but the social and socio-economic significance of owning a well or tubewell, as well as the pursuit of social independence from well or borewell owners, explains the phenomenal increase in the number of such individual infrastructures and the deepening of borewells. Moreover, as groundwater is not regarded as a common resource, those affected by the overdigging of a neighbouring well do not consider it legitimate to lodge a complaint (cases described by Janakarajan and Moench 2002; Shah 1993; Palanisami). All these aspects are to be found in the area studied, where we will focus on three particular points: (i) the social inequity that exists regarding access to water (Chap. 6), (ii) the logic lying behind the quest for underground water, which does not simply follow an economic logic (Chap. 8), (iii) knowledge related to groundwater and its recharge (Chap. 7), this latter being a crucial factor for understanding overexploitation and changes in the relationship between the Tamil agrarian community and its water resource.

What tank irrigation crisis?

Tank irrigation has been undergoing difficulties which are harder to understand, because they are more complex and have earlier origins. Since the eighties they have resulted in a sharp decline in the irrigated area in Tamil Nadu, though this is not the only consequence. This recent decline in areas irrigated by tanks is clearly the consequence of a proliferation of bores and pumps, techniques which compete with collective irrigation. Other environmental and social factors add to this and explain the state of physical disrepair and encroachment of tank systems (Janakarajan 1993, Palanisami *et al.* 1998, Vaidyanathan 2001). Nevertheless, the reasons for this decline in tank irrigated areas is a controversial topic.

For some, notably environmental activists and NGOs (Agarwal *et al.* 1997, publications from Dhan Foundation), the collective lack of interest in these infrastructures, as well as the decline in community practices, was caused by changes introduced in the nineteenth century by British colonisers. The appropriation of the infrastructures by the State, which implies a lack of involvement on the part of rural communities, would be the main reason for this. The recommended solution is therefore communal ownership of infrastructures and above all a revival of tanks and local institutions through the concept of community management (Dhan Foundation 2004).

D. Mosse (1999, 2003) does not agree with this explanation, both regarding the historical interpretation and the envisaged solution. In his analysis of tanks in the Ramnad region in south-east Tamil Nadu he places the societal and environmental context in the past: 'the production system [of tanks] (...) is inherently unstable, changeable, fluid and subject to calamity. [There], local and regional institutions not only respond to the inherent uncertainty of water resources, but also actually contribute to or amplify this uncertainty' (2003: 7). He thus protests about the image of autonomous villages managing the resource in a stable, equitable way, and he tells stories of vulnerable systems dependent on unreliable investments by warrior overlords. Water management has to be analysed from an angle which should not isolate the communities from the political context and whereby State and society are mutually associated. Thus, 'the impact of colonial governance on the water commons (…) has more to do with changing systems of State than the erosion of village tradition' (p.11). Moreover, so-called traditional rules were, in fact, often defined initially by the government (p.10). By the end of the nineteenth century, writings by colonial officials were already pointing a finger at the collapse of village traditions. D. Mosse analyses the image of the autonomous village community as a colonial construction that answered the colonial administration's need for tank maintenance and repair (pp. 247-254). And indeed, there are numerous points of convergence between the colonial point of view and the current notion of community management (Mosse 1999: 319-324). In the same way, "environmentalist narratives about community tradition, state power and irrigation decline" must be understood as a "strategic representation of community and state": they themselves are part of the politics of resource use, not a historical reality (2003: 12). As for the community management model, this alone cannot ensure the success of collective action or of the institution itself for the simple reason that the long-term survival of Water Users Associations (WUAs) has more to do with legal legitimacy, effective links with the government and forging productive, functional relationships with the State Department rather than internal characteristics of the model or of the user communities themselves (pp. 291-292, referring to Singh's work in Hooja *et al.* 2002). Added to these criticisms are those about the "myth of community" (Guijt *et al.* 1998), which tends to idealize a community's aptitude to cooperate and to forge harmonious relationships, as well as to underestimate social heterogeneity and conflicts of interests.

Works by Janakarajan and Meinzen-Dick in the eighties have shown that local institutions were not all defunct, contrary to what could be understood from the description of the decline in tanks and of the weakness of institutions that contributed to an 'under-recognition of village systems resource use' (Mosse, 1999: 321). Besides, some of these institutions may still be effective in tank management. Nevertheless, the changing societal context—mostly the change in land ownership patterns in terms of class, caste, and the number of farmers, changes in the mode of production—has also influenced the disintegration of local institutions in recent years (Janakarajan 1993).

To sum up, significant changes therefore either directly or indirectly affected tank management over the colonisation period, and this has been the case ever since the Green Revolution. The centralisation of common resource management and the appropriation by the State of hydraulic infrastructures has for a long time affected the relationship between state bodies, village communities and the water resource, which has notable repercussions on the upkeep of tanks, and therefore, their storage capacity and their (technical) performance (defined by Palanisami *et al.* (2000) as the ratio irrigated area/ recorded command area). However, as shown in Chapters 9 and 10, the impact of these changes in terms of social relations, equity, poverty alleviations, is far from negligible. More recently, access to groundwater and the change in land tenure and ownership patterns has transformed the balance of power at village level.

For some years now, a new wave of change has emerged, linked to the government's sudden interest in tanks, which they had neglected for decades. This trend is in keeping with the phase initiated in the eighties of declining investments in major water projects and of improving already existing infrastructures. At the same time, militant environmentalists have campaigned for the revival of traditional structures better adapted to local situations than huge infrastructures. Government development policies have consequently benefited from this trend to win acceptance of their tank rehabilitation programmes. Since the end of the nineties, various local NGOs, often operating on outside funding or government programmes have been involved in the rehabilitation and revival of traditional irrigation systems. Formal user associations have been created, as required by the State of Tamil Nadu for any rehabilitation project. The rights of these associations were lawfully recognized in the year 2000 (Tamil Nadu Farmer Management Irrigation Act) (cf. Chap. 9.2).

These policies on participatory management and on the revival of traditions raise several questions. Some address the way these associations are created and operated: according to what precepts are they created? How can the WUA minimise the resource's exclusion principle, especially in a context of increased competition? How do they address social equity in this highly hierarchical society? Who takes part, on what terms and for what purpose? The case of Puducherry (Chap. 9 and 10) will prove interesting in this respect, because the rules for selecting members of tank associations are laid out by a local NGO advocating equity and the empowerment of women and other socially deprived people, in tank affairs. The challenge is not so much the revival of local "traditional" management as the invention of new methods of resource management, local and government regulations, and the integration of local (technical and social) and administrative constraints. Other issues challenge the principle of creating *ad hoc* associations: in what way do these WUAs enable a better management of water? What ensures their effectiveness, especially in comparison with local institutions described as "falling apart"? How does the "community" vision championed by international institutions or NGOs limit the water management of complex and interlinked irrigation systems? How do associations maintain and perceive their relationship with government bodies and their role in resource management, and what fault do associations find with tank rehabilitation programmes? Lastly, another issue concerns the contradiction in rehabilitating old infrastructures that were criticized in the seventies as being ill-adapted to the system of production advocated by Green Revolution policies and since then forsaken by many users. Some studies suggest that beyond a certain population density threshold (170 pers/km^2) tanks, which take up a lot of space, seem to be ill-adapted; demographic pressure in rural areas tends to bring about effects that reduce operational effectiveness and the extent of tank irrigation systems (Von Oppen and Rao 1987: 22). For some researchers, the answer therefore lies in designing some tanks not as irrigation infrastructures, but as ponds for groundwater recharge when village institutions lose total interest in the irrigation function of the tank (Palanisami *et al.* 2000). In many places there may be speculation that the State's interest in tank rehabilitation is guided more by this groundwater recharge function—which would again be a means of giving priority to groundwater (GW)— than by a more effective and fair share in surface water. The same question may be asked with regard to members of the associations, given that a tank has multiple functions (see Chap. 10 & 11).

Tank irrigation is thus subject to contradictory tendencies; on the one hand, a decline in the rank it holds in irrigated agriculture due to physical and management reasons, as well as to the recent boom in GW irrigation. However, this decline takes different forms because of the diversity of local situations, ranging from heavy dependency on the tank to abandoning it totally, from active collective management to almost non-existent management, from a relatively good availability of water in the tank to a situation of considerable siltation and encroachment; on the other hand, the government is attempting to counter this tendency and promote a "revival of village traditions". Its renewed interest in these infrastructures takes the form of a participatory management policy, the spending of important sums of money to rehabilitate tanks, and the creation of formal associations which, as we shall see in the cases studied, show no interest in better management of irrigation water.

Interplay of various forms of irrigation

Even though the crises affecting groundwater and tank irrigation are not the result of the same management problems, they are not totally dissociated, especially in the study area where both types of infrastructure coexist on the same (village but also regional) territory. It therefore seems difficult to talk of water management without referring to the interdependent management dynamics of these two types of irrigation. We maintain that these crises cannot be solved if both types of resource are separated systematically as far as their management method is concerned. Due to this difference in management and in resource property, to the importance of the balance of power in water management, and to the increasing share of groundwater in irrigated areas, we have seen cases where vested interests prevail in water management as well as that of tanks. Is it not disproportionate to invest so much effort in participatory management policies focusing exclusively on surface water, while no limitations whatsoever are imposed on groundwater which supplies more than half the irrigated surface area? Is there not a contradiction in recognising that collective infrastructures recharge groundwater, whereas access to this is associated with private property? In other words, water in tanks is considered to be common property, yet it recharges water for individual appropriation (see also Chap. 10). Talking of recharge implies recognising the hydraulic interaction that exists between surface water and groundwater. Yet the way in which this interaction is perceived and interpreted by users themselves and how it influences their practices is not well documented. What explains the visibly frantic pumping and absence of any economy or limitation in water use? In order to fully understand customary usage, this has to be compared with the knowledge imparted by irrigators, which we shall attempt to do in Chapter 7. In Part 1 and 2 we will discuss the role of tanks in groundwater recharge, an idea that is fraught with practical and conceptual repercussions.

The influence of tubewell irrigation on the individual's lack of interest in tank irrigation has already been demonstrated (Palanisami *et al.* 1998, Janakarajan 1993). However, the influence groundwater has—as a favourite irrigation method—on the degree of concern some farmers show towards tank-use is under-estimated in literature. In our study region we encountered two separate cases that deserve to be presented here: these cases prompt serious discussion about the notion of groundwater recharge and about the solutions found by farmers to combine surface water usage with groundwater usage, and efforts to situate this combination with regard to surface water (Chap. 7).

Another aspect to be taken into consideration in analysing the complementarity/competitivity of irrigation by pumping and by tanks is the type of crop to be irrigated. Indeed, tanks are intimately and historically associated with rice farming. This link is still strong because the land in the tanks' command area is very often too clayey to allow crop diversity other than sugarcane and banana.[8] Although supplying a culturally vital crop, i.e. rice, tank irrigation may seem a limited technique in terms of cropping possibilities, contrary to groundwater irrigation of formerly rainfed land which offers further alternatives. Thus, on the one hand, the cropping pattern in tank command areas is limited (rice, sugarcane, banana or vegetables) and, on the other hand, most of these crops are heavy water consumers[9] (yet at the same time recharge the upper aquifer). In many places in the alluvial area, the tank is never used or only very rarely for irrigation (see Chap. 4), and farmers only rely on deep aquifers for irrigation. In such places, it is preferable to encourage surface water in combination with groundwater if necessary (even if surface water is also subject to competition), in order to limit the use of groundwater from deep aquifers. Research into rice cropping practices using less water needs to be pursued in order to increase tank performance. The introduction of shorter-cycle varieties (100 days instead of 150) from the mid-seventies onwards, during the second phase of the Green Revolution, led to a double crop of rice in some parts of Tamil Nadu (Trébuil *et al.* 2004: 88-89), yet also increased tank performance. Indeed, according to farmers in our study area, this made tank irrigated rice farming less risky as far as water availability is concerned, since they needed water for a shorter period of time. Upgrading tanks and their effectiveness is one option in making South Indian farming, especially rice farming, less dependent on groundwater.

More generally, it is only through a global and integrated approach to using tank water and groundwater that the local dynamics of water management can be addressed (Chap. 4, 5 and 6). One of the problems we face, therefore, is the diversity of the situations and factors to be taken into consideration to grasp this diversity. Diversity concerns both the tanks (size, source of water supply, dependency on rain, extent of use) and the combination of using surface water and groundwater on a village scale. How are we to understand that a tank is no longer used, even though the neighbouring village, situated in the same environmental conditions and with similar access to water, still uses it? Is it only local social, historical features that help us to understand the dynamics of water management or do regional environmental features help define a general tendency which manifests itself in various ways depending on local conditions? One of the major drawbacks of statistical work on regional studies about tanks is indeed that it does not clearly specify the differences in environmental features, such as in Vaidyanathan (2001) or Janakarajan (1993, 1999). Other studies, such as the one on tanks in different regions by Palanisami *et al.* (2003) are carried out according to tank-specific criteria, not to the more global context of groundwater access. It is regrettable that this latest study, which nevertheless integrates numerous tank characteristics, includes tanks in Vellore and Sivaganga districts in the same "tank intensive region" category, while these districts could be characterised by double irrigation and by tank irrigation respectively. A geographical approach to irrigation has been adopted in this book to demonstrate the interaction between irrigation by surface water and irrigation by groundwater, as well as the need to adopt different scales for water management analysis.

Choice of the study area

The study area is in the hinterland of Puducherry. We deliberately chose not to limit ourselves to watershed or administrative limits. In the first case, this is due to a completely anthropised and modified hydrographic network, since the water from a river basin is notably transported to another basin (cf. Chap. 4). The administrative division does not stem from any logic that could be of use to us since, on the contrary, it conceals environmental diversities. Yet to disregard the administrative division has a major drawback in that the statistical data for the area would be incomplete. The whole of Puducherry district is nevertheless included.

To determine the study area, we favoured a random division according to a rectangular demarcation which was moreover perfectly adapted to work using satellite imagery. The choice of limits within this Puducherry hinterland did at least take account of hydraulic logic, encompassing all the irrigation systems downstream from Tirukoilur deviation dam (*anicut*) on the Ponniyar. On paper these systems represent a command area of nearly 18,000 ha and their size is such that they can be found in the 1989 Irrigation Atlas on a scale of 1:1,000,000 (Kundu 1989 vol.1). The western limit of the study area is therefore situated at Tirukoilur and the southern limit is near the Godilam that flows into the sea near Cuddalore (southern limit of the irrigation systems represented in the atlas). The eastern limit is the sea. As for the northern limit, it was defined by the town of Tindivanam, which allowed the Nallavur watershed (also called the Kaluvelly watershed by development agents) to be included. There are various advantages to selecting this case study area: we were able to use the results of previous work on irrigation conducted in this study area (Bossy 1985, Oswald 1989, Marius Gnanou 1991, Druesnes 1997, some TRPP[10] reports); a visit was made of one of the famous pilot tanks mentioned in grey literature on irrigation, Kedar, rehabilitated by the EEC project in Villupuram district; and a reliable description of the living conditions of outcasts was gathered from Viramma's account (1998).

This geographical demarcation shows a common characteristic: as already mentioned, every village is equipped with a tank and wells, which allows us to study the various dynamics for tank and groundwater irrigation.

Research collaboration and methodology

This book combines, on the one hand, results from a research project carried out at the French Institute of Pondicherry (IFP) over a three-year period in the Puducherry hinterland and, on the other hand, witness accounts from development experts (NGO members and an international expert) involved in research and development projects in the same region, notably in Kaluvelly watershed and Puducherry district respectively (cf. Chap. 5 and 10).

The project run by IFP starting in mid-2005 stemmed from a preliminary study carried out in 2000. It included Indian, French and Spanish researchers, as well as students of various nationalities. Some were on short field internships: one week for five students from the "Sustainable Environment" Masters Course at the Puducherry University, two weeks for ten students specialising in "water management" from SupaAgro, Montpellier, France, as well as for G. Savin and W. Méas from the Université de Provence (however, the latter worked throughout the year on the data under the supervision of P. Oliva). Some other students were on longer field internships, ranging from four to six months, such as F. Cesselin (2000), L. Caillot (2005), N. Cayon-Glayère (2006)—co-supervised by M. Roesch, IFP, O. Lefebvre (2006), P. Ariza, E. Galán and T. Serrano (2007) supervised by V. Reyes-Garcia. This project also benefited from worthwhile exchanges with students who were carrying out work at the same time for their Masters Course on a similar topic: B. Bentz (2000) was working in the region with the NGO Harvest; N. Coste and N. Ploumpidis (2007), C. Rognant (2005) were working on neighbouring regions, supervised by K. Anupama (IFP) and Y. Gunnell (former researcher at IFP).

To grasp the diversity of the situations, we worked on different scales. It was possible to cover the whole study area using already existing maps ranging from a scale of 1: 50,000 to 1: 200,000 (Chap. 2 to 4) and five satellite pictures ranging from a resolution of 23.5 to 79 m, dating from 1977, 1991, 2000 and 2004 (some results in Chap. 4). A very rich GIS database was also available that associates data from the 1991 census with each village represented by a point on a cartographic background. This GIS database, designed by a demographer, is of particular interest for our purposes since it contains irrigation data.[11] Thus, for the thousand or so villages in the study area, we have the net irrigated areas by type of irrigation.

Yet the fact that the study area covers two different state territories (Tamil Nadu and Puducherry) complicates statistical acquisition as well as the analysis of geographical phenomena, given that Puducherry district is not an unbroken geographical area. This last point favours using a GIS with localised data, which combines the various localised data available. This is done in order to demarcate homogeneous areas from a water usage point of view. We then also compared the statistical data with surveys (1 or 2 questionnaires per village) carried out in more than 200 villages (Chap. 4).

In Part 2 and 3, a more in-depth analysis of seven villages (see Fig. 1 Box 6) introduces the social aspects of groundwater access and tank management institutions.

Structure of the book

We based our work on the principle that a change in scale (plot-village-region) is needed in order to grasp the local situation: a regional analysis enables the place to be integrated into a set of wider geographical features and a finer scale provides an understanding of local specificities and diversities. Reference to another scale of analysis is therefore regularly made using village examples to illustrate a regional characteristic, by referring to regional features and to intra-village diversities when village examples are taken into consideration. For the sake of clarity, the first part focuses on describing regional characteristics (Puducherry hinterland and Kalluvelly watershed), whereas Parts 2 and 3 are based on village case studies.

As irrigation has undergone major changes since Independence, a brief description of the changes is essential in order to understand the current situation and the challenges it faces (Chap. 1). The first part (Chap. 2 to 5) therefore sets out to define the environmental factors that come into play in the spatial distribution of the use of irrigation infrastructures (tank, well and tubewell)—without overlooking the fact that social factors must be taken into account. Thus the geographical characteristics of irrigation are presented, whereby regions with the same irrigation characteristics may be defined (Chap. 3 and 4) and an integrated approach to addressing the problem of the drop in water tables may be adopted (Chap. 5). Parts 2 and 3 (Chap. 6 to 11) refer to perceptions, knowledge and institutions. Part 2 focuses on groundwater, on the one hand, by analysing how it has developed in various villages, showing that diverse scenarii exist depending on historical and environmental conditions (Chap. 6), and that economic logic alone cannot explain changes in infrastructures (Chap. 8) while, on the other hand, studying local knowledge about groundwater when placed in the political context of groundwater recharge (Chap. 7). Finally, the third part is devoted to the institutions managing tanks, whether informal or formal, with a detailed description in Chapter 9 of past and current institutions in charge of tank management. Chapters 10 and 11 question the participatory management model through a two-angle approach: the gender study applied to the TRPP project (Chap. 10) and the concept of an ecosystem through multiple uses of two tanks (Chap. 11).

NOTES

1. Translated from the French.
2. A similar idea is raised by Kumar *et al.* (2008) regarding the negative impact downstream of action upstream in closed basins and water-scarce regions, where, they say, small water harvesting structures do not improve water supplies on a basin scale, and even have an adverse impact on downstream use, basically because of "Peter taking Paul's water".
3. As of September 2006, Puducherry is the new name of the City, District and the Territory (composed of four scattered districts) previously known as Pondicherry.
4. At MIDS (Madras Institute of Development Studies), Chennai, the main researchers working on water issues were Vaidyanathan, Janakarajan, Sivasubramaniyam and Sengupta and Rajugopal; Palanisami was head of the research programmes at TNAU (Tamil Nadu Agricultural University) when their work was published.
5. Indeed, canal irrigation represents 27 per cent of the net irrigated area in Tamil Nadu as well as in India (2004 data, from NSSO 2007). The main issues regarding canal irrigation are the lack of efficacy of large irrigation systems, the elite seizing power during the transfer of management to water user associations, poor drainage causing overlogging and salinity problems, a lack of water supply in downstream areas due to an increase in water diversion in upstream areas, etc. A lot of work has been published about this type of irrigation, the most recent, significant research concerning South India being: Joshi & Hooja 2000, Hooja *et al.* 2002, Gulati *et al.* 2005.
6. Same in Gujarat: Moench 1992 a; Shah 1993; contributions in Agarwal 1997.
7. The mid-term report of the 10th Plan (GoI 2005); numerous associations such as that of V. Shiva.
8. In Tamil Nadu, in 1996, 71 per cent of the surface area irrigated by tanks was planted with rice (Palanisami *et al.* 2001: 45). For our study region, we do not have any statistics, though tanks seem to be mainly used to irrigate paddy-fields.
9. In Asia, paddy-fields use 90 per cent of irrigation water (Trébuil *et al.* 2004: 134). In India, there is a lesser proportion, since paddy-fields only represent 55 per cent of irrigated land versus 80 per cent in some countries in South-East Asia. Paddy water requirement = 12,000m^3 per ha or 2,500 *l* per kg produced, Sugarcane = 18,000 m^3/ha. .
10. Tank Rehabilitation Project Pondicherry, see Chapter 9 and 10.
11. C. Z. Guilmoto designed this GIS database first at IFP for Tamil Nadu, then for Southern India through the EMIS programme (*Espace et mesure en Inde du Sud*). See Guilmoto *et al.* (2004) for the mapping and GIS methodology used, and Guilmoto (2002) for the description of irrigation data. I produced

the maps presented in Chapter 1.1 and 4 during the EMIS joint research programme.

REFERENCES

Agarwal, A. Editor. 1997. *The challenge of the balance: environmental economics in India. Proceedings of the National Environment and Economics Meeting, January 1994*. New Delhi: Centre for Science and Environment.

Agarwal, A., and S. Narain. Editors. 1997. *Dying wisdom. Rise, fall and potential of India's traditional harvesting systems*. New Delhi: Centre for Science and Environment (State of India's Environment, 4, a citizens' report).

Ariza, P., E. Galan, and T. Serrano. 2007. Water tanks as ecosystems. Ecosystemic perception of local population for integral management of water tanks in Tamil Nadu, South India. Final report for Environmental Sciences Degree, Universitat Autònoma de Barcelona.

Bentz, B. 2001. Histoire sociale - Histoire technique: les tanks du bassin versant du Kalluvelli (Inde), Mémoire du Master of sciences DAT (Développement Agricole Tropical) du CNEARC, Montpellier.

Bossy, A. 1985. *Evaluation des ressources en eau et de leur gestion. Cas d'un village en zone tropicale sèche (Tamil Nadu-Inde)*. Ecole Nationale du Génie rural des Eaux et Forêts (GREF), Diplôme d'Etudes Approfondies.

Caillot, L. 2005. Accès à l'eau et vulnérabilité en zone rurale. Rapport de Master de l'Ecole Supérieure Européenne, IFP, ESE-IER.

Cayon-Glayère, N. 2006. Rentabilité de l'agriculture irriguée dans un village de Pondichéry, Mémoire en vue de l'obtention du Diplôme d'Ingénieur en Agronomie Tropicale, CNEARC, Montpellier.

Cesselin, F. 2001. *Irrigation par tank et motopompes à Tiruvenainallur au Tamil Nadu: compétition ou complémentarité de techniques ?* Mémoire du Master of sciences DAT (Développement Agricole Tropical) du CNEARC, Montpellier.

Coste, N., and N. Ploumpidis. 2007. "Bittersweet water. Irrigation practices and modern challenges in South India. Trilingual DVD (Eaux douces, Eaux amères-Thaneer: inippum kacappum). Co-authors : K. Anupama, O. Aubriot, Y. Gunnell," Puducherry: French Institute of Pondicherry (Publications Hors Série 5).

Cullet, P. 2006. Water law reforms. Analysis of recent developments. *Journal of the Indian Law Institute* 48 (2):206-231.

Druesnes, N. 1997. *Révolution doublement verte. Les biovillages à Pondichéry*. CIRAD (Programmes Economie, Politiques et Marchés-doc. de travail n°29).

Dubash, N.K. 2002. *Tubewell capitalism: groundwater development and agrarian change in Gujarat*. Delhi: Oxford University Press.

Government of India. 2007. *Ground water management and ownership (report of the expert group)*. Planning commission.

Guijt, I., and M.K. Shah. 1998. *The Myth of Community, Gender Issues in Participatory Development*. London: Intermediate Technology Publications.

Guilmoto, C.Z. 2002. Irrigation and the great India rural database. Vignettes from South India. *Economic and Political Weekly* 37 (13):1223-1228.

Guilmoto, C.Z., S. Oliveau, V. Chasles, R. Delage, and S. Vella. Editors. 2004. *Mapping out Social Change in South India. A Geographic information system and its applications*. Pondicherry: French Institute of Pondicherry.

Gulati, A., M.-D. Ruth, and K.V. Raju. 2005. *Institutional Reforms in Indian Irrigation*. New Delhi: Sage Publications and IFPRI.

Hooja, R., G. Pangare, and K.V. Raju. 2002. *Users in water management. The Andhra model and its replicability in India*. Jaipur and New Delhi: Rawat Publications.

Iyer, R.R. 2007. *Towards Water Wisdom. Limits, Justice, Harmony*. New Delhi: Sage Publications.

Jairath, J., and V. Ballabh. Editors. 2008. *Droughts and Integrated Water Resource in South Asia*. Vol. 2. *Water in South Asia*. New Delhi: Sage Publications India.

Janakarajan, S. 1993. In search of tanks: some hidden facts. *Economic and Political Weekly* 28:A53-A60 (Review of Agriculture).

——. 1994. *Trading in groundwater: a source of power and accumulation*. Madras Institute of Development Studies (Working Paper No. 123).

——. 1997. Consequences of aquifer overexploitation: prosperity and deprivation. *Review of Development and Change* 2.

——. 1999. "Conflicts over the invisible resource in Tamil Nadu. Is there a way out?," in *Rethinking the mosaic.* Edited by M. Moench, E. Caspari, and A. Dixit, pp. 123-159. Kathmandu: Nepal Water Conservation Foundation.

Janakarajan, S., and M. Moench. 2002. *Are wells a potential threat to farmers' wellbeing? The case of deteriorating groundwater irrigation in Tamil Nadu.* Madras Institute of Development Studies (MIDS working paper No. 174).

Joshi, L.K., and R. Hooja. 2000. *Participation Irrigation Management. Paradigm for the 21st century.* New Delhi: Rawat Publications.

Kemper, K.E. 2003. "Rethinking groundwater management," in *Rethinking water management. Innovative approaches to contemporary issues.* Edited by C.M. Figuères, C. Tortajada, and J. Rockström, pp. 120-143. London: Earthscan Publications.

Kumar, M.D. 2007. *Groundwater Management in India. Physical, Institutional and Policy Alternatives*. New Delhi: Sage Publications.

Kumar, M.D., A. Patel, R. Ravindranath, and O.P. Singh. 2008. Chasing a mirage: water harvesting and artificial recharge in naturally water-scarce regions. *Economic and Political Weekly* 44:61-71.

Kundu, A.K. Editor. 1989 (1st edition 1972). *Irrigation atlas of India*. Vol. 1 & 2. Calcutta: National Atlas and Thematic Mapping Organisation, Government of India.

Lefebvre, O. 2006. Dynamiques de gestion de l'eau dans un contexte de surexploitation des eaux souterraines. La région de Pondichéry et de Villupuram au Tamil Nadu, Inde du Sud. Mémoire présenté en vue de l'obtention du Mastère Spécialisé Développement Agricole Tropical, CNEARC, Montpellier.

Marius-Gnanou, K. 1991. "L'irrigation et les mutations socio-économiques récentes dans la région de Pondichéry (Inde)," in *Eau et aménagement dans les régions inter-tropicales, T.2 (Espaces tropicaux, n°3)*. vol. 2. Edited by P. Vennetier, pp. 133-160. Talence: CEGET-CNRS.

Méas, W. 2007. *Les "tanks" dans l'arrière-pays de Pondichéry (Inde): distribution spatiale et corrélation avec les données démographiques et socio-économiques (Census of India), intégration SIG*. Mémoire de Master 1 "Sciences géographiques", Université de Provence, Centre de traitement de l'information géographique.

Moench, M. 1992a. Chasing the Watertable. Equity and sustainability in groundwater management. *Economic and Political Weekly* 27 (51-52): A171-A177.

Molden, D. Editor. 2007. *Water for Food, Water for Life: A Comprehensive Assessment of Water Management in Agriculture*. London & Colombo: Earthscan & International Water Management Institute.

Mosse, D. 1999. Colonial and contemporary ideologies of 'community management': the case of tank irrigation development in South India. *Modern Asian Studies* 33:303-338.

——. 2003. *The Rule of Water. Statecraft, Ecology and Collective Action in South India*. New Delhi: Oxford University Press.

Narayanamoorthy, A. 2006. State of India's farmers. *Economic and Political Weekly* 41:471-473.

Narayanamoorthy, A., and R.S. Deshpande. 2005. *Where water seeps! Towards a new phase in India's Irrigation Reforms*. New Delhi: Academic Foundation.

Oswald, O. 1989. An expert system for the diagnosis of tank irrigated systems: A feasibility study. Ph.D., Anna University.

Palanisami, K., and K.W. Easter. 2000. *Tank irrigation in the 21st century. What next?* New Delhi: Discovery Publishing House.

Palanisami, K., and R. Balasubramanian. 1998. Common property and private prosperity: tanks vs private wells in Tamil Nadu. *Indian Journal of Agric. Economy* 53:600-613.

Palanisami, K., P. Paramasivam, B. Chandrasekaran, and S. Karunanithi. 2003. *Participatory irrigation management in Tamil Nadu.* Coimbatore: Water Technology Centre, Tamil Nadu Agricultural University.

Palanisami, K., P. Paramasivam, C. Karthikeyan, and A. Rajagopal. 2001. *Sustainability of tank irrigation systems in South India.* Coimbatore: Water Technology Centre, Tamil Nadu Agricultural University.

Rajagopal, A., and S. Janakarajan. 2001. "Water-rights and participatory irrigation management in India: the case of surface water sector in Tamil Nadu state," www.water-2001.de/datenbank/ 546076235.41368.14/ WATER%20RIGHT%20PAP.doc.

Ramasamy, C., K. Palanisami, K. Lokanadhan, and M. Anjugam. 1999. *Coping behaviour to growing groundwater scarcity in Tamil Nadu.* Coimbatore: Water Technology Centre, Tamil Nadu Agricultural University.

Rognant, C. 2005. Evolution de la gestion de la ressource en eau: irrigation traditionnelle, milieu et société rurale à Melavalavu, Tamil Nadu - Inde. Maîtrise, Université Paris I - Panthéon Sorbonne.

Savin, G. 2007. Les modes d'irrigation par tank dans l'arrière pays de Pondichéry: organisation et évolution entre 1977 et 2007. Approche par télédétection et SIG., Mémoire de Master 1 "Sciences géographiques", Université de Provence, Centre de traitement de l'information géographique.

Shah, T. 1993. *Groundwater markets and irrigation development. Political Economy and Practical Policy.* Bombay: Oxford University Press.

Trébuil, G., and M. Hossain. 2004. *Le riz. Enjeux écologiques et économiques.* Paris: Belin (Mappemonde Coll.).

Vaidyanathan, A. 1996. Depletion of groundwater: some issues. *Indian Journal of Agricultural Economics* 51:184-192.

——. Editor. 2001. *Tanks of South India.* Delhi: Centre for Science and Environment.

Van der Leeuw, S. 2006. "Crises vécues, crises perçues," in *Temps et espaces des crises de l'environnement.* Edited by C. Beck, Y. Luginbühl, and T. Muxart, pp. 351-368. Paris: Quae éditions (coll. Indisciplines).

Venot, J.P. 2008. Entre immobilisme et adaptabilité. Trajectoire d'évolution du bassin versant de la Krishna, Inde du sud. PhD, Université de ParisX-Nanterre.

Villalba, B. 2006. "Dire la crise? L'injonction contradictoire de l'idée de crise écologique dans le champ politique," in *Temps et espaces des crises de l'environnement.* Edited by C. Beck, Y. Luginbühl, and T. Muxart, pp. 383-391. Paris: Editions Quae.

Viramma. 1998. *Life of an Untouchable (Texts collected, translated and annotated by Josiane Racine and Jean-Luc Racine.).* Paris, New York: Unesco Press.

Von Oppen, M., and K.V.S. Rao. Editors. 1987. *Tank irrigation in semi-arid tropical India. Economic evaluation and alternatives for improvement. Research Bulletin No.10.* Patancheru, Andhra Pradesh: ICRISAT.

CHAPTER 1

IRRIGATION IN SOUTH INDIA: A COMBINATION OF TRADITION AND MODERNITY

Olivia Aubriot

Even though India today occupies first place as the world's most irrigated area, it is not only due to its efforts to develop modern irrigation techniques since the colonial period, indeed showing significant progress since Independence. This is also linked to the country's great irrigation tradition, to various methods and to their spreading throughout the country[1]. Many of these methods are still in use, though some have been modernized (especially diversion weirs and canals), while others, such as tanks, suffer competition from new techniques (pumps for groundwater irrigation). Tank irrigation is a collective irrigation system typical of India and Sri Lanka.

Table 1.1: Proportion of net and gross irrigated area in India and in Tamil Nadu in 2003-04

	India *million hectares**	*% irrigation*	Tamil Nadu *million hectares***	% irrigation
Net sown area	140.9		4.69	
Net irrigated area	55.1	39	2.15	45
Gross sown area	190.64		5.32	
Gross irrigated area	76.8	40	2.48	46

* From NSSO data; ** From Tamil Nadu Season and Crop Report 2003-04.

1.1. A Brief History of Irrigation

In order to grasp the current situation (Table 1.1) and the changes affecting the importance of tanks, a brief presentation of the main stages in the history of irrigation in South India and Tamil Nadu is called for. It will become clear how the various political choices have influenced the development of some specific techniques over time.

(a) *Irrigation in India: A Tradition*

There are many literary testimonies attesting to the development of irrigation in India and Tamil Nadu in particular, during the first centuries of the country's history (Adicéam, 1966[2]: 160). The three main types of irrigation found in India—canal, tank and well irrigation—have been in existence for centuries.

Canal irrigation consists in diverting water from a river into a canal. This book will only briefly touch upon the subject, as this technique is not used in the study area. Merely a few dozen kilometres to the south is the Cauvery delta which has been Tamil Nadu's rice-bowl for centuries. Here, the *Grand Anicut,* located on the river at the entrance to the delta to divert water to canals, was built in the second century AD (Agarwal *et al.*, 1997: 255).

Groundwater irrigation using wells is also most probably an old technique, as there are fourteen names for "well" in the Vedic lexicon (*Nighantu*, 16th to 5th century BC), while the grammarian Panini (5th BC) mentions two types of well, and other texts attest to the early presence of such a technique in India, as well as in the Tamil countryside ([A.]: 156-160). The Pallava charts dating from the 7th century AD recall that digging them was nevertheless subject to regulations and could only be undertaken when authorized by sovereigns, no doubt in order to control water distribution depending on the requirements, rather than to limit their number ([A.]: 161).

Tank irrigation is characterised by the presence of a man-made lake-reservoir or "tank"[3] which stores water during the rainy season and releases it to paddy fields cultivated during the same season or even later (*see* Box 1). Even if some canals divert water from a river to channel it into tanks (thus called system tanks), it is classified as tank irrigation, not as canal irrigation. The tank technique dates back a long way, since Third Sangam literature (300 BC to 200 AD) mentions the cropping of irrigated paddy and describes the tanks' infrastructure, while an inscription dated 200 BC describes the construction of a tank in Sivaganga district which still functions (Ratnavel *et al.*, 2006: 10-11). Yet the development of tanks in North Tamil Nadu is mainly due to two dynasties; that of the Pallava (6th to 10th centuries AD) and of the Cholas (10th to 13th centuries AD). Most tanks were built before the 18th century

AD ([A]: 162). One important technical advance was developing the sluice-weir technique between the 6th to 8th centuries that made gravity irrigation possible as opposed to the earlier method of lifting irrigation from ponds (Nandi, 2000 cited by E. Shah, 2003: 35). Inscriptions reveal that the construction of tanks required knowledge of hydrology and hydraulics ([A.]: 154-168; Ratnavel, 2006: 16-20). However, the layout of tanks in a chain or cascade formation, as we see today, does not come from a predefined programme but from various adaptations developed by trial and error, and from various additions over time ([A]: 41; Rotzer, 1984: 152; Ludden, 1989: 53). Tanks are not a static technology, on the contrary, they "have been modified in a variety of ways in response to changing state-society relations and agrarian transformations" (E. Shah, 2003: 26).

The importance of irrigation in Tamil history has been acknowledged by all authors. "Tamil people built their civilization around irrigation" (Ludden, 1989: 20). "Their recorded history begins only after they learned to grow [wet] paddy" [which requires water in Tamil Nadu conditions] (...) Medieval generations transformed South Indian village society by bringing nature under fuller control then ever before and by making places sacred" (*ibid.*: 15-20).

However, the history of the development, management and maintenance of tanks before 1800 is subject to controversy. The role of kings in building them and the autonomy of villages in tank governance are the main sources of discord. Any merit gained by building a tank from the Sangam period onward is one argument for explaining why individuals and local head-men became involved in tank development. Various actors participated at different levels in the irrigation development process: temples, private individuals, merchants, village assemblies, ministers of state and the king's government (Mosse, 2003). Some records prove that elected village assemblies existed and controlled tank maintenance—such as in Bahour, located in the study area, and where 8th-century plates mention a Sanskrit University and a 10th-century text informs us of tank maintenance rules: a village assembly collected the tank tax from cultivators, and each villager, whether 10 or 80 years old, except for untouchables, had to take part in tank desilting work every year; otherwise, the assembly made them pay fines[4] (Vijayavenugopal, 2006: 23). Yet D. Mosse, who more virulently criticizes the colonial and contemporary assumption that tanks were community-controlled systems, suggests that this idea stemmed from documents issued from medieval Brahmadeya villages—villages offered by rulers to Brahman communities for learning and ritual purposes—but these documents refer to a minority of privileged settlements and cannot be accepted as a general model (2003: 58). Moreover, "during the period of the earliest tank inscriptions (9th-11th centuries) the basic economic and political unit (...) was not the self-governing village (*ur*) but the micro-regional *natu*" (*ibid.*, referring to Stein, 1980). The fragmented historical sources and their specific political and environmental context prompt caution about generalizing here as if there were just one history behind tank development. E. Shah (2003) has even shown from cases studied in Karnataka that "tank technology is shaped as a result of the social relation to power in particular historical, agro-climatic and agrarian contexts".

Box 1: What is a tank?

A tank (*eri* in Tamil) is a partially embanked artificial lake or reservoir, of shallow depth where water is stored during the rainy season to be released to paddy-fields by gravity through sluices and irrigation canals. Water distribution takes place mainly during the same rainy season, and also in some cases just afterwards, while the rest of the year the tank remains empty. Thus, a tank is both (i) a centuries-old water harvesting system which collects rainwater, runoff water or in some cases water which has been deviated from a river, (ii) a collectively managed irrigation system—a function for which it has primarily been studied, and (iii) a collective multiple use structure, depending on the different resources its ecosystem provides (water, fish, trees, grass, silt, soil, *see* Chap. 11). *See* Fig. 11.0 in Chap 11.

Tanks are made of a bund (*karai*) of earth, a manmade embankment which blocks the water flow and allows its storage upstream in the tank-bed, also called the waterspread area of the tank. The bund is equipped with a surplus weir (*kalungu*) from which surplus water flows into a surplus course, and with one or several sluices (*maduwu*) which deliver water to irrigation canals supplying the fields. The area of the tank-bed whose level is below that of the sluice is the dead storage (*arangani*) used to keep more water, particularly for fish farming. The area registered as irrigable is called the command area or *ayacut* by the administration, though this last term is scarcely used or known by farmers. Thus, four major components constitute a tank irrigation system: (i) the source catchment area and the supply channels, (ii) the tank itself with its bund, water spread area, surplus weir and sluices, (iii) water irrigation distribution channels and irrigated fields, and (iv) surplus course (Ratnavel *et al.*, 2006: 5).

Tanks are rarely isolated infrastructures. Drawing supplies from the stream draining the watershed, they can either be organized in cascade or in a chain, with the surplus of the upstream tank feeding tanks downstream or they can be linked—in this case they are called system tanks—to a common channel fed by a diversion weir (*anicut*) or reservoir across a river. A unique feature of the surplus weir construction is that one or more vents provide water at a lower level than the crest of the weir, in order to release water into a lower tank in the chain before this tank fills up (Ratnavel *et al.*, 2006: 6). Tanks are, therefore, interconnected to a network of different types of canal crisscrossing the landscape (Fig. 1.1): a supply channel (*odai* or *kolakkal*) collects surface run-off from the catchment area of (cultivated or uncultivated) land upstream; a canal connecting two tanks

(*Contd.*)

Box 1 (*Contd.*)

collects the surplus water from the upper one to bring it to the one downstream; the surplus course carrying water to another canal is used as a drainage canal but also, in some cases, as a supply canal for downstream tanks; the main channel of the network of system tanks diverts water from the river; the supply canal for system tanks is fed by the previous main channel.

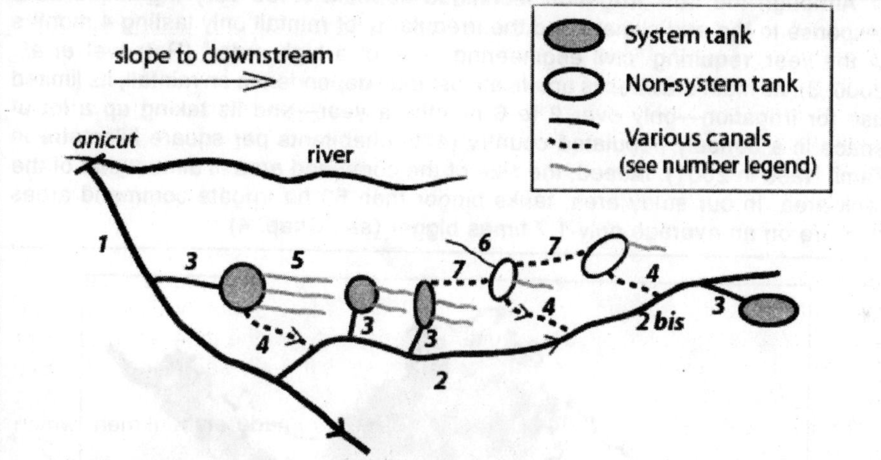

Canal Number Legend

—*Canals for system tanks*—1: main channel of the ST network; 2:branch of supply channel of the ST network 1; 2bis: same channel, as drainage and supply channel; 3: supply channel of a ST—*Canals for all tanks*—4: surplus water canal; 5: irrigation canals—*Canals for non-system tanks*—6: supply channel (collecting run-off); 7: canal connecting two tanks: surplus water channel for the upstream tank and supply channel for the downstream tank.

Fig. 1.1: Schematic Drawing of the Various Types of Canal

Tanks have been classified, since British rule in the old Madras Presidency (nowadays Tamil Nadu and parts of Andhra Pradesh), as "system" and "non-system" tanks according to their supply (Vaidyanathan, 2001: 15). A system tank draws its supply from a river, in addition to the yield of its own catchment area, whereas a non-system tank is rainfed. From the 39,202 tanks in Tamil Nadu, only less than 10 per cent are system tanks, serving 21 per cent of the total *ayacut* of the tanks (*ibid.*).

They are also classified as Panchayat Union or Public Works Department (PWD) tanks depending on the management authority in-charge of their maintenance; Panchayat Union tanks are rainfed tanks with a command area of less than 40 ha whereas PWD tanks are rainfed with a command area of more than 40 ha or system tanks (whatever the command area). Ex-*zamin* tanks were constructed by *zamindars* (landlords) during the British period. After abolition of the *zamindari* system by the State government in 1957, they were transferred to Panchayat Union or PWD according to the size of their command area (Palanisami

(*Contd.*)

Box 1 (*Contd.*)

et al., 2000: 11). The number of PWD tanks amounts to 8,939 in Tamil Nadu (23 per cent of all tanks) but serve 75 per cent of the entire *ayacut*. In Puducherry, however, this distinction does not exist and the 84 tanks in Puducherry district are under PWD management, whatever their size or their water supply source (half have an ayacut below 40 ha, of which only 18 are non-system tanks).

Although the tank irrigation technique appears to be very ingenious as a response to the arid climate and the irregularity of rainfall only lasting 4 months of the year, requiring "civil engineering skills of a high order" (Ratnavel *et al.*, 2006: 3), its main drawbacks are its almost total dependence on rainfall, its limited use for irrigation—only over 2 to 6 months a year—and its taking up a lot of space in a densely populated country (479 inhabitants per square kilometre in Tamil Nadu in 2001). Indeed, the size of the command area is almost that of the tank area. In our study area, tanks bigger than 50 ha irrigate command areas that are on an average only 1.7 times bigger (*see* Chap. 4).

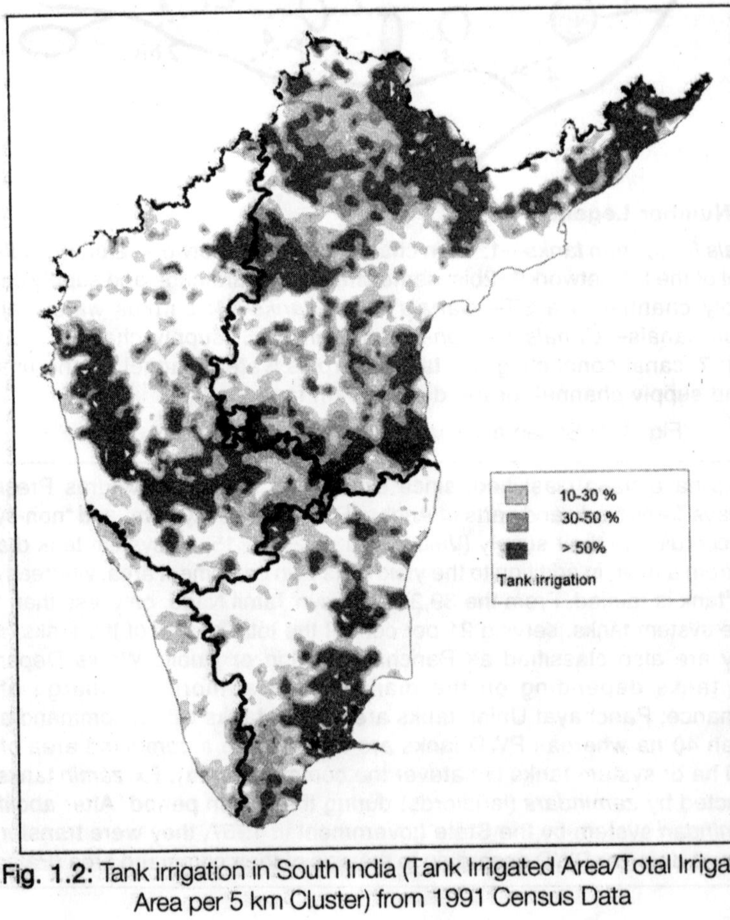

Fig. 1.2: Tank irrigation in South India (Tank Irrigated Area/Total Irrigated Area per 5 km Cluster) from 1991 Census Data

Why are tanks always referred to as the traditional irrigation type in South India?[5] There are indeed many of them, they are an important source of irrigation and they represent (ed) a key element in the socio-economic activities as well as the social life of the village. Let us give a few figures: they number 159,000 in South India (Vaidyanathan, 2001: 17-19) for 65,000 villages, and in Tamil Nadu itself there are about 39,000 for 15,000 villages—villages accounted for in the census, i.e. revenue villages[6]. Hence in regions where there is a greater concentration of tanks—mostly in coastal areas of Tamil Nadu, south and north coastal Andhra, and parts of Telangana[7]—nearly each hamlet has its own tank and each revenue village has two to three tanks. With regard to the irrigated area, until the 1960s, tanks irrigated more land (38%) than canals (36%) and wells (24%) in Tamil Nadu. Indeed, they have shaped the landscape of Tamil Nadu countryside, covering the equivalent of 15 per cent of cultivated land. Yet tanks are not simply an irrigation technique: they have multiple uses and functions within the village; they contribute to maintaining the environment (*see* Box 1 and Chap. 11): they form "part of the 'public domain' through which social relations are articulated, reproduced and challenged" (Mosse, 1997, 2003). Therefore, they hold a privileged place in rural life.

Tanks are not the only technique for harvesting and storing water, but the main one for irrigation. Indeed, in each village in Tamil Nadu, one can find square temple tanks, ponds (*kulam* also called *oorani* around Madurai) used for domestic purposes (and sometimes irrigation), and small ponds *(kuttai)* located in fields and used for cattle to drink from. South Indian people have an old tradition of harvesting rain and runoff by channelling it into water bodies.

What are the factors then to explain their presence, specifically in India? Three main factors are often given (Palanisami *et al.*, 2000; Agarwal *et al.*, 1997): (i) the rainfall regime which is very erratic, unpredictable, and generally limited to a few months during the monsoon, sometimes with huge amounts of rain over just a few days: *see* the graphs of the Puducherry area in Chap. 2 as one example of the annual, pluri-annual and geographical irregularity of rainfall, and Fig. 1.3 for the geographical distribution of the mean rainfall in South India; (ii) the hydrography, with there not being many perennial rivers (only Cauvery and Tambraparni for Tamil Nadu), so that canal irrigation could not easily be extended; (iii) the geology, which does not facilitate groundwater recharge. The hard granite gneisses, which represents the geological formation of most of the Southern Peninsular "serves as a barrier to the deep percolation from the tanks built above" (Ratnavel *et al.*, 2006: 7), and the clay could be used to increase the watertightness. Due to these environmental features, a solution had to be found to store water: either rain water, or the runoff or the river flood water is captured in tanks and used mainly for (wet) rice cultivation[8] which requires great amounts of water for its irrigation in this semi-arid climate where rainfall is irregular during the monsoon. Other factors come into play to explain the spatial distribution of tanks. The flat relief with a small slope towards the east helps to drain and store water ([A.]: 318). K. Rotzer (1984: 171) has shown that in North Karnataka, where the soil is permeable because of its basaltic origin, the tank construction technique might have been improved and the tank developed as a large-scale irrigation system when Muslims arrived in this region (14th AD) with their knowledge of lime mortar that helped to weatherproof the reservoir bund. On such basaltic ground, the "Muslim" tank required two to three times less material and labour than the "Indian" tank (*ibid.*: 194).

One difference stands out in the spatial organisation of tanks, which shows the expertise developed to use rainfall water as much as possible. In non-system tanks (see Box 1 for definition), each tank receives runoff water from its catchment area and the surplus water from the upper tank, allowing the cascade effect of potential surplus water from upstream areas of the watershed to be used downstream; this is necessary as rain can be very localised and heavy. "In the South Indian upland States of Andhra Pradesh and Karnataka, upstream tanks are usually smaller in size, their purpose being to trap sediment so that water can be supplied to the larger tail-end tanks with greater efficiency. This extraordinary indigenous technology reveals a deep understanding of topography, soil systems, and runoff flow paths in an environment subject to variability and irregularity" (Gunnell *et al.*, 2003; same notion in Krishnaswamy, 1947 cited by Mosse, 2006). In system tanks, the complex organisation of the canals and tanks reflects the will to share the available water (diverted from the river) among all the tanks ([A.]: 318). The whole technique and layout of the tank networks might have been thought out with the purpose of sharing water as equally as possible. The tank system can then be seen as a technique of storing *and* sharing water.

(b) Colonial Period: Centralizing Water Management and Construction of Modern Dams

The development of 'modern' irrigation started in the mid-19th century, during the (British and French) colonial period, with the introduction of new techniques and reforms. The British launched the construction of big dams, the strengthening of diversion dams on rivers and the building of long canals. The first Irrigation Commission (1901-1903) was set up, following a severe famine at the end of the 19th century and as a response to the report of the Famine Commission of 1898. The returns on investment were treated as a major issue of debate, so that only the schemes likely to produce some revenue received immediate attention. The focus was then on 'productive' irrigation (not a 'protective' one) and large irrigation schemes. In British India, from 1878 to 1947, irrigated land was multiplied by 2.4, going from 11.8 million hectares (M ha) to 28.2 M ha; yet 80 per cent of the cultivated area was still at the mercy of the monsoon (Narayanamoorthy *et al.*, 2005: 25-27).

The British did not lose complete interest in the tanks but they barely invested in their maintenance. On the contrary, they enacted reforms that brought important changes in the management of tanks: the colonial government took on the ownership and the maintenance of the tanks; a centralized institution (Public Works Department—PWD) was then created in 1836 to take charge of their maintenance; the tenure taxation system changed, the State became the 'universal landlord' and land revenue became its primary objective, thus increasing the land tax that proved too high for farmers (Vani in Agarwal *et al.,* 1997: 300-301; Mukundan, 1988: 10, 13). However, the colonial administration soon realized that it had neither the human nor financial capacity to maintain the 47,000 tanks of the Madras Presidency and the 39,000 of Mysore State[9]. It required some contribution on the part of villagers, and in 1858 it passed a law—to no effect—whereby village communities were forced to participate in work to maintain and protect the hydraulic infrastructure (Agarwal *et al.,* 1997: 251; 301). Successive laws and amendments have brought no significant changes (Rajagopal *et al.,* 2001: 6). Forced labour, as imposed by the British, was based on collective practices that existed before colonization and were called *kudimaramat*. Today the traditional and customary aspects of *kudimaramat* provoke fierce controversy. According to some authors (Mukundan, 1988; Mosse, 1999, 2003), *kudimaramat*[10] was carried out under certain conditions that are often overlooked; the idea of voluntary participation by villagers to meet the administration's irrigation needs is a complete fabrication on the part of the British: the colonial government needed autonomous village institutions, and therefore, invented this so-called "village tradition". According to others, such as Vaidyanathan (1999), the practice of *kudimaramat* declined with the change in taxation imposed by the British. In Puducherry, the French set up formal tank associations, called "Caisses Communes" whose objective was to maintain irrigation canals (tanks being the responsibility of the colonial government), which they rarely did (*see* Box 8).

Many authors credit the British with the decline in traditional village water management. But for Mosse (2003: 11), the identification of the collapse "is an impossible task involving a seemingly endless journey back in time". Similarly, Subbarayalu summarizes in Coste *et al.* (2007) that "the British did not destroy everything that previously existed, but actually changed what was already under way. It has been an ongoing process of change over the last ten centuries". Nevertheless, we can deduce that the centralization of common resources and the appropriation of hydraulic infrastructures by the State in the 19th century altered the relationship between village communities on the one hand and the relationship between State and villagers on the other hand.

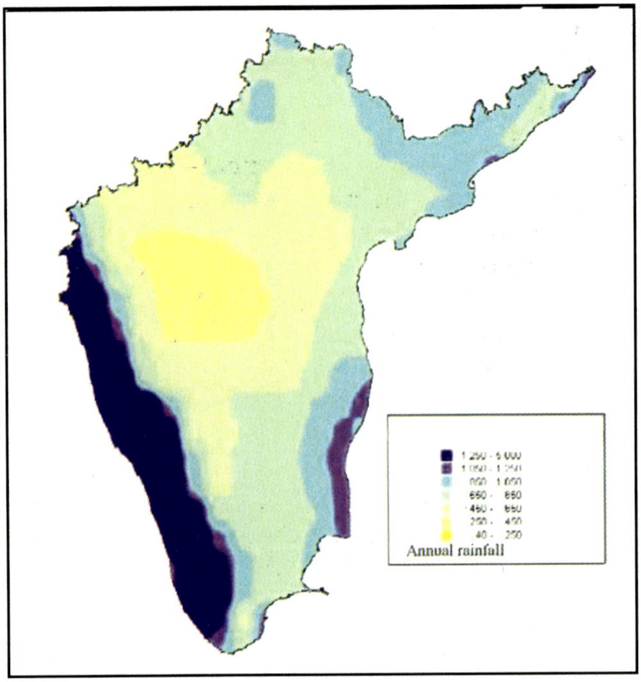

Fig. 1.3: Geographical distribution of the annual mean rainfall in South India

At the time of Independence, the policy for developing large infrastructures, as initiated by the British, was pursued and intensified (see Fig. 1.4 and Fig. 1.5, where we see that the higher densities of irrigated land are today still situated below dams, with canal irrigation). The policy was more a question of encouraging multi-purpose irrigation projects with emphasis on hydro-electricity, with the demand for energy (electricity) being a major concern in developing the irrigation system over this period (Narayanamoorthy *et al.*, 2005: 26-27, 38). Dams have without a doubt helped India to increase its grain production; they store a large volume of water and help regulate the flow of major rivers. Irrigated areas, especially upstream ones, dispose of a theoretically secured water supply which is also available during the dry season. This double (spatial and temporal) improvement provided for by dams, in addition to the associated electricity production potential, explains the interest shown in these structures in the Indian territory land settlement (Durand-Dastès, 1995: 318). These characteristics also justify their name as "temples of modern India" by Pandit Jawaharlal Nehru in the first years after Independence.

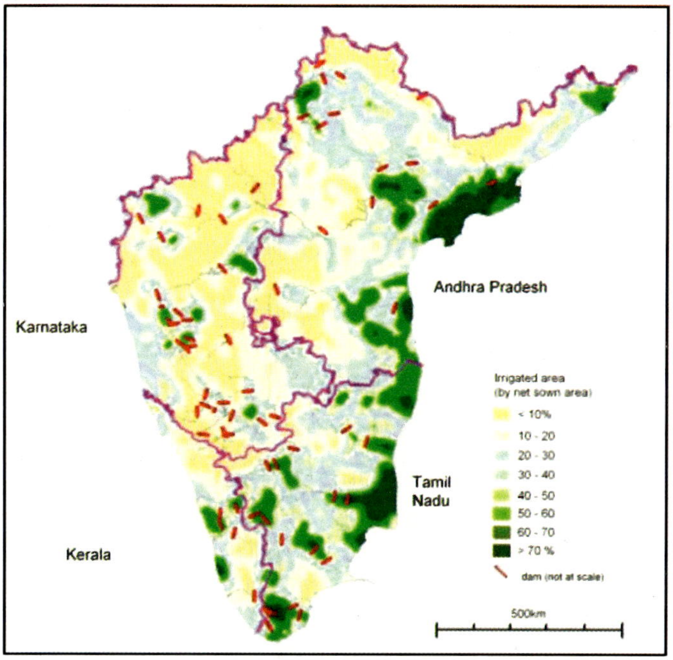

Fig. 1.4: Irrigated Area and Main Dams in South India (1991 Census Data)

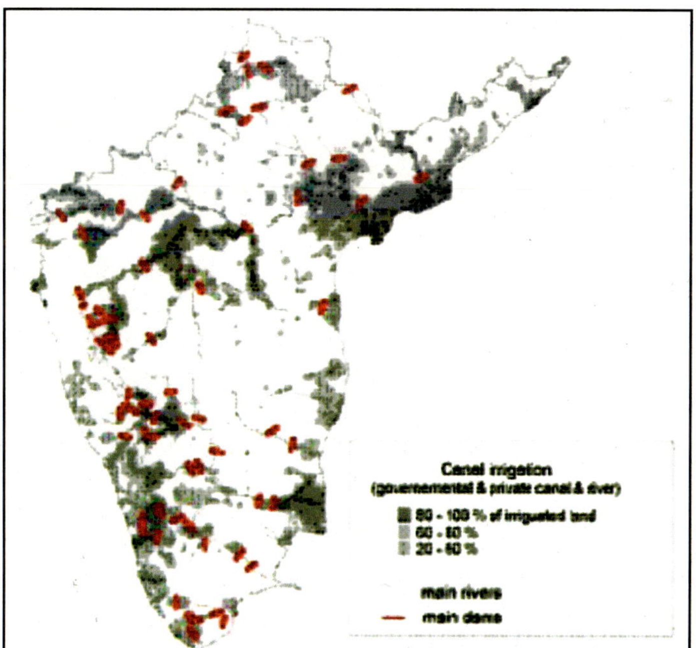

Fig. 1.5: Canal irrigated areas in South India (1991 Census Data)

(c) Green Revolution: Rise of Groundwater Irrigation

The second main step in the modern irrigation development process in India is the Green Revolution which started in the mid-sixties, one of the objectives of which was to increase agricultural production through the use of fertilizers, improved varieties and the development of irrigation. Meanwhile, priority was given to modern techniques, the banking sector was developed in rural areas and trade infrastructures were implemented. Thus, major irrigation projects followed their course. The motorized use of groundwater was encouraged: subsidies for purchasing pumps, the electrification of the countryside, electricity subsidies for agricultural purposes—and even free electricity since the end of the 1980s in some States, such as Tamil Nadu, were incentives that led to the proliferation of wells, of motorized pumps (more effective than human or animal pumping techniques used at the time) and of deep tubewells, which were the result of private initiatives (Fig. 1.6).

Farmers rapidly adopted pumping for their irrigation as it presented many advantages. From a technical point of view, for example, tubewells relieved farmers more or less of climatic constraints and of the constraints of collective irrigation, and in particular regarding the presence of water in the canal for down streamers. Farmers could therefore irrigate when they saw fit, especially during the dry season or before the first rains. This was accompanied by the fact that the tapped water belonged to the owner of the soil who was free to use it as they liked—as long as water and electricity were available. Thanks to this policy, this technique spread very widely over the Indian territory and it was possible to irrigate land previously deprived of water or where little water was available.

An important stage in groundwater irrigation is the introduction in the late 1970s (in Tamil Nadu) of powerful submersible pumps, used alongside tubewells, which allowed access to greater depths (cf. Table 1.2). In Tamil Nadu there are fewer tubewells than dug wells (Fig. 1.8) due to the geological features, since hard rock covers a large part of the State.

During the entire Green Revolution period (mid-sixties to the eighties), tanks continued to be neglected by the public authorities, whereas pumping irrigation flourished. Though both these types of irrigation often exist in the same areas, the rapid development of borewells undoubtedly affected tank irrigation. Here is one of the most retold scenario: only the wealthiest owners have been able to invest in pumps, now using their free access to groundwater to sell it to the poorest and thus maintain their local influence. Besides, these owners are often influential people, who play an important role in taking collective decisions, especially those concerning the management of water tanks. However, since groundwater has freed them of the constraints of collective

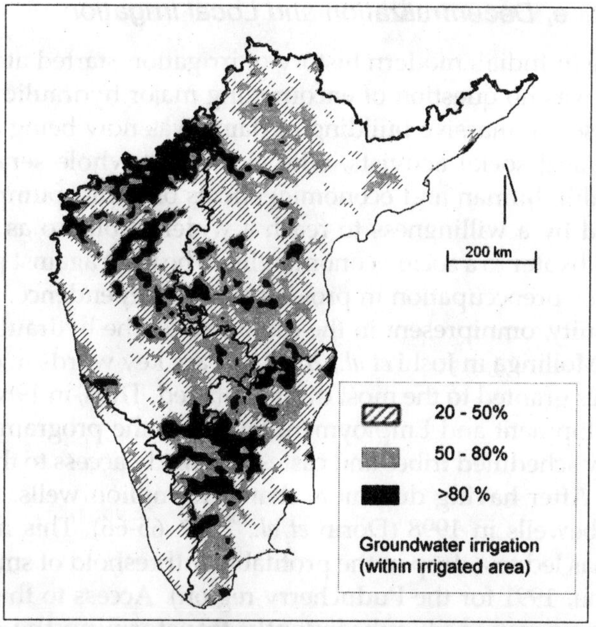

Fig. 1.6: Groundwater irrigation in South India

irrigation and has provided them with local economic power, these peasants have often become disinterested in collective irrigation, resulting in a steep decline in its management (Palanisami *et al.*, 1998; Janakarajan, 1993). Many authors therefore hold the massive proliferation of motorized pumps responsible for the drop in tank management (Agarwal *et al.*, 1997; Palanisami *et al.*, 1998). For others, it is not the prime cause of the decline, but it did serve as a catalyst to spark off a phenomenon that had already begun (Mosse, 2003). In addition to this, other factors have come into play since the sixties, such as the land transfer from upper castes to lower castes and land distribution between castes, which affected land tenure, the quantity of land owned, and castes' control over the means of production such as water and land (Janakarajan, 1993).

Table 1.2: Change in the number of wells and tubewells in Tamil Nadu

	Number of dug wells	*Number of tubewells*
1850[1]	220,000	
1902[1]	442,700	
1960[1]	909,700	1,683
1980[2]	1,607,449	49,104
1993[3]	1,533,839	144,123
2005[4]	1,608,524	251,853

[1] *from Adicéam 1966;* [2] *from Season and Crop Report 1979-80;* [3] *Minor Irrigation Census, 2007;* [4]*from Season and Crop Report 2004-2005*

(d) Social Justice, Decentralization and Local Irrigation

The third phase in India's modern history of irrigation started at the end of the eighties. There was no question of encouraging major hydraulic projects, as in the past, because the massive building of dams was now being challenged by environmental and social activists, who exposed a whole series of harmful (ecological, health, human and economic) effects of dams; pumping irrigation was still backed by a willingness to reach a wider public so as not to restrict access to groundwater to a socio-economic elite. The fight against poverty, which has been a major preoccupation in projects since Independence, and actions in the name of equity, omnipresent in the discourse on the hydraulic policy since the 1970s-80s (Mollinga in Joshi *et al.*, 2000: 74), are key words in explaining the special subsidies granted to the most disadvantaged. Thus, in 1988, the Minister for Rural Development and Employment launched the programme "a Million Wells" whereby scheduled tribes and castes had ready access to the free digging of open wells. After having dug more than 1.15 million wells, the scope was widened to tubewells in 1998 (Dorin *et al.*, 2002: 65-66). This massive use of groundwater has led to a drop in the profitability threshold of small farmsteads (Marius Gnanou, 1991 for the Puducherry region). Access to this resource has therefore been encouraged, subsequently transforming the geography of irrigation, which is no longer limited to collective systems (canals or tanks). Thus, Fig. 1.6 shows that the geography of groundwater use is more diffuse than the other two types of irrigation, except for the areas densely irrigated by canals.

Due to all these measures, groundwater has become the main source of irrigation water in India as in Tamil Nadu (cf. Fig. 1.7). In India, from 1950 to 2005, the number of motors rose from less than 100,000 to almost 19 million (most wells are now fitted with a motor) while the area irrigated by groundwater increased from 6 to 46 million hectares (GoI 2007; NSSO 2007). Yet these measures have also contributed to the overexploitation of water tables whose level has dropped dramatically since the 1990s. Fig. 1.7 shows that during the three years of drought, 2001 to 2003, the only area to have suffered no decrease at all is the area irrigated by tubewells, which on the contrary has continued to expand.

In terms of surface water use, it was agreed to focus on small-scale infrastructures and to enhance the productivity of existing systems by upgrading infrastructures and reforming management processes. It was also decided to increase the capacity of recovering and storing rain water ("rain water harvesting"), by building new structures or improving old ones. Tanks—structures for irrigation and for storing water—fully meet both these requirements. In addition, the State plays on quite a different aspect to get its tank rehabilitation policy accepted: it takes advantage of the school of thought launched back in the 1980s by environmentalists whereby old traditional structures, better adapted to local situations, must be renovated. Various local NGOs (operating on foreign but also local funding) are taking part in the rehabilitation and the renovation of traditional irrigation systems. Yet for any rehabilitation project, Central Government advocates some form of contribution by peasants in creating Water Users Associations (WUA) according to World Bank guidelines, as will be discussed in Chapters 9 and 10. "Our sentiment is that participation is primarily perceived by the State as instrumental, i.e. as a convenient means of increasing the efficacy of its action by transferring to local actors certain project management costs, without fundamentally challenging the usual standards of operation. Decentralization might, therefore, prove to be a convenient way of finding solutions to the State's financial crisis, and to the Administration's legitimacy crisis, rather than a tool for resource management which corresponds to a real development policy."[11] (Landy and Bon, 2005).

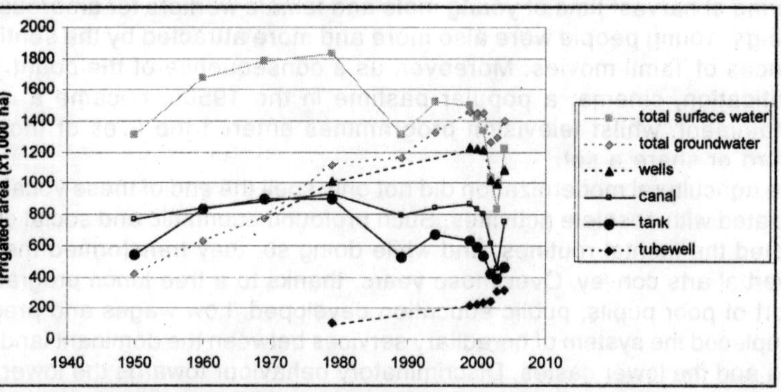

Fig. 1.7: Change in the Irrigated Area According to Type of Irrigation, in Tamil Nadu

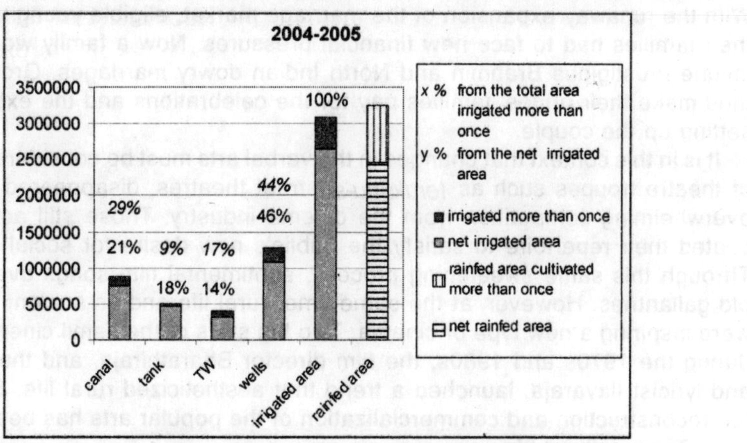

Fig. 1.8: Net and gross irrigated and rainfed agriculture, in Tamil Nadu (from Season and Crop Report 2004-2005)

Box 2: The Green Revolution in Tamil Nadu: Its Social and Cultural Impact by Martine Gestin

During the two decades of the "green revolution", the 1970s and 1980s, the rural population of Tamil Nadu entered a phase of radical modernization which overthrew their oral traditions, world representation and verbal arts. During this period, a large number of work songs lost their original purpose to become mere folklore. Due to the development of public bus transportations all over the State, long family journeys in convoys of zebu drawn carts (*mâṭṭu vaṇṭi*) disappeared. Wagoners no longer had the opportunity to sing alternating songs, one driving behind the other in single file, for their convoy's delight. In many places, the "picottah", a rather perilous irrigation system, was replaced by electric pumps. Since workers no longer had to action the swinging motion of the water-drawing poles, the *ẹrra pāṭṭu-s* irrigation songs especially meant for swinging, became scarce. New farming calendars discouraged the periodic gathering at harvest time of young male and female workers for amorous verbal sparrings. Young people were also more and more attracted by the sentimental romances of Tamil movies. Moreover, as a consequence of the countryside's electrification, cinema, a popular pastime in the 1950s, became a massive entertainment, whilst television programmes entered the lives of those able to afford or share a set.

The agricultural modernization did not only spell the end of these vocal genres associated with obsolete activities. Such profound economic and social changes disrupted the mental routines, and while doing so, they transformed the values that verbal arts convey. Over those years, thanks to a free lunch programme in support of poor pupils, public education developed. Low wages and precarious jobs replaced the system of hereditary services between the dominant landowning castes and the lower castes. Discriminatory behaviour towards the lower castes was reduced, at least in its most visible forms. Money-based calculations and monetary transactions came into use everywhere, including amongst the poorest. With the runaway expansion of the marriage market, eligible young women and their families had to face new financial pressures. Now a family would start to imitate prestigious Brahmin and North Indian dowry marriages. Grooms would thus make their brides' families pay for the celebrations and the extra costs of setting up the couple.

It is in this context that changes in the verbal arts must be considered. Legions of theatre troupes such as *terukūttu-s;* street theatres, disappeared faced with overwhelming competition from the cinema industry. Those still active sugar-coated their repertoire to satisfy the public's new desire for social distinction. Through this same sweetening process, sentimental film songs overthrew the old gallantries. However, at the same time, rural life and its ancient verbal arts were inspiring a new type of cinema. Two big stars of the Tamil cinema industry during the 1970s and 1980s, the film director Bharathiraja, and the composer and lyricist Ilayaraja, launched a trend that aestheticized rural life. Since then, the reconstruction and commercialization of the popular arts has been imposed upon Tamil cultural life.

1.2. Challenges in the Agricultural Water Sector

Among the various challenges in the agricultural water sector, we will first of all present some related to the economic and socio-economic context, then others linked to water management. Here we will not address those related to the quality of water (either due to pollution or to soil salinity in coastal areas or on waterlogged land) or to competition with other sectors (industry, drinking water).

(a) Irrigation Development and Agricultural Production

All political and economic investments made by successive governments since Independence and aimed at developing irrigation and the agricultural sector (hybrid varieties, chemical fertilizer, mechanization, credit, commercialization, electrification of the countryside) have had obvious repercussions in terms of food production. With the efforts realized before[12] and after the Green Revolution, the country achieved food self-sufficiency as of 1973. India increased its cereal production over a fifty-year period (1960 to 2007). Paddy production was multiplied by 2.8, that of wheat was sevenfold (NSSO data[13]). Meanwhile the population grew nearly threefold, with this addition amounting to more than 650 million consumers.

However, this all too positive picture does have a dark side, that is the less satisfactory aspects of agricultural development: in 2001, 300 million Indians lived below the poverty line (Dorin and Landy, 2002: 9); the Green Revolution has failed to expunge the disparity that exists in the development process across regions and cultivators, since it was based mainly on the development of irrigated crops, such as wheat and rice (improved varieties, minimum loan level, market structure), and at the same time left the production of other cereals, which nevertheless consume less water, to stagnate. This agricultural policy has had the effect of accentuating the relative drawback of such crops and of heightening inequalities between farmers with irrigation and those with no access to water. Thus, according to the UNESCO report dated 2003 (cited by Landy *et al.*, 2004) "in India, 69 per cent of people in non-irrigated areas are poor; in irrigated areas this figure falls to 2 per cent." Wheat and paddy yields have not increased any further, and since the 1990s the rise in agricultural production remains steady at 1.7 per cent, that is less than the population's rate of growth (1.8) (Landy, 2006: 24). Another problem highlighted nowadays in the context of climatic changes is the need for research into varieties capable of withstanding hotter temperatures.

In Tamil Nadu, irrigation is relatively dependent upon rainfall, since tanks and wells are mainly filled by rain or by the aquifer that rain recharges (Fig. 1.7). Thus, during the dry period 2001 to 2003, the production of food grain nearly reached that of the 1950s (according to data from Season and Crop Reports), showing how vulnerable irrigated agriculture is.

In view of the increasing demand for water—whether for agricultural, domestic or industrial uses—and for agricultural products due to the population increase, one of the challenges for agriculture is to continue to increase food production and to improve water productivity. The latter can be improved either in irrigated agriculture, on which much research has so far concentrated, or in rainfed agriculture, by increasing the productivity of green water[14] which presents a larger scope (Molden, 2007). Irrigated agriculture could save water either through changes in irrigation techniques or/and practices (such as micro-irrigation or the System of Rice Intensification: SRI) or through changes in the cropping pattern, since paddy and sugarcane, which consume a lot of water, occupy 81 per cent of the irrigated food crop area and 66 per cent of the gross net irrigated area in Tamil Nadu (figures from Season and Crop Report 2001-2002). Water savings in rice and sugarcane cultivation are an issue that needs to be addressed. Technical changes have already been encouraged: SRI that can reduce water consumption by 30 per cent is recommended[15] and micro-irrigation is subsidized.

For sugarcane, which is necessarily cultivated using ground water (GW), drip irrigation is actually a way of reducing the quantity of extracted GW. However, this technique is not suitable for paddy and does not necessarily prove to be a panacea, since it has a reverse effect on GW exploitation, because farmers may increase the irrigated area and with it, the total use of GW which would lead to a more rapid depletion of the aquifer (Kemper 2003: 126). India, and Asia in general, will have to devise a way of producing more rice using less water (Trébuil *et al.*, 2004).

Changing the cropping pattern is therefore an added solution. It is particularly appropriate in places where paddy is cropped during the very hot season as a third crop, using only GW, as can be seen in the coastal area or in the alluvial area around Puducherry: this cropping comes across as an ecological absurdity since evapotranspiration is at its highest. It is as if deep GW was tapped in order for it then to be made to evaporate immediately. In this respect, Puducherry government has an interesting response, granting subsidies during this period for cropping other than paddy in order to use less water. Reducing the rice cropping area is, of course, a difficult political choice, since it implies making important changes at State or national level at least in eating habits (to introduce more diversified food, whereas, on the contrary, the diet tends to depend more on rice than ever before), and in the economic sphere since importations of paddy from other countries with a better paddy yield have to be planned in advance. For Tamil farmers who have access to irrigation, the idea of purchasing paddy, instead of growing it, even if it is not profitable, is not yet conceivable (*see* also Chap. 8). And in the rice field itself, changing over from paddy cultivation to another crop is not an easy task, as it requires changing the nature of the clayey soil. Moreover, rice must not be seen merely as a greedy crop, as it is well adapted to local environmental conditions such as large amounts of rain during the monsoon and submersion irrigation through canals and tanks. Better water management on the scale of a field is not enough: it also has to be conceived within an irrigation network or on a regional scale, which is rarely the case.

Many economists and political experts recommend that agriculture evolve towards a private-public partnership which would provide better technical monitoring of farmers and lead to the development of agro-based industries: the Green Revolution already included peasant farming in commercial agriculture; the aim of the private-public partnership is to modernize the latter and to offer a better response to market demand (Rao, 2005). Such views encourage industrial development as a solution to rural development and particularly conceive agriculture in terms of modern techniques. Thus, the place of traditional techniques, such as seasonal ponds (tanks) in such an economy is threatened. If this collective irrigation system has helped to explain a historically high population density in a semi-arid climate, is it still adapted to current needs regarding agricultural production? How can this heritage from the past, fully steeped in tradition, face the changes taking place in society and current constraints of water management? In what way is this heritage from the past a burden or, on the contrary, a framework offering new potential to local communities and to governments in charge of defining water policies?

(b) Farmers' Socio-Economic Conditions and Social Changes

Farmers in Tamil Nadu, and in Villupuram district in particular, have on average a small land property (most farmers have less than 2 acres, see figures in Chap. 8), so that their capacity to invest is low. The agricultural sector is often portrayed as being non-profitable, since with the rising cost of chemical inputs farmers have fallen into debt; a survey carried out by the NSSO confirms the pathetic conditions of Indian farmers and reveals the very low income of Tamil Nadu farmers (Narayanamoorthy, 2006). Marginal farmers or landless agricultural workers regularly migrate on a seasonal basis to other places in search of work (either in sugarcane producing areas, or in towns as unskilled labourers on building sites), so that pluriactivity is quite common.

Some of the biggest land owners, who seized the opportunity of shifting their activity to other sectors, moved to town. "The space made vacant by them in the agrarian structure is being filled in by sections of the other backward classes and at times even by the scheduled castes" (Rao and Nair, 2003: 3352). Within the space of fifty years, land holding has therefore evolved, with some land being transferred from upper castes to lower castes, with the fragmentation of land holding and an increase in the number of holdings (Janakarajan, 1993). In Alapakkam in Vellore district, over a period of 80 years prior to 2000,

"Brahmin and Mudaliar dominance (in land ownership) has been replaced by (high castes) Pillai and Reddy communities" while lower castes and SCs have also increased their landownership (Sivasubramaniyam, 2002). The impact of such change in landholding is visible in the decline in the number of landowners and the position of authority of upper castes (*ibid.*, Janakarajan, 1993). In our study area, it is not rare to hear that big landowners were from the Reddy or the Brahmin caste, and only one or two families are still in the village today with reduced landownership compared to before; the last big wave of departure to the cities was in the early 1990s.

Another characteristic of the rural area is that the agricultural sector no longer attracts young people who prefer to go to town, especially those who have minimum schooling. This attitude raises questions relative to the socio-economic status (and thus the power of investment), adaptation and decision-making capacities of farmers-to-be.

(c) Management of GW Overexploitation

Many places in India suffer from ecological degradation, mostly due to the Green Revolution which has led to the non-sustainable use of land and water since water logging, groundwater depletion, water pollution, water salinity, soil fertility loss and salinity are the main issues encountered in many parts of the country (*see* for example the articles in the *Indian Journal of Agricultural Economics*, dated July-September 1991 about the Green Revolution; the agro-ecological crisis in the rural district of North Arcot, Tamil Nadu, described in Harris-White *et al.* 2004). Groundwater plays a large part in this ecological degradation since the resource supplies more than 50 per cent of the irrigated area and its use has surged since the 1960s. This GW withdrawal in India is unequalled in the rest of the world: it has increased from 25km^3 in the 1960s to 230 km^3 in the 2,000s, whereas other countries' GW withdrawal ranges from 50 to 110 km^3 maximum (*see* Fig. 10.1 in Shah, 2007). It has led to GW overexploitation in more than 15 per cent of the blocks of the country and in more than half of the blocks in major agricultural States, including Tamil Nadu (GoI, 2007: 7).

In sum, the main factors explaining groundwater overexploitation are: open access to GW to millions of bore owners, highly subsidized energy for pumping (in the case of electric pumps) and even a free supply in Tamil Nadu and Punjab, a very limited regulatory mechanism for the exploitation of the resource, along with technical, agronomic and social advantages already cited, which explain farmers' frenzied quest for GW. The main solutions used so far by the Indian Government to address the GW issue refer to legal, economic, and institutional spheres, the economic one operating through three classes of intervention: direct regulation such as the licensing of new wells; indirect regulation through taxes and subsidies, especially for wells and pumps, a power supply and credit supply; and public investments in productive infrastructures, such as the State tubewell programme, and recharge infrastructures (*see* Shah, 1993). In order to compensate for the scarcity of water India has indeed chosen to give priority to rain water harvesting and GW recharge: the State is still playing the role of water provider through modern techniques (dams, pumps, tubewells) and the GW recharge programme, when it "could play a useful role as a protector of water resources" (Rajagopal *et al.*, 2001: 15). The country is still in resource development mode (Moench, 1992a) or supply-side in its response to GW depletion (Shah, 2007: 28), even if since the 1990s scientists' discourse has prompted a shift to resource management mode, or proactive demand management (Moench 1992a, Shah 2007c, Kumar 2007). Generally speaking, very few measures are aimed at limiting water withdrawal or at collective management of the underground resource. Besides, politicians do not dare take any restrictive measures which are too unpopular.

Yet moving over to a water management approach is one of the recommendations laid out in the report by a group of experts (GoI, 2007: 46-48), since "experience with regulation in some states shows that it is not possible to have significant control over the use of GW through legal provisions because of difficulties in enforcement" (*ibid.*: 47), particularly because of "logistical problems of regulating a large number of small, dispersed users" (Shah, 2007b: 27). However, the way this move needs to be made is subject to controversy. The main points of dissension concern: the legal regime of water ownership and water use; electricity pricing and supply as a form of indirect regulation; management institutions (see below). Some of these points of dissension are due to a fundamental difference in points of view: whether to keep GW separate from surface water and regulate them both separately, or to have an integrated approach to water management. The latter requires closer ties with the scientific community (especially in hydrological, hydro-geological, environmental and ecological sciences) [Narasimhan's criticism (2008) of the report by the group of experts about its simplistic conclusions from an earth-science perspective], educating the community at large about water, which implies circulating data (Moench, 1992b; Narasimhan, 2008), and indeed more reliable data[16].

In this short description of the challenges related to GW management we will not present the solutions already developed by the States (*see* Shah, 1993; Moench, 1992a; Bhatia, 1992) or the various solutions proposed in the literature (*see* for example, Kumar, 2007; Mukherji, 2007; Dubash, 2007; Shah, 2007c; Narendranath *et al.*, 2005; Kemper, 2003; Moench, 1992b), as they make reference to a huge amount of literature and would have to be developed in length, with arguments, lines of reasoning, etc. Let us sum up the points that seem to be of particular interest to us as far as limitations for integrated water management are concerned.

Data Related to GW

Government data do exist, but experts often express some doubt about their reliability. Data are indeed organized on a block basis, not according to a hydrological unit (reflecting the importance of such data for financing development); data collection and their interpretation may be influenced by external factors, since data play an irrefutable role in determining financial allocations for groundwater development; the recharge component is estimated using the 'water-table fluctuation method' which in itself contains much scientific conjecture, due in particular to the small number of measurements per year (only 2 to 4); the key process of lateral inflows and outflows are not captured in the water balance (FAO, 2003: 26-36).

Moreover, data are difficult to access, especially those related to an aquifer other than the upper one. Data concerning deep aquifers are not clearly exploited in reports: thus in Tamil Nadu GW perspective reports for Villupuram district (GOTN, 2002) and the Puducherry report about water resource management (GOP, 1999), the recharge process of deep aquifers is not mentioned; estimates of the gross GW recharge are given without specifying the proportion concerning upper and deep aquifers, the confined and unconfined aquifers; for Villupuram district, only the variations in the upper aquifer water table are disclosed. Yet authors of these reports are aware of the necessity of limiting the utilization of groundwater in order to curb GW depletion, which creates numerous socio-economic problems: "Regulation is the need of the hour", "steps have to be taken" (GOTN, 2002: 64, 65), "utmost care is necessary" (GOP, 1999). The lack of transparency about such data is certainly linked to the political sensitivity of the subject. Yet, as we will see in Chap. 7, the absence of such information about deep aquifers—reinforced by the complexity of the hydro geological functioning—leads to various interpretations of the phenomena, with some villagers manipulating local knowledge in their own self-interests.

Since groundwater recharge programmes are implemented in many places throughout the country, it is not rare in seminars to see NGOs or even researchers demonstrating the variations in the water table over a short period and thus alluding to the success of the programme: yet a short period for measurements does not give any scientific proof. The political impact of such data on funding development programmes distorts the objective analysis of the decline in the water table.

Thus the challenge is to produce reliable data—knowing that some data analysis requires years of data, which are not always available, in order to be reliable—, to provide access to these data with some objective analysis, and to clearly distinguish the various aquifers according to their recharge dynamics.

Legal Aspects Regarding GW Ownership and Use

Let us first discuss the ambiguity of and the controversy about the GW ownership regime and then speak about its regulation by law.

"Under the Easement Act of 1882, groundwater is considered an easement connected to land ownership. Ownership of groundwater thus falls to the landowner, who is free to extract and use it as she/he deems fit" (GoI, 2005: 228). In India, ambiguity surrounding the rights that govern groundwater ownership blurs the distinction between the water ownership regime (associated with the landed property regime) and the user rights (not clearly defined, thus considered unlimited). Many associations are pushing for a change in the GW property regime and for water to be recognized as a common resource managed by communities in order to compensate for privatization of water throughout the country which allows all landowners, including the State, to overexploit the resource (Shiva, 2002). The ambiguity concerning ownership rights explains the confusion made here between the nature of the resource (water is in any case common property) and the ownership regime (here implicitly a common property regime). The Commission's mid-term report on the 10th Plan suggests that groundwater ownership should be the subject of public debate, with two solutions being put forward: either a landowner is permitted to use only the quantum required for his use, while the balance available should be a community resource, or the shallow part of the aquifer is associated with the land while the deep aquifer should be a community resource (GoI, 2005: 229). By contrast, the report drafted by experts in 2007 first of all clearly separates the property rights regime from water management regulation, with the State having to intervene in the latter, and secondly, suggests that "no change in the basic legal regime relating to GW seems necessary". Yet "extraction rights can and should be curbed by the State if the use of GW is considered 'excessive'" (GoI, 2007: 41). The notion of "excessiveness" is nevertheless vague and does not help to define water user rights. Furthermore, as Narasimhan (2008) points out, maintaining a situation where the legal regime is based on the Easement Act, where GW is treated as an entity totally distinct from surface water and as immutable property, is a way of refusing to recognize modern scientific knowledge, which associates surface water with ground water.

Some other legal aspects relate to the direct regulation laws on GW use, passed by each State, since water is a State matter. The Central State merely presents bills, as it did in 1970, 1992, 1996 and 2005. However, very few States have enacted their own law. It was the case of Tamil Nadu in 2003, yet with still no implementation decree. Nevertheless, regulating water extraction seems indispensable, since "the GW situation in the State is among the worst in the country" (GoI, 2007: 25). By law, new wells have to be registered and in areas identified as "critical", a water user permit is required, specifying the location of the well, the number of pumping hours and the period of use.[17] Establishing new borewells or excessive deepening of existing wells in areas classed as "overexploited" is not to be permitted. To be effective, this regulation needs to be monitored by an inspection agency. Yet this is not the role of the Central Ground Water Authority created in 1997, whose mission is to declare which areas are to be protected and to run educational programmes on harvesting rain water (GoI, 2005: 229).

As we have already pointed out, the approach using purely regulatory control has been heavily criticized, as it has so far had little real impact. In Tamil Nadu, for example, a minimum distance between wells is required for those seeking a government loan, though as most farmers have recourse to non-institutional credit, these rules are not respected (Palanisami, 2002: 9). Andhra Pradesh passed a law in 2002 regulating groundwater exploitation, but once the dry season arrives, the law is no longer applied (Narendranath *et al.*, 2005: 5565). "Even where legislation relating to groundwater has been enacted, it has remained largely an exercise on paper only" (GoI, 2005: 228). Such top-down regulatory approaches would hardly "be effective with numerous tiny users in rural areas and very high heterogeneity in the resource conditions" (Kumar, 2007: 289).

Water Rights and Water Markets

Water rights is a concept which is rather ambiguous as it refers either to property rights (which are composed of a set of rights including the right of access, of use, to manage, to exclude, to alienate water according to Schlager *et al.*, 1992), or to a more specific notion, the share of water which is generally a quantified right of use. Water rights can be either individual or collective.

Currently in India, groundwater rights are linked to land ownership, they are unlimited and unregulated. The extracted water is thus perceived as an individual property by bore owners, who tap GW without being aware of the impact on adjacent wells[18]. These features of GW rights have led to the emergence of water markets: in the case of rural-rural water markets, well owners sell water extracted from their well to other farmers for cash payment or a share of the crop. These informal markets are widespread in India and supply 30 per cent to 60 per cent of the GW irrigated area according to estimates (Kumar, 2007: 63-64). Dubash (2002) has shown, through two in-depth case studies, that water markets are not uniform and that their 'thickness' in terms of water exchange depends on environmental conditions, local history, the social and socio-economic context and thus cannot be induced simply by policy intervention—as opposed to neoclassical economics. As far as the problem of depletion is concerned, they may exacerbate it by broadening access to GW. "In the absence of formal property rights over GW, water markets do nothing to address the problem of depletion" (p. 248).

Nevertheless, water markets do have many advantages[19] (*see* Mukherji, 2007), including providing access to GW irrigation to farmers who cannot afford their own well and pumping equipment (Shah, 1993). Interestingly enough, this advantage is also present in the local discourse in the area studied, because Tamil farmers do not say that GW is sold or bought, but that well owners "give" water to help others who do not otherwise have access to GW. The price of this "help" is nonetheless quite expensive (often one-third of crop production for paddy). The implications of water markets on the equity issue are thus one of the notions currently under debate among researchers. Some see these markets as a solution to supply irrigation water to a large part of the population and to render access to water fairer. Thus they advocate their extension, since competitive water markets should lead to more effective use of water and more equitable outcomes (Shah, 1993). Others point out that water markets are a source of power and accumulation, of more dependence on the upper classes for those lacking resources and marginal farmers, especially when the water markets shrink due to some resource availability problem (Janakarajan and Moench, 2006; Prakash, 2005). This shrink in water markets consequent to the reduction of the electricity supply and power subsidy has been confirmed in Uttar Pradesh as well as in Gujarat with the implementation in 2004 of the Jyotirgram scheme; with marginal and landless farmers being the ones who have to bear the burden of this limitation (Shah *et al.*, 2008, 2009). Kumar (2007: 288 and following pages) proposes the promotion of water markets with tradable water rights at community level, that is to allocate individual water rights even to landless people, so that big land owners would have to buy water if they want to irrigate their land fully. According to him, such water markets would encourage the economically effective use of water, and would also integrate concerns of equity and sustainability.

The allocation of water rights often appears as one of the steps in water sharing. Water rights are indeed one of the issues present in most of the water reform proposals over the world. Two approaches to this concept exists (Molle, 2004)—the concept being used here in the restrictive sense since "water rights remain an abstraction if they are not defined in terms of quantity, timing and even quality"(p. 211): (i) the establishment of formal rights by prescription through a bureaucratic process; but only a few conclusive experiences are available, including those in developing countries such as Chile, Mexico, South Africa and Tanzania (p. 210) and (ii) flexible allocation rules designed through a gradual and continuous process of negotiation (the terms right and allotment are used to distinguish these two types of water rights). Both types face many difficulties in their implementation, including the necessity to address fluctuation in supply and curtail water rights accordingly. Moreover, "bureaucracies often oppose such processes and resist the redistribution of roles perceived as a threat to their power, and therefore, tend to favour a definition of rights established under the auspices of the administration" (p. 213).

Regulation Through Electricity Supply and Price

As most pumps are electrical (Shah *et al.*, 2004, Fig. 6), a way of indirectly regulating water use[20] is, therefore, to act on electricity, through its supply and its price. The Electricity Board does indeed have a great role to play in groundwater demand management, though the Water Secretary can do little to manage the water demand directly (Shah, 2007c). Tamil Nadu opted for free electricity in 1991 and rationing schedules (10-hour supply of three-phase electricity per 24 hour cycle among which 6 hours at day-time in 2008). It is commonly recognized that free electricity encourages ineffective water use and even water wastage. The impact of the price of electricity on the effectiveness of water use and on the sustainable patterns of GW development is among the most controversial points of discussion related to state regulation of GW use. On the one hand, Tamil Nadu is not really concerned by this debate since politicians do not dare to modify farmers' free electricity privileges[21], as this would be an unpopular measure that could lead to losing elections; yet on the other hand, it is a recognized fact by many authors that this free power generates a waste of water, and it has also been demonstrated by Janakarajan and Moench (2006) that most existing power subsidies benefited wealthy farmers: "This is a clear case where policy reform is required" (*ibid.*).

In States where electricity is highly subsidized (but not free), some authors are all for the solution of making electricity more expensive while ensuring the quality of the supply, which would favour a more effective use of water (Kumar, 2002), whereas others have come to the conclusion that a policy based solely on the price of energy will not have the expected result: rationing hours of electricity supply, fixing differentiated tariffs, improving the effectiveness of pumps and working on public awareness of the effects of overexploiting groundwater would have a greater impact (Narayanamoorthy *et al.*, 2005: 242). Narendranath *et al.* (2004: 5563), farmers from Andhra Pradesh, explain why farmers are reluctant to pay electricity: after spending so much on sinking a borewell and installing a pump set, farmers cannot sustain another financial burden, especially when compared to canal irrigation for which the government covers the capital cost. Moreover, pump set farming is very risky since the duration of pump set work is greatly reduced, because of the water decline and of motors burning out due to a poor electricity supply. Some economists suggest that rich farmers should pay for the power consumed, but those advocating the opposing situation argue that such a measure would concern very few farmers and thus would not be effective (in Chittoor district in Andhra Pradesh, for example, only 10 per cent of farmers would be concerned according to Narendranath *et al.* 2004: 5565). In their proposals set out in the article entitled "To free or not to free power", these authors only put forward three conclusions regarding electricity, suggesting that electricity should remain highly subsidized and be seen as an incentive to help with productive investment. Shah *et al.* (2004) propose an 'intelligent' management of the electricity supply, taking into account agronomic conditions and crop choices, water requirement periods for which supply should be guaranteed, while the total annual electricity supply should be reduced. However, as evoked above, any reduction in the power supply will be borne by marginal farmers and water buyers. In Uttar Pradesh, "the energy-squeeze is inducing small-holders to adapt/respond in myriad ways, from shifting to water-saving crops, to water and energy-saving irrigation technologies (...) [but] it is also gradually driving small-holders out of irrigation, and increasingly, from farming itself "(Shah *et al.*, 2009). As the cases from Punjab and Uttar Pradesh demonstrate, restricting the electricity supply and subsidy inevitably comes with some unfair social and socio-economic consequences that politicians have to face.

Ground Water Management Institutions

Mitigating the issue of over-exploitation of groundwater cannot be achieved by applying only one measure: a combination of political, legal, economic, technical and institutional changes is a need. Among the institutional changes, groundwater management by user groups and collective action is an alternative solution proposed since the 1990s by scholars (*see* for example Moench, 1992a). The odd scheme or local initiative has been implemented or has emerged at village level, such as government tube wells, *pani panchayat* created in 1972 (Narayanamoorthy, 2005), or tube well cooperatives, but these actions represent sporadic efforts with too small an impact to Shah's liking (1997); in most places GW is still managed individually.

Developing institutions to better manage the boom is complicated because of the physical properties of the resource and their interaction with social systems (Giordano *et al.*, 2007: 4). Due to the invisibility of the resource, its complex hydrogeological functioning making it difficult to measure, its geographical heterogeneity affecting its availability, and also to variations "in the socio-economic and political strengths, weaknesses and desires of the societies involved", any GW management solution has to be a local solution: "groundwater management approaches that work in one country or region may not do so in another" (*ibid.*). "GW through user-groups faces a variety of challenges from the perspective of common property resource management experiences", with the main issues to be addressed being: delimitation of the user group and of the resource boundary, resource availability and user information, control of free riders (i.e. those who consume more than their fair share of the public resource, who benefit from the collective resource due to the non-respect of the rules) and management scale (Moench, 1992a). For community-based local GW management, Kumar (2007: 305 etc.) therefore proposes including three levels in the hierarchy: (i) village institutions, (ii) watershed institutions (that would particularly help to avoid some villages from acting as free riders and to settle conflicts between villages), and (iii) management institutions at the aquifer level, to address the depletion problem.

As Kemper (2003: 139) explains as a general feature, but which is really in keeping with the Indian case, such collective action implies that groundwater users have to be "recognized as true stakeholders entitled to information about the resource on which they are so dependent. For many agencies this implies a significant shift from being centralized agencies that keep the information about water availability to themselves and take decisions without the participation of other stakeholders. (...) GW users also need a voice in decision-making", a point of view shared by Molle (2004: 212) about "allotments [which] require a real political commitment to supporting a process of empowerment of users in the decision making of allocation, at both the scheme and the basin level". As a realist, Kemper concludes: "given the alarming rate at which GW aquifers are being overdrafted and polluted in so many regions, the question is whether there is time for water users"; the various processes may take decades, and in many cases, it may be too late for this process (*ibid.*: 141). Some scholars, therefore, propose setting up user committees at electricity transformer level (Narendranath *et al.*, 2005).

Nevertheless, an increasing number of scholars suggest that groundwater user-groups be implemented (Dubash, 2007; Kumar, 2007), since "the key will be in getting the local farm community to 'own' the problem (rather than forcing a solution on it)" (Narendranath *et al.*, 2005). Even the expert group recognizes that "there is a need to rely less on control mechanisms and more on decentralization and cooperative action. The State Act must oblige the State Government to involve the *panchayati raj* institutions and facilitate the creation and effective functioning of water users groups" (GoI, 2007: 41). However, "developing collective action institutions for GW requires additional care and is best undertaken through a pilot testing approach" (p. 45). Nevertheless, organizing villagers into water user-groups and involving them in the management of local resources must be seen as part of the water reforms: participatory management is indeed an integral part of the new governance model.

(d) Reforms in the Water Sector

There have been significant water sector reforms in India though they differ from one State to another, due to the degree of individual State application of the guidelines provided by the Central Government through Bills (first National Water Policy in 1987, last one in 2002; GW model Bill first introduced in 1970, last Bill in 2005). "The process of water law reform is now picking up for a variety of reasons which include a strong policy push from international aid agencies" (Cullet, 2006). By analysing water law reforms on an Indian scale, Cullet describes three main types of legal intervention—the introduction of independent water regulatory authorities, the introduction of legislation and schemes to encourage the participation of users in water service delivery, and lastly the introduction of GW laws to dispense with the principles of control over and access to water linked to land ownership.

However, as far as Tamil Nadu is concerned, the Tamil Nadu GW Act (2003) which in particular provides for the establishment of the Tamil Nadu GW Authority, has no implementation decree yet, and very few River Basin Boards have been formed since 2001. Thus, until now Tamil Nadu has mainly been involved in legislation and schemes concerning user participation: firstly, through the Tamil Nadu Farmers Management Irrigation System Act (2000), which only concerns surface water[22] and provides a legal framework for water user associations (WUAs); and secondly, through the "democratization of water management" experiment for drinking water, launched in 2004 in 472 villages (*see* TWAD website, Tamil Nadu Water Supply and Drainage being the government agency with the mandate to supply water to the entire State, except for Chennai city). The latter focuses on consultation, information and public awareness, a somewhat exceptional, novel and innovative process for it to be mentioned. It is, therefore, fully in keeping with new governance perspectives: forums are organized with communities during which discussions are held on the responsibility villagers may have in the water crisis and in any eventual solutions.

However, the State is progressively pulling out from any funding, the principle underlying such a policy being that water is considered to be economic goods (Cullet, 2006). The aim is for drinking water to be managed by local authorities, with the risk of not having sufficient financial means and of the State relinquishing its responsibility that all this entails. The same goes for irrigation, operation and maintenance costs which should be covered by users, this goal being part of the national water law. Some WUAs have thus been created by the government agency, the Water Resources Organisation, in some tank irrigation systems. Moreover, any tank rehabilitation programme or watershed projects, implemented through a NGO or a government Agency or Ministry, must also involve villagers organized into user associations.

Involving users is the basis of decentralization and participation principles, one of the spearheads of water sector reforms throughout the world. "This is meant to provide a framework for decentralizing decision-making to the lowest level and to allow 'beneficiaries and other stakeholders' to be involved from the project planning stage. (...) the state is thus called upon to change its role from that of a service provider to that of a regulator. (...) In principle, participation is conceived as an umbrella term that covers participation from policy planning and project design to the management of water infrastructures. In practice, the focus is on participation at the tail end of the process. In fact, the word participation is some sort of a misnomer. On the one hand, what is envisaged is not so much the possibility for farmers and users to participate in taking decisions affecting them but the blanket imposition of a new system of local water use and control scheme based on commercial principles even where there may be successful systems of water governance already in place. On the other hand, the participation which is envisaged at the local level is not the participation of everyone using water. Water user associations schemes that have been set up in recent years generally provide that their members are land owners and land occupiers. The focus on land ownership and occupation as a basis for governing the use and control of water is likely to reinforce inequalities in access to water between people who have access to land and all others" (Cullet, 2006). Besides, traditional irrigation institutions face difficulties in managing tanks and tank irrigation, due to certain compelling socio-economic, technological and institutional factors (Janakarajan, 1993). "There has been a huge accumulation of neglected repairs over a long period which has paralyzed irrigation management in many systems". How can an Act solve this age-old problem of neglected maintenance? (Janakarajan, 2003).

Summary

In this brief description we have seen how irrigation in India is a combination of tradition and modernity, whether as far as techniques or institutions and laws are concerned. Moreover, it is characterised by a spectacular upsurge in groundwater irrigation, which has emerged bringing with it a negative socio-economic and environmental impact. However, the State has intervened to a disproportionate degree in surface water and GW management regulations. Users of the latter indeed enjoy the freedom to use it, which partly explains the overexploitation of the resource. Conversely, tank users face greater constraints due to the fact that tanks and canal networks are government structures, water is a State resource, and irrigation is collectively owned. Even if the country is engaged in decentralization and participatory management reforms, not all the necessary means are provided for new formal water users associations to fully participate in the decision-making process concerning water management at village level or even at irrigation network level. These participatory management reforms may be interpreted as a way of concealing the fact that, of the two main economic trends (water as a market resource *versus* water as a resource collectively owned and managed), the former is gradually imposing its principles (water as an economic good is, for example, one of the principles guiding the reform process, Cullet, 2006).

Changes in water management are crucial for mitigating the agricultural water crisis; changes that should take place from plot level to irrigation network or even basin level. This issue is particularly at stake here, as there will be a greater demand for water (due to population growth and the consequent challenge regarding farm production), competition with other sectors will also become tougher and climate change may reduce the availability of water and alter the climatic pattern.

NOTES

1. See Agarwal *et al.*, 1997 for a description of many of these techniques throughout India.
2. Adicéam (1966), often used here as a reference, is referred to as [A.] in this chapter.
3. The term "tank" comes from the Portuguese term "tanque", water reservoir ([A.]: 301).
4. The translation into English of this part of the text was done by Venkatasubramanyam, IFP.
5. Tanks also exist in the rest of the country, but they are fewer. However, there are quite a large number of them in Sri Lanka.
6. A revenue village is the limit of a village recognized by the administration for land records and taxation. It can be made up of a village, generally divided into two zones in Tamil Nadu, the *ur* (settlement area for castes) and the colony (settlement area for outcasts), it may be made up of a village and hamlets, or of hamlet-village, i.e. village from a social point of view, but included within a revenue village.
7. See maps in Vaidyanathan, 2001 for South India and that of Von Oppen and Rao, 1980 for the whole of India, both on a district scale. Fig. 1.2 represents the spatial distribution of tank irrigation at a 5 km cluster level, based on village data from the 1991 census, does not really give the distribution of tanks: it underestimates the importance of tanks in the landscape in areas where groundwater is massively used, and it overestimates it in areas where rainfed agriculture is dominant such as in Karnataka and the western part of Andhra Pradesh (*see* Fig. 1.4). See Guilmoto *et al.* (2004) for the mapping and GIS methodology used in maps using census data.

8. Wet rice is rice which grows in plane fields which are steeped in water, as opposed to dry or rainfed rice which is grown in any field, and relies only on rain for its watering. Rainfed rice is however very rare in Tamil Nadu.
9. On the whole, Mysore State corresponds to the current Karnataka State. However, its western coast belongs to the Madras Presidency which itself encompasses most of the current States of Tamil Nadu and Andhra Pradesh.
10. Adicéam (1966: 176) suggests that the *kudimarat* was created due to the lack of land availability for workers. Indeed, in villages where there was not enough funding or no rich landowner, the village assembly granted common land to those who could repair the tank. When, in due course, common cultivable land became scarce, cultivators had to share the cost of tank maintenance, that is *kudimarat* [A: 176].
11. Our translation from French.
12. The agricultural progress made during the years leading up to the Green Revolution is noteworthy and often underestimated: irrigated surfaces grew by 26 per cent, the use of fertilizer by 625 per cent and agricultural production increased considerably (Dorin and Landy, 2002: 55).
13. For 2007, these data have been updated with the third advance estimates for crop production from Directorate of Economics and Statistics.
14. Green water refers to the soil moisture generated from rainfall that infiltrates the soil and is available for uptake by plants. It constitutes the main water resource in rainfed agriculture (Molden, 2007: 43).
15. Prasad *et al.* (2005) trace back over the history of how this practice spread, as well as the initial resistance it met on the part of Tamil scientists in 2002, whereas on the contrary NGOs favoured this practice. It has only been recommended by the Department of Agriculture in Tamil Nadu and in the Union Territory of Puducherry since 2004.
16. The calculation error in the assessment of utilizable water resources on a nationwide scale as pointed out by Garg and Hassan (2007) particularly concerns the inclusion of the interaction of surface and groundwater in the form of regenerated groundwater in one institution's estimates, whereas another institution took these figures and added the return flows to the surface water: the regenerated GW was, therefore, integrated in the calculation twice, leading to an overestimation of the utilizable water resources in various studies, ranging from 66 per cent to 88 per cent.
17. http://wrmin.nic.in/cgwb_modelbill.pdf.
18. See the striking examples in Janakarajan *et al.*, 2002. See also, regarding competition in the industrial-agricultural sectors, the case of Coca Cola which was referred to the High Court of Kerala on the issue of excessive exploitation of GW (GoI, 2007: 17), with the plant being accused of drying up and polluting neighbouring wells (Bijoy, 2006). Sugarcane factories

are also said to be the main actor behind the reversal of the hydraulic gradient and the 36 m drop over a period of two decades in Puducherry (GOP, 1999: 7).

19. Positive and negative impacts of water markets are also discussed in Chap. 6 of this volume.
20. Other types of indirect influence on water use have been seen above: change in the cropping pattern, innovation in irrigation practices (such as SRI), and subsidies for modern irrigation facilities. The latter are more or less developed and effective, with Puducherry district, for example, being well equipped in underground pipes but not in drip irrigation facilities.
21. The shift to the flat-rate tariff and then to free electricity was an indulgence in populist politics to cultivate and seize an important and powerful vote bank (Dubash, 2007). "A brief attempt at making the farmers pay at least partly for the power met with an electoral rebuff for the ruling party which then quickly reversed its stand" (Narendranath *et al.*, 2005: 5566).
22. This Act is similar to that of Andhra Pradesh. The main difference between these two States with regard to water management is that water is free of charge in Tamil Nadu, either for surface water—in large irrigation systems Telugu farmers have to pay a water charge, see for example, Gulati *et al.*, 2005 for discussions about the reforms in large canal irrigation—or for GW since electricity is free of charge in Tamil Nadu.

REFERENCES

Adicéam, E. 1966. *La géographie de l'irrigation au Tamil Nadu*. Paris: Ecole Française d'Extrême Orient.

Agarwal, A., and S. Narain. Editors. 1997. *Dying wisdom. Rise, fall and potential of India's traditional harvesting systems.* New Delhi: Centre for Science and Environment (State of India's Environment, 4, a citizens' report).

Bhatia, B. 1992. Lush fields and parched throats. Political economy of groundwater in Gujarat. *Economic and Political Weekly* 27: A142-169.

Bijoy, C. R. 2006. Kerala's Plachimada Struggle. *Economic and Political Weekly* 41: 4332-4339.

Bon, E., and F. Landy. 2005. "L'irrigation en Inde rurale à l'épreuve de la décentralisation "in *Développement durable et territoires [on line]*.

Coste, N., and N. Ploumpidis. 2007. "Bittersweet water. Irrigation practices and modern challenges in South India. Trilingual DVD (Eaux douces, Eaux amères-Thaneer: inippum kacappum). Co-authors: K. Anupama, O. Aubriot, Y. Gunnell," Puducherry: French Institute of Pondicherry (Publications Hors Série 5).

Cullet, P. 2006. Water law reforms. Analysis of recent developments. *Journal of the Indian Law Institute* 48 (2): 206-231.

Dorin, B., and F. Landy. 2002. *Agriculture et alimentation de l'Inde. Les vertes années* (1947-2001). Paris: INRA Editions.

Dubash, N. K. 2002. *Tubewell capitalism: groundwater development and agrarian change in Gujarat.* Delhi: Oxford University Press.

— —. 2007. The electricity-groundwater conundrum: case for a political solution to a political problem. *Economic and Political Weekly* 42: 45-55.

Durand-Dastès, F. 1995. "Le monde indien," in *Afrique du Nord, Moyen Orient, Monde indien.* Edited by G. Mutin and F. Durand-Dastès, pp. 245-480. Paris: Belin, Reclus (Géographie Universelle, sous la direction de Roger Brunet).

FAO, 2003. "Groundwater Management—The Search for Practical Approaches," in *Water reports* 25. Edited by FAO. Rome, 43 p.: http://www.fao.org/DOCREP/005/Y4502E/Y4502E00.HTM.

Garg, N. K., and Q. Hassan. 2007. Alarming scarcity of water in India. *Current Science* 93: 932-941.

Giordano, M., and K. G. Villholth. 2007. "The agricultural groundwater revolution: setting the stage," in *Community-based water law and water resource management reform in developing countries.* Edited by B. van Koppen, M. Giordano, and J. Butterworth, pp. 1-4. Oxford: CABI (Comprehensive assessment of water management in Agriculture series 3).

Government of India. 2005. *Mid-term Appraisal of 10th Five Year Plan (2002-2007).* Planning Commission (http://planningcommission.nic.in/midterm/midtermapp.html).

— —. 2007. *Ground water management and ownership (report of the expert group).* Planning Commission.

Government of Pondicherry. 1999. *Water resources and their management, Pondicherry Region (water for life, save water, save Pondicherry).* Department of Agriculture.

Government of Tamil Nadu. 2002. *Groundwater perspectives. A profile of Villupuram district,* Tamil Nadu. State Ground and Surface Water Resources Data Centre, Water Resources Organisation.

Guilmoto, C. Z., S. Oliveau, V. Chasles, R. Delage, and S. Vella. Editors. 2004. *Mapping out Social Change in South India. A Geographic Information System and its applications.* Pondicherry: French Insitute of Pondicherry.

Gulati, A., M.D. Ruth, and K. V. Raju. 2005. *Institutional Reforms in Indian Irrigation.* New Delhi: Sage Publications and IFPRI.

Gunnell, Y., and K. Anupama. 2003. Past and present status of runoff harvesting systems in dryland peninsular India: a critical review. *Ambio* 32: 320-324.

Harris-White, B., and S. Janakarajan. 2004. *Rural India Facing the 21st Century: Essays on Long Term Village Change and Recent Development Policy.* London: Anthem Press.

Janakarajan, S. 1993. In search of tanks: some hidden facts. *Economic and Political Weekly* 28: A53-A60 (Review of Agriculture).

——. 1994. *Trading in groundwater: a source of power and accumulation*. Madras Institute of Development Studies (Working Paper No. 123).

Janakarajan, S., and M. Moench. 2006. Are Wells a Potential Threat to Farmers' Well-being? *Economic and Political Weekly* 41: 3977-3987.

Joshi, L. K., and R. Hooja. 2000. *Participation irrigation management. Paradigm for the 21st century*. New Delhi: Rawat Publications.

Kemper, K. E. 2003. "Rethinking groundwater management," in *Rethinking water management. Innovative approaches to comptemporary issues*. Edited by C. M. Figueres, C. Tortajada, and J. Rockström, pp. 120-143. London: Earthscan Publications.

Krishnaswamy, S. Y. 1947. *Rural Problems in Madras*. Madras: Superintendent Government Press.

Kumar, M. D. 2007. *Groundwater Management in India. Physical, Institutional and Policy Alternatives*. New Delhi: Sage Publications.

Landy, F. 2006. Décentralisation et gestion participative en Inde. Introduction. *BAGF (Bulletin de l'Association de Géographes Français)* 83: 145-149.

Landy, F., and E. Bon. 2004. Irrigation in rural India and the decentralization test. *Poverty and Environment Times, UNEP, March 2004*. (http://www.grida.no/publications/et/ep2/page/2494.aspx).

——. 2005. "Irrigation in rural India and the decentralization test," UNEP, GRID Apendum (www.grida.no/publications/et/ep2/page/2494.aspx).

Ludden, D. 1989. *Peasant history in South India*. Delhi: Oxford University Press.

Moench, M. 1992a. Chasing the Watertable. Equity and sustainability in groundwater management. *Economic and Political Weekly* 27 (51-52): A171-A177.

——. 1992b. Drawing down the Buffer. Science and Politics of Ground Water Management in India. *Economic and Political Weekly* 27: A7-A13.

Molden, D. Editor. 2007. *Water for Food, Water for Life: A Comprehensive Assessment of Water Management*. London and Colombo: Earthscan and IWMI.

Molle, F. 2004. Defining water rights: by prescription or negociation? *Water Policy* 6: 207-227.

Mosse, D. 1997. The Symbolic Making of a Common Property Resource: History, Ecology and Locality in a Tank-irrigated Landscape in South India. *Development and Change* 28: 467-504.

——. 1999. Colonial and contemporary ideologies of 'community management': the case of tank irrigation development in South India. *Modern Asian Studies* 33: 303-338.

——. 2003. *The rule of water. Statecraft, Ecology and Collective Action in South India*. New Delhi: Oxford University Press.

——. 2006. Rule and representation: transformations in the governance of the water commons in British South India. *The Journal of Asian Studies* 65: 61-90.

Mukherji, A. 2007. Implications of alternative institutional arrangements in groundwater sharing. Evidence from West Bengal. *Economic and Political Weekly* 42: 2543-2551.

Mukundan, T. M. 1988. The very systems of south India. *PPST Bulletin*: 37 p.

Nandi, R. N. 2000. *State Formation, Agrarian Growth and Social Change in Feudal South India A.D.* 600-1200. Delhi: Manohar.

Narasimhan, T. N. 2008. Groundwater management and ownership. *Economic and Political Weekly* 43: 21-27.

Narayanamoorthy, A. 2006. State of India's farmers. *Economic and Political Weekly* 41:471-473.

Narayanamoorthy, A., and R. S. Deshpande. 2005. *Where water seeps! Towards a new phase in India's Irrigation Reforms*. New Delhi: Academic Foundation.

Narendranath, G., S. Uma, and K. R. Reddy. 2005. To free or not to free power. Understanding the context of free power to agriculture. *Economic and Political Weekly* 40: 5561-5570.

Palanisami, K. 2002. "Techno-Economic Feasibility of Groundwater Over-Exploitation in Tamil Nadu." Annual Partners' meet 2002 of IWMI-TATA water policy research programme, Colombo, 2002.

Palanisami, K., and R. Balasubramanian. 1998. Common property and private prosperity: tanks Vs. private wells in Tamil Nadu. *Indian Journal of Agric. Economy* 53: 600-613.

Palanisami, K., and K. W. Easter. 2000. *Tank irrigation in the 21st century. What next?* New Delhi: Discovery Publishing House.

Prakash, A. 2005. *The Dark Zone. Groundwater irrigation, politics and social power in North Gujarat*. New Delhi: Orient Longman.

Prasad, C. S., P. K. Basu, and A. Hall. 2005. Assessing SRI as a Process: Evidence from India. Water Policy Research (http://www.iwmi.org/iwmi-tata) 22:7p.

Rajagopal, A., and S. Janakarajan. 2001. "Water-rights and participatory irrigation management in India: the case of surface water sector in Tamil Nadu state." www.water-2001.de/datenbank/546076235.41368.14/WATER per cent 20RIGHT per cent 20PAP.doc.

Rao, G. N., and K. N. Nair. 2003. Change and transformation in rural South India. Findings from village studies. *Economic and Political Weekly* 38:3349-3354.

Rao, N. J. Editor. 2005. *Indian agriculture. Issues and perspectives*. Hyderabad (India): The ICFAI University Press.

Ratnavel, S. M., and P. Gomathinayagam. 2006. *In search of Ancient Wisdom—Irrigation tanks*. Madurai: DHAN Foundation.

Rotzer, K. 1984. Bijapur: alimentation en eau d'une ville musulmane du dekkan aux XVIe-XVIIe siècles. *Bulletin de l'Ecole Française d'Extrême-Orient* 73: 125-195+28 PL photos.

Schlager, E., and E. Ostrom. 1992. Property rights regimes and natural resources: a conceptual analysis. *Land Economics* 68: 249-62.

Shah, E. 2003. *Social Designs. Tank irrigation technology and agrarian transformation in Karnataka, South India. Wageningen University Water Resources series.* New Delhi: Orient Longman.

Shah, T. 1993. *Groundwater Markets and Irrigation Development. Political Economy and Practical Policy.* Bombay: Oxford University Press.

— —. 1997. "Groundwater management in India," in *The challenge of the balance: environmental economics in India. Proceeding of the National Environment and Economics Meeting, January 1994.* Edited by A. Anil, pp. 299-304. Delhi: Centre for Science and Environment.

— —. 2007a. "Groundwater: a global assessment of scale and significance," in *Water for Food, Water for Life.* Edited by D. Molden, pp. 395-423. Colombo: IWMI.

— —. 2007b. "The groundwater economy of South Asia: an assessment of size, significance and socio-ecological impacts," in *The agricultural groundwater revolution: opportunities and threats to development.* Edited by B. Van Koppen, M. Giordano, and K. G. Villholth. Oxford: CABI (Comprehensive assessment of water management in Agriculture series 3).

— —. 2007c. "Issues in reforming informal water economies of low-income countries: examples from India and elsewhere," in *The agricultural groundwater revolution: opportunities and threats to development.* Edited by B. Van Koppen, M. Giordano, and K. G. Villholth. Oxford: CABI (Comprehensive assessment of water management in Agriculture series 3).

Shah, T., M. U. Hassan, M. Z. Khattak, P. S. Banerjee, O. P. Singh, and S. U. Rehman. 2009. Is irrigation water free? A reality check in the Indo-Gangetic Basin. *World Development* 37: 422-434.

Shah, T., C. Scott, A. Kishore, and A. Sharma. 2004 "Energy-irrigation nexus in South Asia: Improving groundwater conservation and power sector viability. [Second (Revised) Edition]," in *Research Report* 70. Colombo, Sri Lanka: International Water Management Institute.

Shah, T., and S. Verma. 2008. Co-management of electricity and groundwater: an assessment of Gujarat's Jyotirgram scheme. *Economic and Political Weekly* 43(7): 59-66.

Shiva, V. 2002. *Water Wars. Privatization, Pollution and Profit.* New Delhi: India Research Press.

Sivasubramanyam, K. 2002. *Alapakkam: a resurvey. Working Paper No. 173:* Madras Institute of Development Studies.

Stein, B. 1980. *Peasant State Society in Medieval South India.* Delhi: Oxford University Press.

Trébuil, G., and M. Hossain. 2004. *Le riz. Enjeux écologiques et économiques*. Paris: Berlin (Mappemonde Coll.).

Vaidyanathan, A. 1999. *Water Resource Management. Institutions and Irrigation Development in India*. New Delhi: Oxford University Press.

——. Editor. 2001. *Tanks of South India*. Delhi: Centre for Science and Environment.

Vijayavenugopal, G. Editor. 2006. *Pondicherry Inscriptions*. Pondicherry: Institut français de Pondichéry, and Ecole française d'Extrême-Orient.

Von Oppen, M., and K. V. Subba Rao. 1980. *Tank Irrigation in Semi-arid Tropical India: Historical Development and Spatial Distribution*. International Crops Research Institute for the Semi Arid Tropics (ICRISAT).

PART 1

REGIONAL GEOGRAPHY OF IRRIGATION

CHAPTER

2

PRESENTATION OF THE STUDY AREA, HINTERLAND OF PUDUCHERRY AND VILLUPURAM

Olivia Aubriot

The study area covers the Puducherry hinterland which is characterised by a high population density (more than 700 inhab/km², i.e. 50 per cent higher than the Tamil Nadu average), developed irrigated farming, yet with marked differences between various parts of the area: from half to fully irrigated, from rain-dependent irrigation to irrigation independent of rainfall; and varying degrees of tank use. The study area thus raises the question of the diversity of irrigation situations in an area where there are tanks and pumps in every village. We have set out to show how environmental factors in accessing water are crucial for understanding some features of water management dynamics, but some of the differences may be explained by the fact that the study area covers two distinct States. Indeed, water (as well as land and agriculture) is a State issue, so that water laws, development programmes and subsidies differ slightly: electricity for farming is free of charge in Tamil Nadu, yet subject to a timetable, whereas it is always available at a very low cost in Puducherry; all 84 tanks in Puducherry district have been rehabilitated under the Tank Rehabilitation Project of Pondicherry (TRPP), whereas in Tamil Nadu the number of tanks is much bigger, the process is more sporadic, and carried out by various means depending on the institution involved in the project [Public Works Department (PWD) or NGO or village Panchayat through the National Rural Employment Guarantee Act 'NREGA' programme]. Colonial policies regarding irrigation were also different. With Puducherry being richer than Tamil Nadu, the socio-economic characteristics are highly contrasted. Moreover,

Puducherry provides more subsidies to farmers who are proportionally much fewer. Puducherry is also a very small State so that many people have political ties with decision makers, consequently readily calling upon their political support when necessary. Such political links seem to be a major interference and an important element in the development process, especially as, up until 2006, no local elections for representatives at village level had been organised for years (since 1968) in Puducherry.

The study area covers 4,357 km² (*see* Fig. 2.1 and Fig. 2.2). As explained in the introduction, it does not coincide with administrative limits: in Tamil Nadu, it covers the eastern half of Villupuram district and a very small part of the northern district of Cuddalore (approximately 1/3 of the former South Arcot district which until 1993 included these 2 districts). However, it encompasses Puducherry district[1] on the territory of the same name (made up of 4 districts). The distinctive feature of Puducherry district is that it is not an unbroken geographical area but, on the contrary, presents an image of scattered parts, interspersed with sections of Villupuram and Cuddalore districts: a village in Puducherry district is very often situated next to a Villupuram village, and it is not rare for farmers to have fields in both States (thus being subjected to different policy and institutional settings, and benefiting from the more favourable policies). There is the unique case of Vadanur village where plots alternate between Villupuram and Puducherry.

To describe this area, we will base our findings on work previously published, especially work in French which, although rather old, may be worthwhile translating into English (Adicéam, 1966; Racine, 1982; Marius Gnanou, 1991).

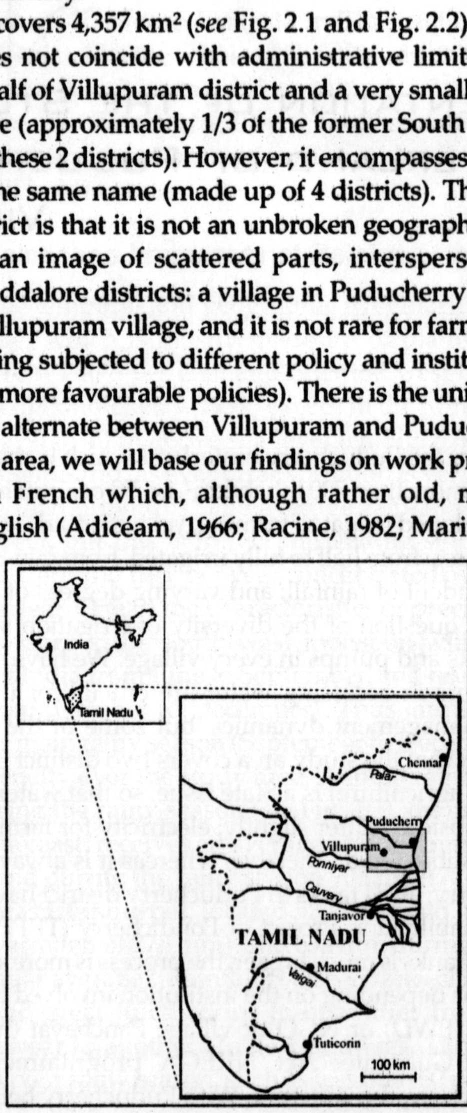

Fig. 2.1: Localisation of the Study Area within Tamil Nadu

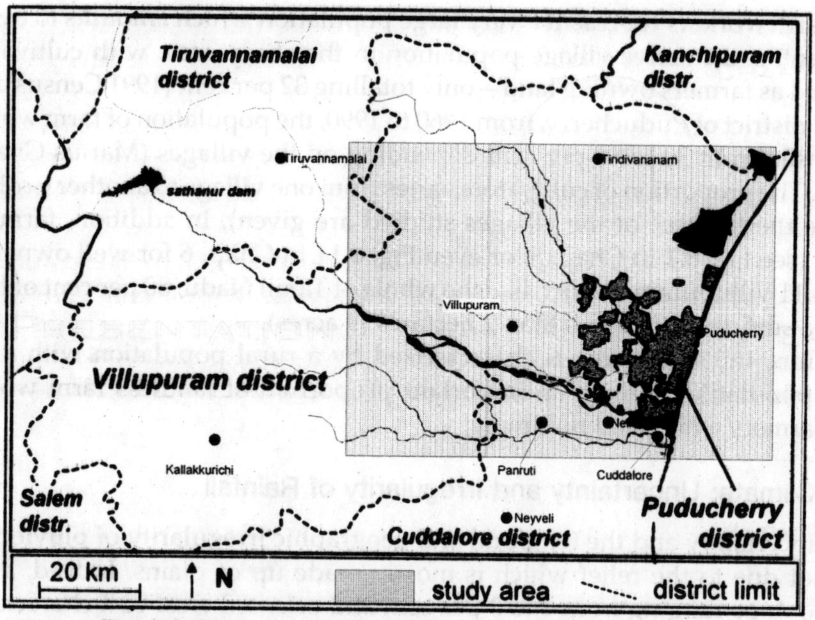

Fig. 2.2: Study Area, Villupuram and Puducherry Districts

2.1. A Rural Farming Population

This study area covers 9 towns and 1,115 villages, about thirty of which are not inhabited since they are Reserved Forests. The population density of the study area is high: more than 710 inhabitants/km^2, which is much higher than Villupuram (406 persons/km^2) or Tamil Nadu (478 inhabitants/km^2). This was calculated using a GIS (geographical information system) linking the map of all revenue villages to the 1991 census database. This population amounted to 2.82 million in 1991, and would in all likelihood have reached 3.1 million in 2001 (if we take a mean of the Puducherry and Villupuram district increase in population between 1991 and 2001). Villupuram is mostly a rural district—the urbanisation rate is only 14 per cent—one of the most backward districts in Tamil Nadu. On the other hand, Puducherry district is highly urbanised (67%) for a population of 0.73 million, with employment in the service sector and industry.

The scheduled caste (SC) population represents a quarter of the total population, a figure higher than the average for TN (19%). The two features (high population density and high proportion of SC) may be linked to the fact that the area was historically well-watered (using tanks), thus supporting higher densities and requiring manpower for rice farming; yet this manpower often comes from the SC population (*see* maps and comments in Oliveau, 2003).

Farm workers represent a very large population which amounts to over 50 per cent of the active village population in the study area, with cultivators–defined as farmers owning land—only totalling 32 per cent (1991 Census data). In the district of Puducherry, from 1960 to 1990, the population of farm workers increased from 30 to 80 per cent depending on the villages (Marius Gnanou, 1991). The proportion of cultivators varies from one village to another (*see* Box 6 where the features of the villages studied are given). In addition, farms are small (see Fig. 8.1 in Chap. 8, or even Fig. 6.11 in Chap. 6 for well owners) in Fig. 6.11 Villupuram district, like the whole of Tamil Nadu, 80 per cent of farms have a surface area of less than 2 hectares (5 acres).

Thus, the study area is characterized by a rural population with a very high population density, an important proportion of landless farm workers and farmers with small holdings.

2.2. Climate: Uncertainty and Irregularity of Rainfall

The uncertainty and the temporal and geographic irregularity of pluviometry are not due to the relief which is mostly made up of plains. Indeed, a very gentle slope ranging from 1 to 3 per cent and oriented west-east characterises this area (Government of Pondicherry, 1999) except in the northern part, where some upland areas are 30 to 40 m above sea level (cf. Fig. 5.6 in Chap. 5). The climate of the region is semi-arid, as evapotranspiration is very high and higher than rainfall for more than 6 months. The average rainfall is relatively high, above 1,000 mm per year, but this is in fact misleading since the number of rainy days is quite small, around fifty per year concentrated over a few months, towards the end of the year. One of the characteristics of precipitation in Tamil Nadu is that it is low during the south-west monsoon which, on the other hand, waters the rest of India from June to September, and much higher during the north-east monsoon[2] (October to December). Thus, around two-thirds of the rain falls during the main rainy season (September to December) with spatial and temporal heterogeneity over the study area. Annual rainfall, as well as the proportion of rain during the rainy season (*see* Fig. 2.3), dwindles from east to west: Cuddalore, on the coast, receives an average of 1,270 mm (with a standard variation of 385 mm) and during the rainy season 903 mm (72 per cent of mean annual rainfall), whereas Ulundurpet, located 53 km inland, receives only 973 mm (with a standard variation of 278 mm)[3] and 634 mm during the rainy season (64 per cent of the annual average). Rainfall is also erratic, characterised by an irregular frequency and great fluctuations in quantities from one month to another and from one year to another (Table 1 and Figs. 1 to 3 in Box 3). The probability of having a poor rainy season, such as less than 60 per cent of average rainfall is rather high (0.42 with 2.3 years as a recurrence interval for Ulundurpet, Lefebvre, 2006: 53).

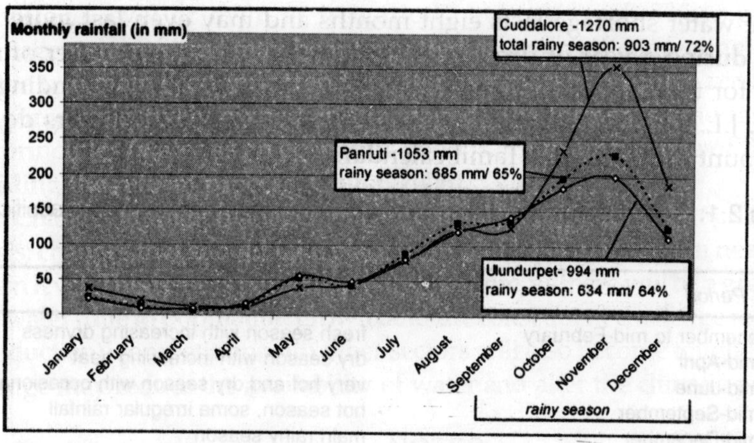

Fig. 2.3: Average monthly rainfall for three stations (calculated for the period 1945-2004)

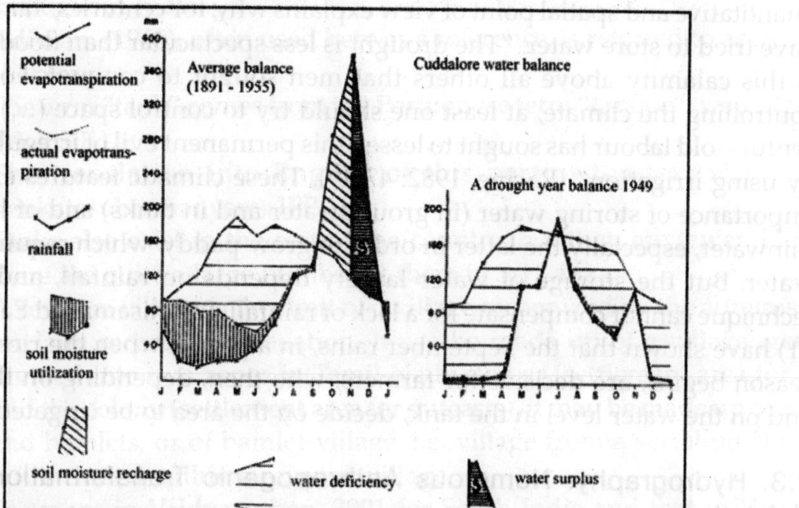

Fig. 2.4: Average and drought year water balance (*source*: Racine 1982, referring to V.P. Subrahmanyam and A.R. Subramaniam, 1964:400)

The water balance analysis (comparison of evapotranspiration and rainfall) reveals various periods (*see* Fig. 2.4): rainfall from September to November renews the soil's humidity; in November-December it creates a water surplus; this reserve of humidity is somewhat rapidly used over January and February when there is a greater water shortage; from March onwards the water shortage becomes more acute, and from May to August, even if some rain does fall, the water balance remains negative. Thus, even if average rainfall is 1270 mm/

year, the water shortage lasts eight months and may even last more than 11 months during a year of drought (Racine, 1982: 36). The number of seasons defined for this semi-arid climate differs from three to six depending on the authors. J.L. Racine (1982: 40-41) defines five seasons, as farmers do, taking into account the wind and Tamil calendar.

Table 2.1: Climatic seasons in the study area, according to Racine's classification (1982), in keeping with that of farmers

Period	*Season*
mid-December to mid-February	fresh season with increasing dryness
up to mid-April	dry season with increasing heat
up to mid-June	very hot and dry season with occasional storm
up to mid-September	hot season, some irregular rainfall
up to mid-December	main rainy season

This great unpredictability of Tamil Nadu's rainfall from a temporal, quantitative and spatial point of view explains why, for centuries, Tamil farmers have tried to store water. "The drought is less spectacular than flooding, but it is this calamity above all others that men sought to conquer. For want of controlling the climate, at least one should try to control space (...) Peasants' century-old labour has sought to lessen this pernanent evil of irregular rainfall by using irrigation" (Racine, 1982: 47-49). These climatic features explain the importance of storing water (in groundwater and in tanks) and of harvesting rainwater, especially the latter in order to grow paddy, which requires a lot of water. But the storage of water largely depends on rainfall, and the tank technique cannot compensate for a lack of rainfall. Palanisami and Easter (2000: 61) have shown that the September rains, in a period when the rice cropping season begins, are decisive for farmers, who then, depending on the rainfall and on the water level in the tank, decide on the area to be irrigated.

2.3. Hydrography: Numerous Anthropogenic Transformations of the Network

As for hydrology, the study area covers the downstream sections of two river basins, that of the Ponniyar, a 320-km long river which starts in Karnataka near the Nandhi hills, and that of the Varahanadi (also named Gingee River) a 80-km long river, covering Nallavur watershed (that of Kalivelli tank) situated in the study area, Fig. 2.5. None of these waterways are perennial, but diversion dams have long been in existence, supplying system tanks, and ever since colonial times they have been renovated to become more modern, stronger structures diverting greater quantities of water. Rivers were equipped with dams at the end of the 1950s to regulate their flow and supply new irrigation networks (Sathanur dam on the Ponniyar in 1958, and Vidur dam in 1960 on the Varahanadi). Since 1999 Sathanur dam has also been used, to generate electricity.

Box 3: A Decrease in Rainfall Over the Last 25 Years?
By Olivier Lefebvre

During discussions with farmers in the hinterland of Puducherry noticed that many of them say that there has been less rain over the last twenty or thirty years. This is why we are proposing this statistical study of the pattern of rainfall in the region, based on the findings of several pluviometric stations. This information comes from the Tamil Nadu Department of Statistics. Numerous stations are present in the zones but the information is often incomplete, which is why we have limited ourselves to three stations for which we have the consecutive months for a hundred years: Cuddalore, Panruti and Vridanchalam.

The first test we were able to make of this information concerned stationariness which lets us know if the rainfall average is constant over time. Two principal results appear. First, we notice that there is a break around 1950 when the rainfall average falls about 10 per cent between the periods 1900-1950 and 1950-2003. This failure of stationariness is more likely to be of human origin (changes in the measurement protocol) than a climatic one. This period does not, in fact, correspond to other periods of broken rhythm observed elsewhere in the world. Consequently, we should study information only from post-1950. We also verified that the series does follow a normal statistical law, that is to say, events are uncertain and independent, repeating a great number of times. Lastly, the rainfall average appeared to remain constant from this year onwards, which prevents us from confirming the claim that there has been less rain per year since the 1980s.

How then are the farmers' statements to be explained? In order to look more closely at this question, we propose to analyze in greater detail, the annual amounts at these three stations since 1950. Earlier, we calculated the average and the standard differentiation of the series between 1950 and 2002. We then related the information to these characteristics (normalized form). Thus 0 indicates a normal year (average), 1 indicates a year of surplus, one time the standard deviation and –1 indicates to a year of deficiency one time the standard deviation. Elsewhere the pluviometric annual total corresponding to a dry or wet year with a statistical occurrence of 5 years: we designate as a dry year one whose pluviometric annual is inferior to that quinquennial limit (and the reverse for wet years). The centred reduced representations and the quinquennial limits are shown on the graph opposite.

The breakdown of dry and wet years for our three stations is synthesized in Table 1 where the periods 1950-1979 may be distinguished on the one hand and 1980-2003 on the other so as to compare the occurrence.

Two main observations are apparent when reading of this table:

- It seems that during the period 1950-1979, the number of wet years is equivalent or even superior to the number of dry ones. During the period 1980-2003, the number of wet years is significantly less.
- The number of dry years has clearly increased in two of the stations.

(*Contd.*)

Box 3 (*Contd.*)

Table 1: Breakdown of Dry and Wet Years Since 1950

	Period 1950-1979			*Period 1980-2003*		
	Number of measures	*Number of dry years*	*Number of humid years*	*Number of Measures*	*Number of Dry years*	*Number of Humid years*
Cuddalore	30	5 [17%]	7	23	9 [39%]	3
Panruti	30	10 [31%]	8	20	4 [21%)	3
Vridanchalam	30	5 [17%]	9	20	6 [31%]	2

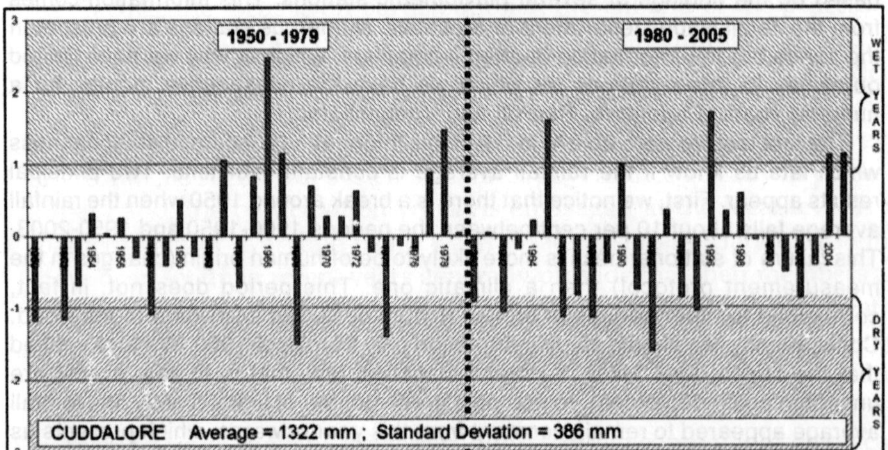

Fig. B 3.1: Normalised Representation of Rains at Cuddalore Station

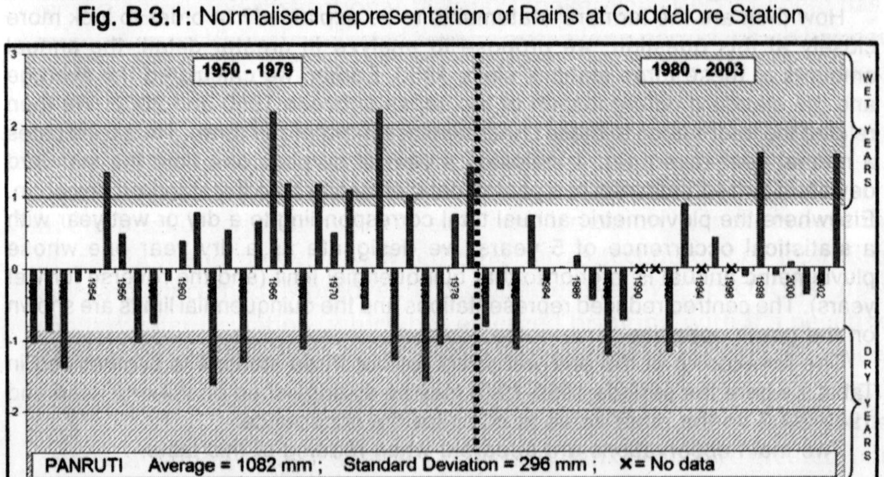

Fig. B 3.2: Normalised Representation of Rains at Panruti Station

(*Contd.*)

Box 3 (*Contd.*)

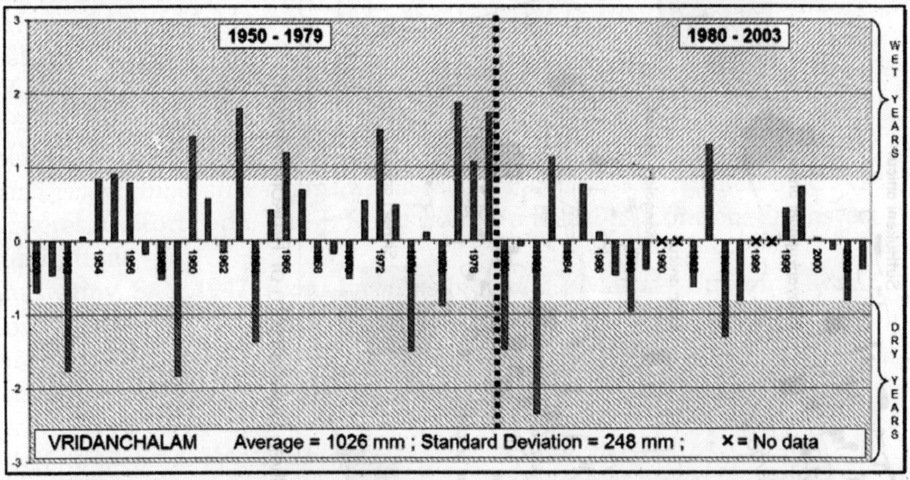

Fig. B 3.3: Normalised Representation of Rains at Vridanchalam Station

These results therefore make it possible for us to understand the farmers' statements. Even though the average rainfall has not decreased the farmers have had to face a greater number of years of little rain and fewer wet years since 1980, with the well-known consequences regarding the filling of tanks and cultivation.

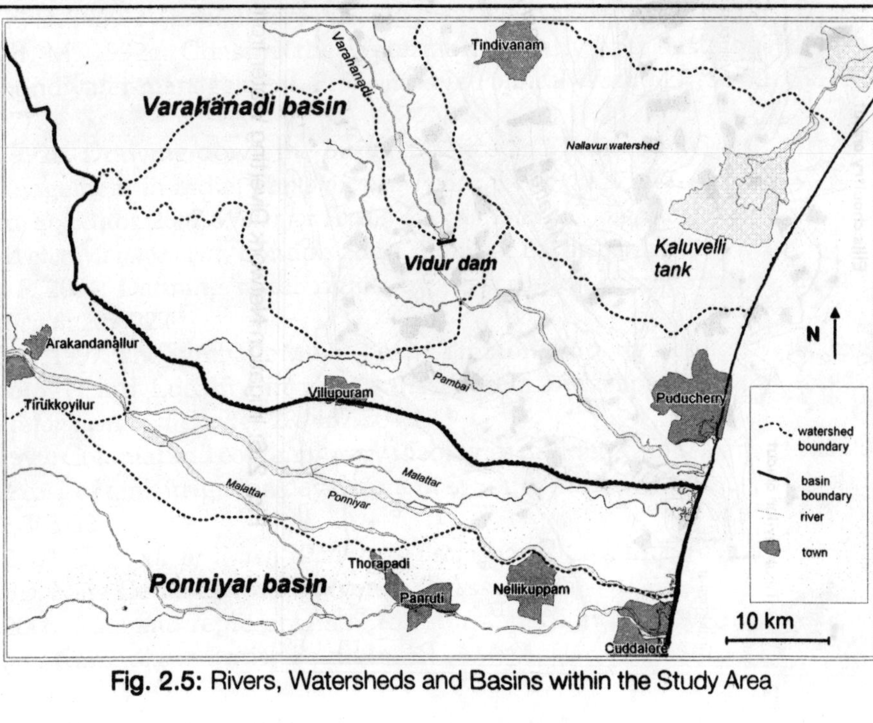

Fig. 2.5: Rivers, Watersheds and Basins within the Study Area

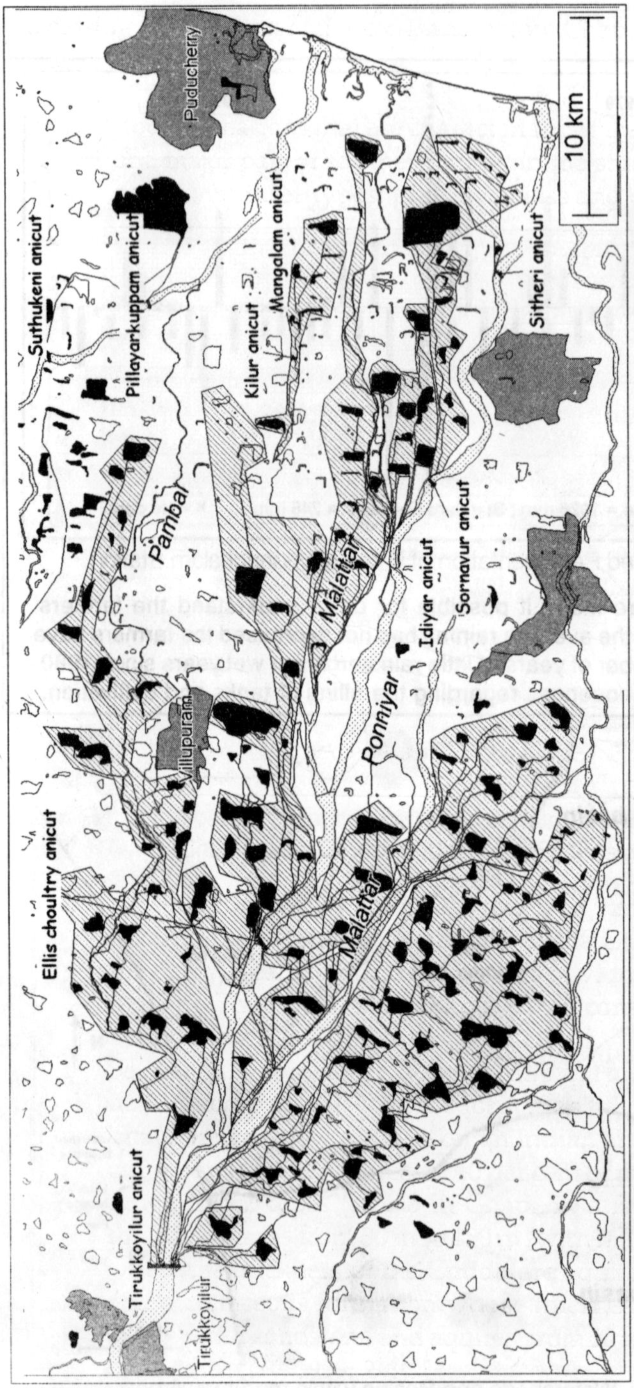

Fig. 2.6: Irrigation Network Diverting Water from the Ponniyar River Downstream of Tirukoilur *Anicut*

Sathanur dam not only supplies the 15,000 ha irrigation network created downstream of the dam, but also much further away, in the study area, a command area dependent on former tank systems and which, on paper, covers almost 18,000 ha. First, 50 km below Sathanur dam, there is a diversion dam called Tirukoilur *Anicut* (9,184 ha) from where four channels fan out (Raghavan channel with a command area of 4,358 ha; Pambai—2,556 ha; Malatar—1,990 ha and Marudhur—280 ha), then Ellis choultry (5,927 ha), Sornavur *Anicut* (3,150 ha) and lastly Sitheri. Sornavur Anicut is the starting point for the Bangaru channel which provides water to Bahur tank, the most famous tank in Puducherry.[4] It supplies water to the rice-bowl area of Puducherry and is the second biggest tank in the district (tank bed of nearly 500 ha). On streams similar to an outflow or a runoff stream, diversion dams may be constructed, such as on the Kuduvayar odai, which comes from the North Malatar, a branch of the Ponniyar: *anicut* of Idiyar, Kilur, Mangalam and Tirujanji. Some diversion dams located on small rivers and which have not been modernised have disappeared, such as one on Pambai odai which more than fifty years ago brought water to Vadanur tank, as a second source of water for the tank. Now the canal is no longer in working order, it is encroached on and not even visible over a distance of 300 m (*see* Fig. 2.6). In the Nallavur watershed, the water from small temporary watercourses is also diverted to tanks and the watercourse catchment area is covered by tanks that harvest the runoff water.

Thus the hydrographic network is characterised, on the one hand, by storage infrastructures with river water being diverted many times to tank networks and it is only recently that dams regulate the flow of the two rivers crossing the study zone; and on the other hand by interlinked waterways with canals bringing water from the watercourse to the tank, from tank to tank, from tank to drainage canal or water course, with streams being used as canals and with the downstream section of the drain functioning as irrigation canals. Everywhere, the diversion and mesh of waterways and water bodies are commonplace. Networks of canals and tanks shape the landscape: the anthropogenic transformation of the hydrographic network is omnipresent. J.L. Racine, who studied the landscape of South Arcot just before the boom in submersible pumps, concluded in 1982 (p. 298), that one of the main features of this regional history is the age-old effort to manage the rural area through technical mastery over water. This concretely means having networks of canals and tanks, and diversion dams on rivers. "Not being able to master the climate, one has attempted to master rivers" (p. 49).

2.4. Geology: Three Main Formations

The study area is characterised by the presence of three main geological formations (Fig. 2.7), which define the greatly contrasting conditions for accessing groundwater (cf. Box 4, and below Chap. 3). These three formations are representative of India:

- Rock formations that are characteristic of the Deccan plateau and hence of a major part of southern India. In the study area they can be broken down into three types: granite, gneiss and charnockite.

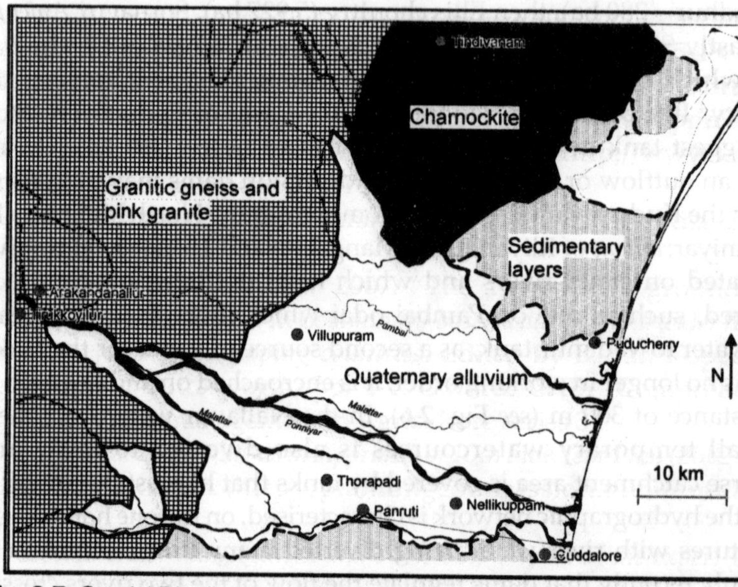

Fig. 2.7: Spatial Distribution of the three Main Geological Formations: Rocky, (Charnockite, Granite and Gneiss) Sedimentary and Alluvial

- Alluvial formations are rarely found in Tamil Nadu, though more so on an Indian scale due especially to the Ganges and Indus basins. In the study area, the thickness of the alluvium deposit varies from 10 m to 55 m. The layers of sand and gravel constitute the alluvial aquifer. Below these are sedimentary formations.
- Sedimentary formations lie below alluvial formations, though they may have outcrop areas (when the formation has some contact with the air). There are very few of these on an Indian and Tamil Nadu scale, yet in our study area they represent 10 per cent of the area. They can be broken down into 2 types (cf. 5.5 in Chap. 5): a Paleocene formation whose thickness varies from 30 m to 450 m (with mainly Cuddalore sandstone, where the thickness of the Cuddalore aquifer varies from 20 to 245 m) and a cretaceous formation whose thickness varies from 50 to 150 m (with Vanur sandstone and aquifer, from 30 to 90 m thick) (Government of Pondicherry, 1999: 2-4).

In the study area, alluvial and sedimentary layers lie, for the most part, parallel to each other, dropping eastward (cf. Fig. 5.7 in Chap. 5), but their thicknesses are not geographically homogeneous, varying according to the localities. Thus, in alluvial formations, three main aquifer systems overlap each other (from the surface to the greatest depths): shallow alluvial aquifers, deep Cuddalore sandstone aquifers, and deep Vanur-Ramanathapuram sandstone aquifers (GoP 1999: 3).

Box 4: Hydrogeology: Some Fundamental Characteristics

First of all, we need to distinguish the upper aquifer from the underneath layers. The upper aquifer is directly supplied by infiltrations from the surface (rainfall, river, lakes, fields, irrigation etc.), and its level progressivey declines during the dry season, rising and falling each year between the rainy and dry season. However, the working of underneath aquifers is more complex and differs according to the geological formation and especially the hydrogeological characteristics.

A second major dichotomy then has to be evoked: rocky areas as opposed to alluvial and sedimentary areas (see Fig. 1).

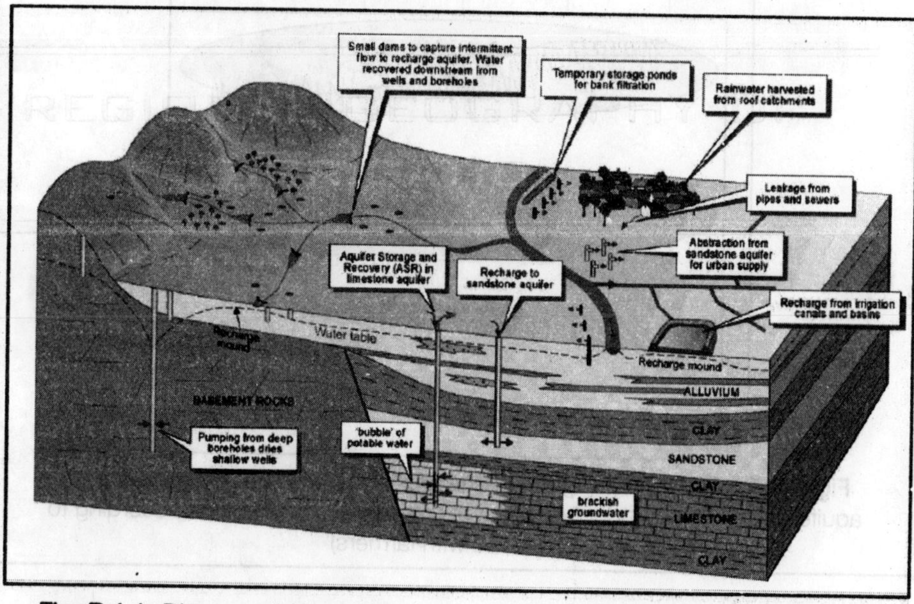

Fig. B 4.1: Diagram of Associated Geological and Hydrogeological Formations
Source: Gale et al., 2002.

- In the rocky areas, below the upper aquifer, water circulates mainly through fissures and weathered areas. Recent studies show that these cracks, many of which are horizontal, are found more in the upper part of the fractured zone. Thus, drilling deeper does not mean a higher probability of

(Contd.)

Box 4 (*Contd.*)

encountering more cracks. (BRGM, 2007: talk at the 3rd International Groundwater Conference—IGC).

- In alluvial and sedimentary formations, water is found primarily in the sandy layers which may be interspersed with more or less impervious layers of clay: hence several aquifers overlap each other, some being confined and others unconfined.

Deeper drilling may therefore correspond to different phenomena: drilling into the same aquifer whose level drops (the aquifer does not refill or the refill is, over the year, lower than the extracted quantity); drilling into a deeper aquifer because the previously tapped aquifer has dried out or is not as productive. Thus, when it is said that the average depth of bores ranges from 30 m to 60 m, it does not necessarily mean that the level of water has dropped continuously between these two depths. This may be due to a drop in productivity of an aquifer and to tapping into a deeper aquifer. The schematic diagram representing the phenomenon observed in the Puducherry region is reproduced in Fig. 2.

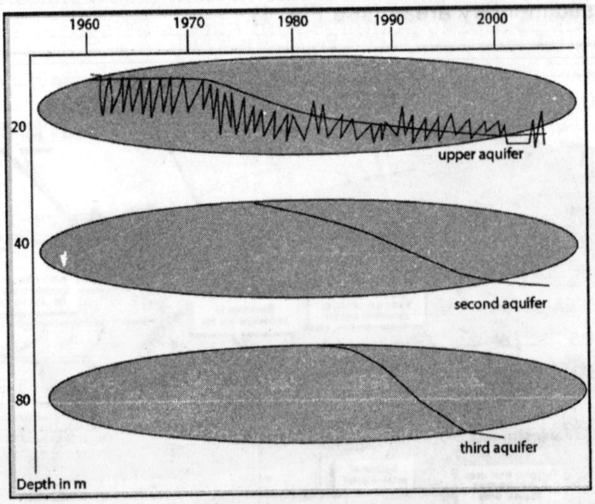

Fig. B 4.2: Schematic diagram of changes in water levels in the three main aquifers used in the Puducherry alluvial area (case of Bahur area according to interviews with farmers)

2.5. Irrigation: Several Features must be taken into Consideration

Irrigation in this study area is marked by several elements that avoid a simplified regionalization of water use. These elements are:

- The co-existence of two types of irrigation infrastructures (groundwater or tank) in each village. Thus 1,900 tanks are distributed over 1,115 revenue villages: tanks and villages were counted using a GIS (see footnote in Introduction) and the tanks mapped out based on topographical maps to a scale of 1:50,000. Tanks vary in size and cover command areas ranging from 4 to 1,600 ha, planted mostly with paddy and sometimes with sugarcane or banana. Wells, which are individual infrastructures for the most part (a few rare government tubewells do exist), irrigate smaller areas of less than 1/2 ha to 7 ha depending on the crops, the demand and water availability. If a lot of wells are situated close to each other, they may form uninterrupted irrigated areas, otherwise land irrigated only by groundwater forms patches in the middle of non-irrigated land.

- A steep decline in tank irrigation around Puducherry, where there are many tubewells spurting out water all year long over the entire cultivated area, suggesting an abundance of water. According to data from the report published by Atkins (2001), a research agency that carried out a survey for TRPP, one quarter of tanks are no longer used at all for irrigation, 30 per cent irrigate less than a quarter of the command area and the remaining 35 per cent irrigate on average less than 50 per cent of the command area. As no detailed data of this type exist for Tamil Nadu, we carried out a field survey to obtain such data.
- A generously irrigated alluvial area.
- Drier areas in the western and northern (rocky and sedimentary) parts of our study area, where groundwater is not as easy to access as in the first case, because of the low groundwater potential and in the second case because of the depth.
- A series of tank systems in the southern part of the study area, contrasting with the rainfed tanks in the northern part of the area.
- Some tanks in Tamil Nadu are run by formal irrigation associations set up by various institutions (Public Works Department, Non-Government Organizations) each with its theories and methods related to the implementation of Water Users Associations (WUAs). WUAs are registered under the TN-FMIS Act, 2000, whereas in Puducherry, Tank Associations were set up within the Tank Rehabilitation Project Pondicherry (TRPP) with no Act been enacted; all the tanks in Puducherry district were managed by these institutions at the time of the study (up to 2007).

For various reasons, irrigation in the district of Puducherry is rather a special case: first of all, given its almost exclusive location in the alluvial area, the district is not representative of what happens elsewhere in Tamil Nadu. Yet the importance of groundwater irrigation—91 per cent of the gross cultivated area is irrigated, 96 per cent of which by GW—and the problems encountered by farmers are similar to those affecting other regions largely irrigated by groundwater, such as Gujarat. Moreover, unlike in Tamil Nadu, the development of groundwater irrigation is not a phenomenon exclusively linked to the Green Revolution. Indeed, statistical analysis (Fig. 2.8) reveals that GW irrigation was already well developed in the sixties, as half of the net irrigated area was irrigated by GW, and tubewells were already being used. The first artesian well was in fact dug in Puducherry by the Savana Textile Plant in 1877, and subsequent wells (up to 100 m) were dug by the Travaux Publics (French Public Works Department) for agricultural purposes. A. Chatterton who visited South Arcot district in 1900 "marvelled that Pondicherry Territory already possessed more than 500 tubewells". Up until the middle of the 20th century, such infrastructures were very rare in Tamil Nadu[5] (Adicéam: 255-256). Thus, irrigation using tubewells in Puducherry dates back further (at first ensured by artesian wells) than in Tamil Nadu (*see* also Table 2.2, where the evolution of well irrigated area in the beginning of 20th can be seen).

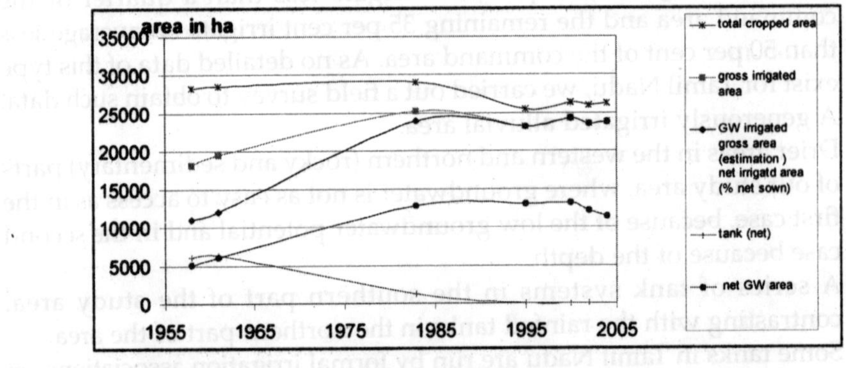

Fig 2.8: Changes in the Irrigated Area in Puducherry District

Table 2.2: Change in the Irrigated area in Puducherry District, with Reference to the Colonial Period (*Data from Season and Crop Report Pondicherry from Each Year; ** from PWD Data)

	Registered tanks ayacut	*Irrigated area*	*Tank irrigated area*	*Non-irrigated area*	*Net sown area*
1824(Bédier)		4052	4052	8991	13043
1958*		12224	6103	7684	19908
1983*		14163	1241	2581	16744
2004*		11403	0	1598	13001
**	6730				

Even if irrigation in Puducherry district is rather a special case, its study is of great interest for various reasons. Firstly, as Fig. 2.9 clearly shows, the area

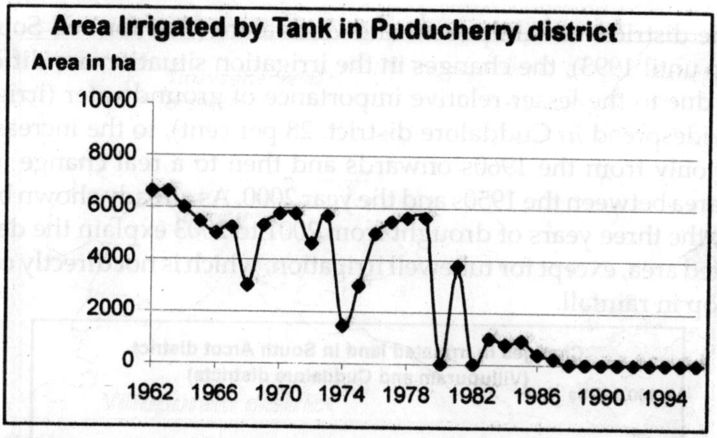

Fig. 2.9: Changes in the Tank Irrigated Area in Puducherry District (from Govt. of Pondicherry Report, 1999, Plate XIX)

irrigated by tanks gradually declined from the beginning of the sixties onwards, and then the surge in submersible pumps in the late seventies put an end to the use of tanks in the district. It is in fact the introduction of powerful pumps, not only the motorization of lifting groundwater, which explains the end of tank irrigation in Puducherry. Secondly, the decline in tank use (only 4 per cent of the net irrigated area is still irrigated by tank water, according to the report by Atkins, 2001) was not mitigated by the rehabilitation project[6] concerning all 84 tanks in the district (even if it was one of the purposes of implementing such a project). Only one small tank has been reused, after 20 years of total disuse, for irrigation (but not without provoking any conflict, Raghunath *et al.*, 2008). Lastly, some lessons may be learnt from tank management institutions. During the colonial period, formal Farmers Associations ("*Syndicates Agricoles*") were set up by the French government and remained in existence until the Independence of Puducherry (1954). Even if many elderly people recognize that these institutions were a good form of intervention by the government (*see* Box 8), such institutions could not exist on their own, showing how fragile formal institutions imposed by governments can be and how reliant on a legal and political context they are. Their status was indeed linked to the colonial legislation in force. Even though recognized as useful by villagers, no endogenous spontaneous organization took over, and 25 years later most tanks had been abandoned. Today, the future of Tank Associations is uncertain, since the *Puducherry Government* announced, at the end of 2008, that small-scale irrigation systems (such as tank irrigation) will be handed over to the *panchayats* and it has decided to suspend the TRPP project for the time being as it will have reached the end of its 10-year period by March 2009.

For the districts of Villupuram and Cuddalore (that formed South Arcot district up until 1993), the changes in the irrigation situation are different (*see* Fig. 2.10) due to the lesser relative importance of groundwater (irrigation by canal is widespread in Cuddalore district: 28 per cent), to the increase in GW irrigation only from the 1960s onwards and then to a real change in the net irrigated area between the 1950s and the year 2000. As already shown by Fig. 1.7 (Chap. 1), the three years of drought from 2001 to 2003 explain the decrease in the irrigated area, except for tubewell irrigation, which is not directly concerned by the drop in rainfall.

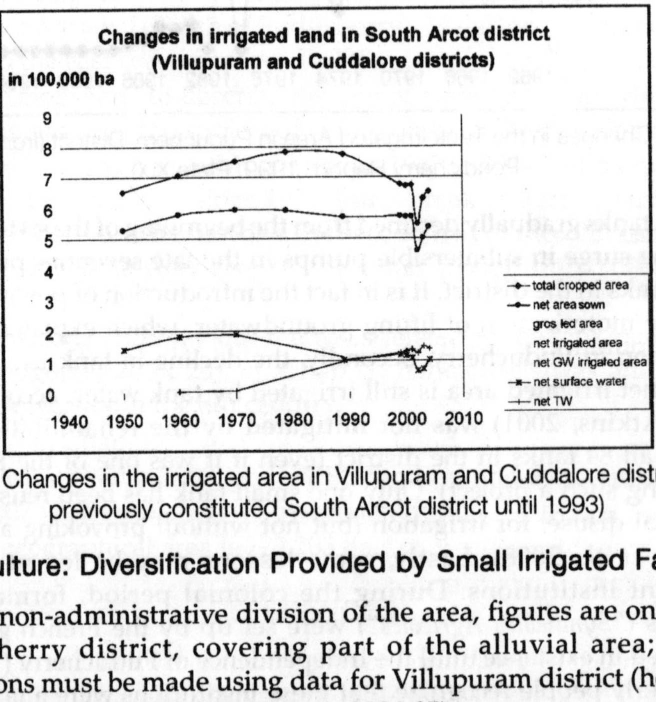

Fig 2.10: Changes in the irrigated area in Villupuram and Cuddalore districts (both previously constituted South Arcot district until 1993)

2.6. Agriculture: Diversification Provided by Small Irrigated Farms

Due to the non-administrative division of the area, figures are only available for Puducherry district, covering part of the alluvial area; otherwise extrapolations must be made using data for Villupuram district (here data are taken from the Season and Crop Report 2004-05).

In the district of Puducherry, agriculture is entirely irrigated (90 per cent of cultivated land) mainly using GW, paddy being the main crop (73 per cent of the gross irrigated area). From 1958 to 2004, the cultivated surface dwindled by a third (Fig. 2.11), due to a move to use cultivable land for non-agricultural purposes (housing) and for the cultivation of casuarinas, a crop classified as "land under miscellaneous tree crops". The gross cultivated area also shrank, but to a lesser extent (7%), though only from the mid-eighties onwards. In contrast, the proportion of irrigated land steadily increased (from 60 per cent to 90 per cent of the cultivated surface) as did the production of cereals which multiplied by 2.3. Hence, thanks to irrigation, the intensified crop cycle initially compensated for the drop in the net cultivated area (hence stagnation of the gross cropped area); then, however, with an already high proportion of irrigated land, the effect of the drop in the net cultivated area was felt on the gross cultivated area, but not on production which was constantly growing (cf. Fig. 2.11).

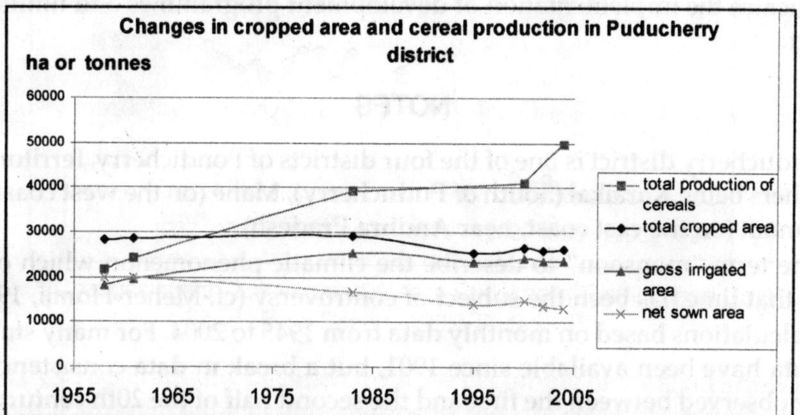

Fig 2.11: Changes in the Cropped Area and Cereal Production in Puducherry

In Villupuram district, the proportion of non-irrigated land is higher (18 per cent of net areas and 40 per cent of gross areas), but this remains low compared to the rest of the State. Villupuram is the district in Tamil Nadu that has the largest gross cultivated area, the largest irrigated area and the largest gross area cultivated with paddy (which represents 44 per cent of the gross cropped area and 61 per cent of the gross irrigated area in the district), even though it is the third biggest district[7]. It is the 2nd rice-producing district in Tamil Nadu, (after Tanjavoor district in the Cauvery Delta whose geographical area is half that of Villupuram) with production amounting to more than 400,000 tonnes (70 per cent of which is grown during the monsoon). Along with Cuddalore district, they are the biggest sugar cane producers, growing 30 per cent of Tamil Nadu's sugarcane between them.

The groundnut covers another 10 per cent of the irrigated area. This crop is irrigated when produced outside the monsoon period, i.e. 42 per cent of its cultivated area. Other irrigated crops are banana, cotton and casuarina (which is not accounted for as a crop in farming statistics, but as a plantation), with tapioca, a kind of millet (*kumbu*).

All this is produced on small farm holdings (most cover less than 2 ha) where farmers seek to diversify their crops. Farming thus relies on small farms, most of which do have access to irrigation and produce rice and even sugarcane. Rice and sugarcane, which are systematically irrigated in this part of India, cover three quarters of irrigated areas in Villupuram and 79 per cent of those in Puducherry. And yet they are large water consumers: making water savings comes down to making changes in cultivation techniques for these crops or to diversifying with less greedy crops. And yet, to be effective, any change in farming practices has to reach thousands of farmers. This numerical feature complicates the implementation of development programmes and limits their impact.

NOTES

1. Puducherry district is one of the four districts of Pondicherry Territory, the others being Karaikal (South of Puducherry), Mahé (on the west coast) and Yanam (on the east coast, near Andhra Pradesh).
2. The term "monsoon" to describe the climatic phenomenon which occurs at that time has been the subject of controversy (cf. Meher-Homji, 1975).
3. Calculations based on monthly data from 1945 to 2004. For many stations, data have been available since 1901, but a break in data consistency can be observed between the first and the second half of the 20th century: this requires using data over one or the other period only (cf. Lefebvre, 2006: 48).
4. Inscriptions dated 1027 lay out the rules related to the tank management (Vijayavenugopal, 2006: 24), cf. Chap. 1. Bahur itself was an important Sanskrit University in the 8th century.
5. cf. Table 1.3 in Chap. 1.
6. TRPP project started in 1999, under EEC funding, and then continued from 2005 to 2007 with Puducherry Government funds.
7. The total geographical area is 722,203 ha.

REFERENCES

2001. *Baseline survey, Tank rehabilitation project Pondicherry—Final report, Vol.II, Appendix A (individual tank summary).* WS Atkins (India) Pvt. Ltd.

Adicéam, E. 1966. *La géographie de l'irrigation au Tamil Nadu*. Paris: Ecole Française d'Extrême Orient.

Gale, I., I. Neumann, R. Callow, and M. Moench. 2002. *The Effectiveness of Artificial Recharge of Groundwater: a review.* British Geological Survey. Commercial Report/02/108N.

Government of Pondicherry. 1999. *Water resources and their management, Pondicherry Region (water for life, save water, save Pondicherry).* Department of Agriculture.

Hooja, R., G. Pangare, and K. V. Raju. 2002. *Users in water management. The Andhra model and its replicability in India.* Jaipur and New Delhi: Rawat Publications.

Lefebvre, O. 2006. Dynamiques de gestion de l'eau dans un contexte de surexploitation des eaux souterraines. La région de Pondichéry et de Villupuram au Tamil Nadu, Inde du Sud, Mémoire présenté en vue de l'obtention du Mastère Spécialisé Développement Agricole Tropical, CNEARC, Montpellier.

Marius-Gnanou, K. 1991. "L'irrigation et les mutations socio-économiques récentes dans la région de Pondichéry (Inde)," in *Eau et l'aménagement dans les régions inter-tropicales, T.2 (Espaces tropicaux, No. 3)*. Vol. 2. Edited by P. Vennetier, pp. 133-160. Talence: CEGET-CNRS.

Meher-Homji, V. M. 1975. On the monsoons of Tamilnad, South India. *The Indian Geographical Journal*, June: 25-30.

Oliveau, S. 2003. "Atlas of South India 1991," cubergéo, http://www.atlasindia.parisgeo.cnrs.fr.

Palanisami, K., and K. W. Easter. 2000. *Tank Irrigation in the 21st Century. What Next?* New Delhi: Discovery Publishing House.

Racine, J. 1982. Milieux, Histoire et Paysage au South Arcot. Fondements et formes traditionnelles de l'occupation de l'espace rural en pays tamoul (Inde du Sud). thèse de Doctorat de troisième cycle, Université de Paris-Sorbonne.

CHAPTER

3

ACCESS TO GROUNDWATER: SPATIAL DISPARITIES

Olivia Aubriot

The motorization of groundwater extraction has made groundwater the primary source of irrigation in the country as well as in Tamil Nadu and even in our study area. Wells, borewells and tubes are indeed commonplace in India. They are so common, that many writers forget to describe what an actual well or a borewell is, and may even write "here we can see wells just like everywhere else in India". Yet wells and bores are not the same everywhere: some are 3 metres in diameter, others 15 metres, some are round, others rectangular; bores can be drilled directly at ground level or inside a well. For those whose immediate field of work does not concern irrigation, there is a certain ignorance about well and borewell techniques. For many, wells and tubes are seen as simple techniques; a well being a hole in the soil and a borewell a tube to raise the water. Then, while looking at the tube coming out of an open well (that means the tube of the pump lifting the water from the open well), a scientist, not an irrigation specialist, talked of borewell! Such considerations are typical of those who do not really know how to "read" the well and tube landscape.

Thus there is, first of all, a need for clarification about the techniques and the terminology. Then, we will discuss some parameters (type of well, density of well, geology, cropping intensity) according to irrigation intensity use and access to groundwater. Can one say, for example, that geological spatial disparities define differences in water access and that the type of infrastructure reflects the availability of the resource? What are the indications of environmental constraints?

3.1. Terminologies and Classifications: What Information do they Provide?

A description is needed so that the reader understands what we are talking about and to avoid any misunderstanding. Let us therefore describe the various types of well we see in the study area according to their physical characteristics, and present the two main types of motor, since water extraction is fully motorized today. We will then discuss the various typologies relative to wells.

First of all, we can distinguish between wells with an open structure and wells where only a tube can be seen: wells *vs.* tubewells (see Fig. 3.1).

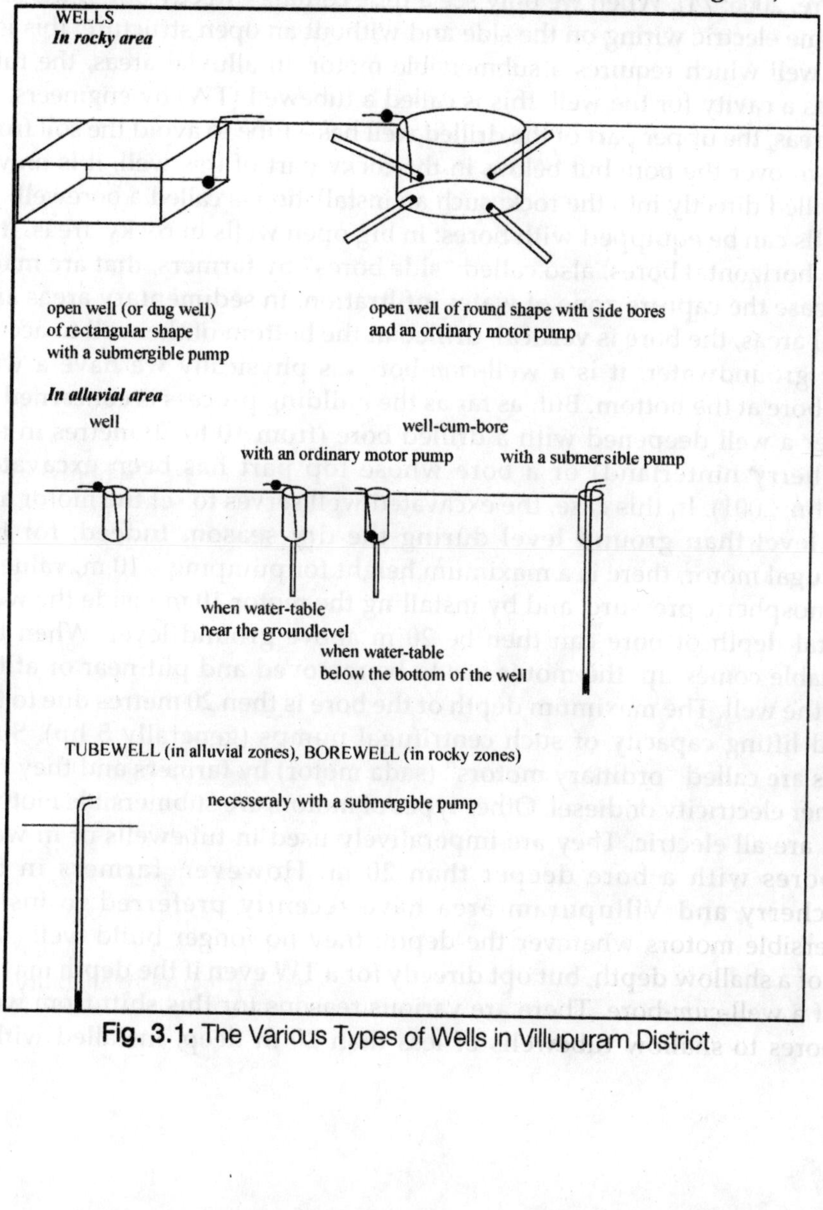

Fig. 3.1: The Various Types of Wells in Villupuram District

Among open wells, various types of well exist according to the soil and subsoil characteristics. Thus, in some rocky or sedimentary zones, open wells are large-size dug-out wells: they may be 10-15 m in diameter (or square-sided) and up to 15-20 m deep; their edge may be reinforced with stones or be directly hewn out of the rock. In alluvial areas, wells are often smaller in size (1 to 3 m in diameter) and with some kind of brick construction to maintain the border of the well. Some larger ones, 4 to 6 m in diameter or even more, did exist in the past in the western part of Pondicherry (Girod, 1937: 32), since farmers from Manamedu near the Ponniyar River still remember their existence (Lefebvre, 2006: 74). When we only see a tube coming directly out of the soil, with some electric wiring on the side and without an open structure: this is a drilled well which requires a submersible motor. In alluvial areas, the tube serves as a cavity for the well: this is called a tubewell (TW) by engineers. In rocky areas, the upper part of the drilled well has a tube to avoid the soil from closing in over the bore but below, in the rocky part of this well, it is only a bore drilled directly into the rock: such an installation is called a borewell.

Wells can be equipped with bores: in big open wells in rocky areas, it is mainly horizontal bores, also called "side bores" by farmers, that are made to increase the capture zone of water infiltration. In sedimentary areas and alluvial areas, the bore is vertical, drilled at the bottom of the well to access deeper groundwater: it is a well-*cum*-bore, as physically we have a well with a bore at the bottom. But, as far as the building process is concerned, it is either a well deepened with a drilled bore (from 10 to 70 metres in the Puducherry hinterland) or a bore whose top part has been excavated (Cesselin, 2001). In this case, the excavated well serves to set the motor at a lower level than ground level during the dry season. Indeed, for the centrifugal motor, there is a maximum height for pumping—10 m, value of the atmospheric pressure, and by installing the motor 10 m inside the well, the total depth of bore can then be 20 m above ground level. When the watertable comes up, the motor has to be removed and put near or at the top of the well. The maximum depth of the bore is then 20 metres due to the limited lifting capacity of such centrifugal pumps (generally 5 hp). Such motors are called "ordinary motors" (sada motor) by farmers and they run on either electricity or diesel. Other types of motors are submersible motors, which are all electric. They are imperatively used in tubewells or in well-*cum*-bores with a bore deeper than 20 m. However, farmers in the Puducherry and Villupuram area have recently preferred to install submersible motors whatever the depth: they no longer build well-*cum*-bores of a shallow depth, but opt directly for a TW even if the depth may be that of a well-*cum*-bore. There are various reasons for this shift from well-*cum*-bores to shallow tubewells of less than 20 m deep, installed with a submersible pump. The first is the cost of drilling, as direct drilling costs less than digging an open well (though the motor is more expensive). Secondly, it is easier to use a submersible pump than an "ordinary motor" which has to be placed at different levels according to the water level in the well. Setting up the latter inside the well is rather tricky and dangerous, and requires at least 2 persons. Lastly, government schemes now exist which grant subsidies to farmers to replace their old pump and buy a submersible pump of less than 10 hp (in Tamil Nadu such a scheme started in 2003, giving 25 per cent subsidies).

The vernacular terminology usually helps to classify techniques or at least it helps to understand the indigenous classification of the environment and techniques. Here, we are confronted with a new technique brought in from outside, and its terminology combines both English and local terms. However, most farmers do not speak English. Does this mean that they use terms borrowed from the technical vocabulary of agricultural experts without really understanding the nuances? Have they reappropriated these terms whilst giving them their own specific meaning? The study tends to prove this and shows the difference between local use of the terms and what can be read in (official, administrative or scientific) literature.

The well as a dug-out structure is known as *kenaru* in Tamil, but where a vertical well is dug and a pipe is used to reach the groundwater, farmers talk of "bore", whether it be a tubewell, a borewell or a well-*cum*-bore, with most farmers not knowing the words "borewell" or "tubewell". A distinction must be made, however, some farmers only use the word "bore" for wells where the pump is immersed (Cesselin, 2001). We see through this dichotomy in the farmer's terminology (*kenaru*/bore) that what is important in this wording is above all the fact that the well has been deepened, that it is deeper than traditional wells, or sometimes that it will draw deeper than so-called "normal" pumps. "Bore" refers implicitly in its two meanings (boring, or more precisely boring with a submersible pump) to recent techniques, drilling and/or submersible pump, i.e. techniques that allow water to be drawn from greater depths. To provide further details or sometimes as a first way of describing the well, another terminology related to water raising techniques is used: "*yatham*", "*erukavalai*", "*suttukavalei*" for old water drawing systems using human strength, animal strength and a bucket wheel (Adicéam, 1966: 257-265), "*sada* motor" (or ordinary motor) and "motor *neer*" (submersible pump, the term "*neer*" meaning "water") respectively. More emphasis is laid on the capacity to raise water by the engine.

We can see that in this farmer's terminology one word may be used to express different types of well (for example, the term "bore"). When we look at sales or layman terminologies, different words are used by different persons for the same installation. Thus the technical distinction made by engineers between "borewell" and "TW" is rarely applied by others, one term being used for the other: see, for example, the numerous websites using these two terms—interestingly they are mostly Indian—interchangeably. These words, which have become "common" terms are not defined as they appear to be too common! Why does this imprecision exist in the use of terms, with the same word defining different types of well? Is this due to the fact that they are English terms? Is it because sales people speak to farmers who do not know the difference? Are these differences so negligible or insignificant that everyday language cannot distinguish the technical differences? On the basis of our studies all these suggestions would seem to be correct.

In this book, we have based our vocabulary on the farmers' terminology which lends importance to the presence of drilling, whenever there is a tube or not. Thus, we use the generic term "well" to define open wells (with no vertical bore) and "bore" or "tubewell" for drilled wells (direct drilling from ground level or bored vertically inside the well). This 'well-TW' dichotomy seems to be commonest in administrative[1] classifications—corresponding more or less to the vernacular one (well/bore)—, even if a glance at the typologies reveals that the Ministry of Water Resources (in its Census of Minor Irrigation Schemes, 2001, online) surprisingly includes the dug-*cum*-borewells in its category of dugwells. A third category exists in this last classification, that of the deep TW defined as being more than 100 m deep.

Most administrative classifications are based on the physical features of wells, such as the presence or not of a cavity for the well, construction work in the cavity, the presence or not presence of a tube, the depth of the bore and sometimes the power of the pump. However, classifications based on morphological characteristics do not—except for large open-wells—help to grasp the real hydrogeological situation facing the farmer: to drill deeper than the depth of traditional wells does not necessarily mean that the well reaches a more productive watertable. Apart from the case of large open-wells that receive water from the upper aquifer, the various well classifications do not therefore tell us much about the characteristics of this irrigation by underground water. It is not so much the depth of the well that is important in understanding the constraints of this irrigation as the type of captured aquifer; data which is rarely explained or available.

3.2. Wells, Geology and Major Environmental Constraints

If we classify the wells in the study area according to their physical features, seven types of well (in a broad sense including wells and bores) can be identified (Table 3.1): large dug wells (also called open wells) with or without side bore, with or without masonry work (type A), small well-*cum*-shallow bores (type B), large and small well-*cum*-deep bores (type C and D), borewells (type E), shallow and deep tubewells (type F and G). It is difficult to determine the limit in distinguishing between a shallow and a deep TW, as it is not a fixed depth, but depends on the geological characteristics. It is more or less 60-70 m and as much as 100 m deep (*see* above for the category used by the Department of Irrigation, in the census of minor irrigation schemes).

Table 3.1: Different Types of well in the Study Area, according to the Geological Characteristics and their Main Physical Features

Type	Geological Characteristics			Depth Feature	Installation Feature
	Rocky Zone	Sedimentary Zone	Alluvial Zone		
A	large open well (+ side bore)	large open well	*(previously small open well)*	*Shallow*	well only
B			open well-*cum*-shallow bore (progressively replaced by shallow TW)		
C		large open well-*cum*-deep bore (few)		*Deep (more than 40-50 m in the case studies)*	we see a well, and a tube/bore deepens the well
D			open well-*cum*-deep bore		
E	borewell				
F		deep tubewell	deep tubewell		tube only
G			shallow tubewell (recent)	*shallow*	

We can see in Table 3.1, that on the one hand, the same type may have different catchment depths while, on the other hand, only the borewell and open well-*cum*-shallow bore are specific to one geological formation, whereas other types may be found in at least two different formations. One type of well is not specific to geological formation, and even less so to the actual hydrogeological situation which the farmer faces. However, in a given geological formation, there will only be a limited number of types of well, open wells being mainly found in rocky areas and well-*cum*-bores and TWs in alluvial and sedimentary areas (cf. schematic map on Fig. 3.2).

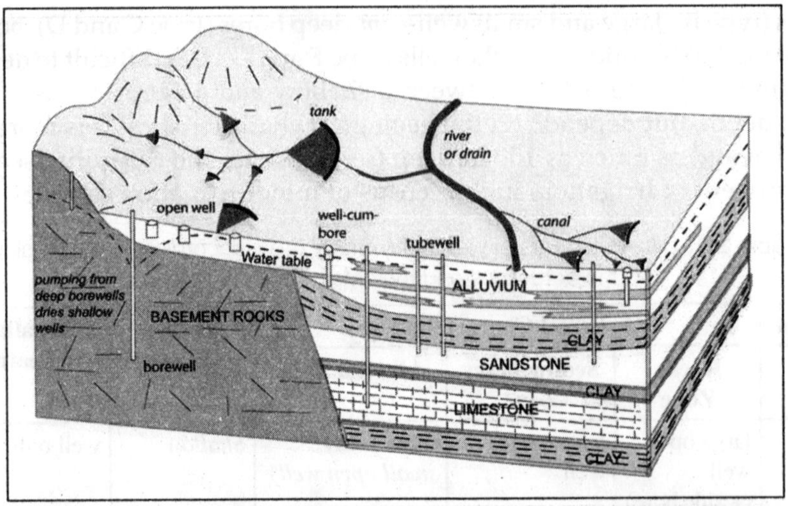

Fig. 3.2: Schematic Cross-section with Reference to Well Types (Adapted from Gale *et al.*, 2002)

The geology determines much of the environmental constraints that farmers face in accessing groundwater. Thus, in rocky areas, the three main constraints are: (i) the hardness and compaction of the rock that makes drilling expensive or risky because of the location of water in cracks (environmental plus economic constraints); (ii) the strong dependence on rainfall to recharge the upper aquifer that supplies the open wells. During the period of drought, from 2001 to 2003, which hit the region, most farmers dug their wells deeper by 2 to 5 metres, hoping to catch water stored deeper; (iii) the flow of such wells dwindles the further we get into the dry season. At Koratur, for example (see location Fig. 1 & 2 in Box 6), the irrigated area during the hot season is 3 to 6 times less than during the rainy season.

In sedimentary areas, open wells which tap groundwater, as in rocky areas, depend on surface water (rain, waterways, irrigation) for any recharge. Otherwise, deep drilling has to be done, since the exploited aquifers in the area are more than 80 m deep. This is expensive and is largely entrusted to the wealthiest farmers. Tubewell owners are indeed real water lords, according to Bentz's description (2001: 67-69): only members of socio-economically dominant castes own such installations and sell the water.

Thus, they maintain that "it is only since we've been selling the water that we've begun to reap profits in agriculture." Besides, according to one of them (*ibid.*), some wealthy owners have built extra wells "simply to sell water to others". The overexploitation of deep groundwater implies a drop in their level and forces farmers to dig even deeper (cf. Chaps. 5 and 6).

In alluvial areas, environmental constraints vary from place to place, depending on geomorphological, geological, hydro-geological (thickness and depth of the aquifer) and the overexploitation of groundwater. We can see on Fig. 3.3 showing the geographical distribution of aquifers in the district of Puducherry, that the most productive aquifers (Cuddalore and Vanur sandstone aquifers, cf. Chap. 2) are at variable depths and over variable thicknesses depending on the villages. The first, located under the upper aquifer, is 245 m thick near Bahour (SW of Puducherry) and 66 m thick west of Puducherry, but exists no longer a dozen kilometres west from there. The second, located under the first (at a distance not specified in the GoP manual, 1999), ranges from 40 to 90 m thick. Furthermore, rather shallow aquifers, which were productive in the past, are less so today, causing farmers to drill deeper to catch another aquifer. But some aquifers, such as that of Ramanathapuram sandstone (below the Vanur aquifer) tapped in Sorapet cannot be used as it contains some hydrogen sulphide, which would explain the rotten smell and the orange colour of the water. Below this is the bedrock, so that the depth of any bore is limited here to 50 m by environmental constraints, while the tapped aquifer is not very productive. The depth of

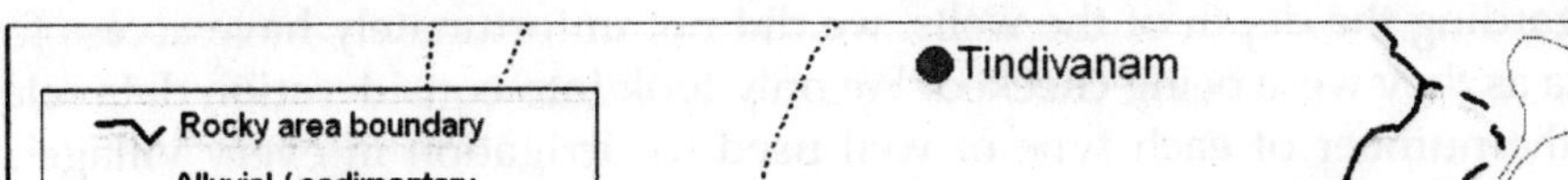

Fig. 3.3: Schematic Representation of the Extent of the Main Aquifers (given that they are not spatially homogeneous)

drilling cannot unfortunately serve as an indicator of the environmental constraints because of the heterogeneity of hydrogeological characteristics in alluvial areas. While a low maximum depth does indeed suggest that a geological limit (rock) prevents drilling deeper, this tells us little of the productive capacity of the captured aquifers. Thus, in another village in the alluvial area (TVN), located near the Ponniyar River, wells are drilled to a maximum depth of 25 metres but they tap a very productive groundwater and allow three crops a year: once again we may conclude that the depth of drilling in no way reflects the hydraulic constraint. The alluvial area is far from being homogeneous in terms of access to water.

3.3. Spatial Distribution and Well Density

We used data from the Well Census conducted in 2004, by the State Ground and Surface Water Resources Data Centre[2], on a village scale. Five types of well are identified: open wells (or dug wells), dug-*cum*-bores, tubewells, borewells and filter points (found only in the Kaveri Delta and made of a small tubewell using pebbles). Types B and D of our typology do not figure in this typology and have to be integrated in type C, even if the depth and the size of wells are different. Regarding the depth of the wells, we did not unfortunately have access to the data as they were being checked. We only took into consideration data relative to the number of each type of well used for irrigation in every village[3]. We combined this data with that of the 1991 Population Census which gives the cultivated area, and then calculated well density per cultivated area[4].

To compare maps, we used the same density classes for all maps. The choice of classification privileged the fact that we had roughly the same number of open wells, which come in larger numbers than TWs and well-*cum*-bores. Moreover, we chose to round off the figures for the limits of classes (10-30-50 wells per 100 ha). We could not take into account the limits reported in literature regarding the influence of the number of wells on the use of a tank, since these figures refer to well density in the tank's command area, whereas here we are calculating these densities over the total cultivated area. Indeed, for a density greater than 20 wells per 100 hectares, Palanisami *et al.* (2000: 74-75) observe a decrease in the use of the tank. Janakarajan (1993: A57) does not give fixed limits for well density but a higher well density is more often linked to a defunct traditional irrigation institution[5].

Our class limits are thus: 10/30/50 wells per 100 ha. Two factors may slightly distort the reading of Figs. 3.5 to 3.7: on the one hand, there are far more villages owning open wells than tubewells while, on the other hand, the density classes do not have a uniform number of records for TW (cf. Fig. 3.4).

It is nonetheless clear from the graph that there are proportionately more high well densities in villages with open wells than in those with TWs. If we

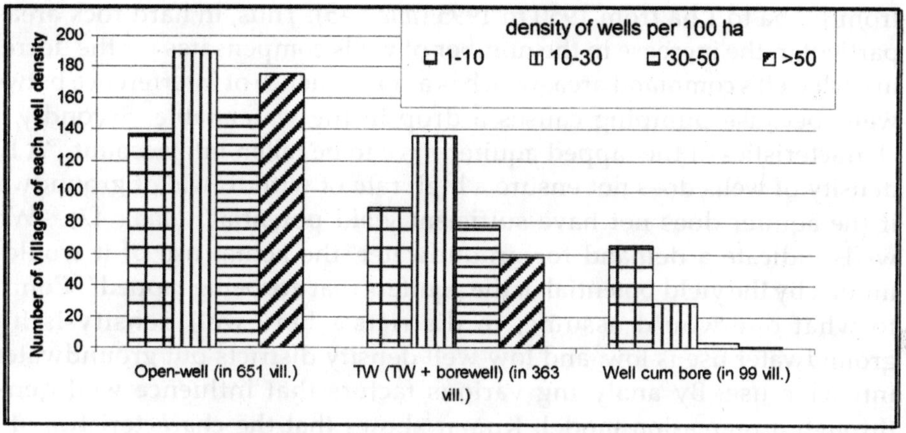

Fig. 3.4: Distribution of well and TW Densities

could correlate these data with the irrigated area, the difference in densities would be even greater because, in a TW area, almost the entire surface area is irrigated, unlike an open well area where there is a lot of rainfed land. This high density of open wells may be explained by the fact that the net surface area irrigated by each well is smaller than the one irrigated by a TW (from 0.4 to 2.8 ha respectively and as much as 8 ha according to our interviews, with a respective average of 0.5 ha and 2.4 ha in Villupuram district as reported by the Minor Irrigation Census, 2000-01). To irrigate the very same area, one needs more open wells than TWs.A greater number of wells does not necessarily mean that more people have access to water, as well owners may have more than one well. Moreover, in the case of open wells, they rarely sell water, unlike tubewell owners who, on the other hand, for the most part sell their water. Depending on the market for water (in particular linked to the availability of the resource, but also to several socio-economic and social factors—cf. Dubash, 2002), all farmers and all land can access water even if there is a relatively small number of wells, whereas in villages with a high well density access may be limited to well owners. Well density alone does not explain the situation regarding water access. This information would have to be combined with the surface area and the percentage of irrigated land, data that we do not have for 2004.

As M.D. Kumar (2007: 42-80) points out, it is crucial not to misunderstand the interpretation of these well densities: the number of wells is not indicative of irrigation intensity[6]. First of all, the growing number of wells over time does not mean that there has been an increase in well irrigation, since some data show that the average area irrigated by a single well has decreased over time. For example, the average command area of a dug-out well in India dropped from 1.7 ha to 1 ha from 1950 to 1995 (*ibid.*: 45). Thus, in hard rock areas in particular, the increase in the number of wells compensates for the decrease in each well's command area, which is a consequence of interference between wells because pumping causes a drop in the water table. Secondly, the characteristics of the tapped aquifer have to be taken into account. "A high density of wells does not ensure a high rate of withdrawal of groundwater if the aquifer does not have sufficient yield potential" (*ibid.*: 47). "While wells indicate a demand for groundwater, the actual use of it would be limited by the yield potential of the aquifers that are being tapped". Contrary to what one would assume, in districts where well density is high, groundwater use is low, and low well density districts put groundwater to intensive use. By analysing various factors that influence well density through a regression model, Kumar shows that the characteristics of the aquifer are more important than other factors—such as the renewable annual recharge, the degree of land fragmentation, the cropped area—to explain well density. Above all, there is no correlation with the population density. By providing this demonstration, Kumar disagrees with Roy and T. Shah's hypothesis (2003) whereby groundwater development is linked to population density and its water demand, not to its physical characteristics. Kumar emphasizes that "any analysis of groundwater socio-ecology without due consideration to physical settings would be incomplete" (p. 80). Such argumentation validates the use of a GIS method which includes local physical characteristics (and breaks down administrative borders). Janakarajan (1991: 46) also specified that well density as such does not mean much as the use of the wells is linked to recharge conditions from surface waters. These remarks reinforce the point that well density does not at all reflect water availability or irrigation intensity.

If we look at the maps (Figs. 3.5 and 3.7) of the density of open wells and tubewells, one is struck by several aspects. Firstly, well density varies greatly from one village to another, which does not allow us to establish homogeneous well density areas. The multitude of factors to be taken into account (as described by Kumar) explains this geographical diversity. Secondly, the geographical dichotomy between wells and bore is not quite as marked as what we were led to think from field trips and the previously shown table. Though wells are mostly found in rocky areas and TWs in sedimentary and alluvial areas, they are not restricted to these areas. In rocky areas, tubewells, or more accurately borewells, can be found in some places in large quantities. Yet the relatively large number of open wells in alluvial areas is what is most surprising. One factor may explain their presence in the alluvial area bordering the rocky area: the geological boundary does not correspond to the geomorphological boundary, and this whole

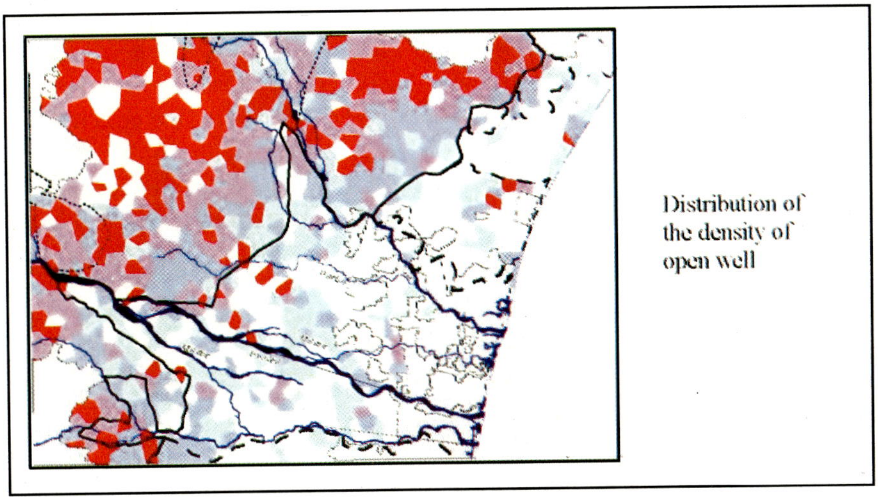

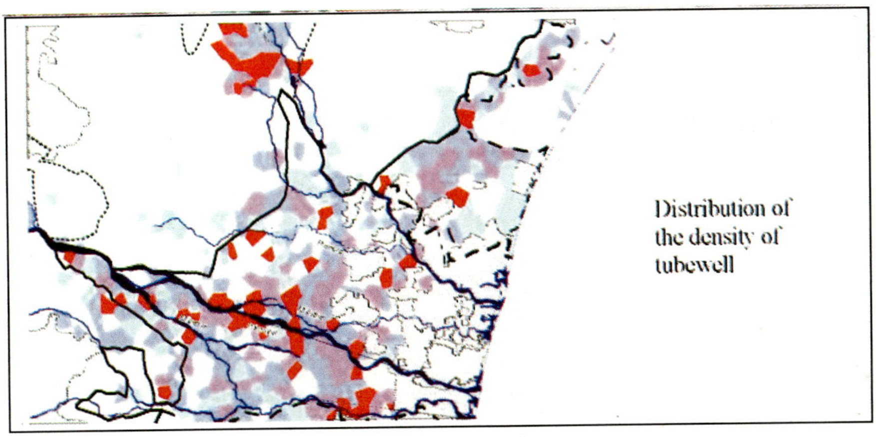

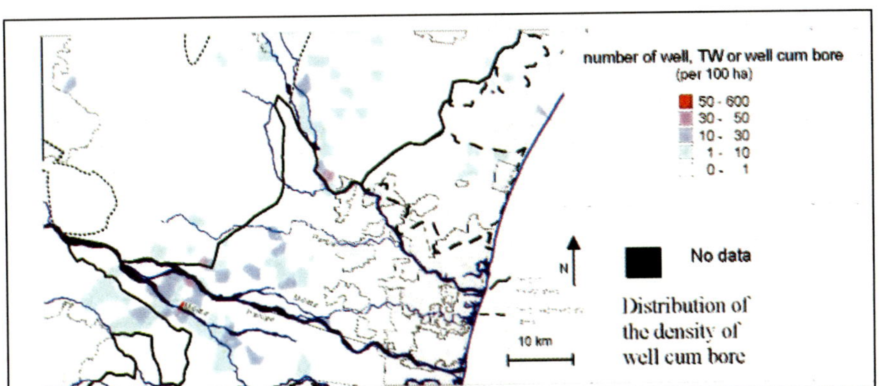

Figs. 3.5, 3.6 and 3.7: Density of open wells, tubewells and borewells, and well-*cum*-bores, on a village scale (the limit of each village is nevertheless fictive)

western part of the alluvial plain has a thin layer of alluvia: the hydro-geological characteristics are similar to those of the rocky area. Thus the geological limit alone must not be taken into consideration, yet the geomorphological limit must be in order to understand the spatial distribution of wells in this area (especially in the North and West of Villupuram town).

Thirdly, the highest well densities (when all open wells and TWs are taken together, Fig. 3.8) are in rocky areas, for the reasons already explained above.

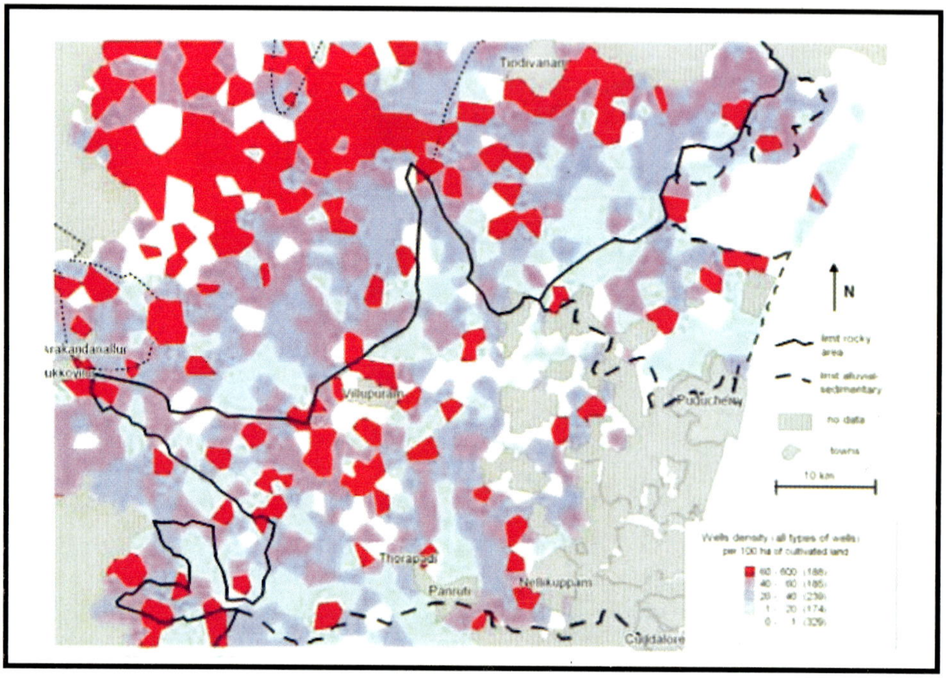

Fig. 3.8: Well Density (all types of well) per village (the limit of each village is nevertheless fictive)

It is interesting to compare the current distribution of wells with the former one prior to the Green Revolution (where there were mostly open wells), which Adicéam's study indeed allows us to do (1966: 222): "The distribution of wells, [on a Tamil Nadu scale] is not only related to the existence of aquifers. Overall, the number of wells is inversely proportional to the number of irrigation channels and tanks: where rivers, channels and tanks are most developed, wells occupy little space. Wells are most common in regions least equipped with channels and tanks." These comments do not hold today because: firstly, easier access to groundwater—either from a technical or an economic point of view, thanks to motorized pumps, the electrification of rural areas and various State subsidies—has helped to spread well and TW technology; secondly, compared to what we have just seen, today environmental conditions for accessing groundwater affect well density more than the presence or absence of a tank does; in this case, there are more wells in rocky areas. Similarly, the places where the irrigation tank was less common in 1991 (proportion of land irrigated in this way, see Fig. 4.4, Chap. 4) are not those where there is now a higher well density, once again because well density does not denote irrigation intensity.

3.4. Do Wells and Bores Reflect a Dichotomy in Accessing Groundwater?

Access to groundwater is related to different elements that can be divided into three main categories: (i) water availability, which can fluctuate over time; (ii) the technical facility with which the resource can be accessed, depending on the depth of the aquifer and the geological features that one comes up against; and (iii) economic and social conditions for using water, such as the financial capital necessary for installing the infrastructure or socio-economic constraints in purchasing water from others.

The question as to whether wells and bores reflect a dichotomy in accessing groundwater implicitly refers to water availability, and to a lesser extent, to its easy access. Wells, in fact, depend on rainfall and surface waters for recharging the upper aquifer they draw water from, whereas bores capture deeper groundwater, often more productive and less reliant on rainfall. Yet this is not necessarily the case and, as we have seen, neither the type of well nor the geological formation nor bore depth gives a clear picture of the hydrogeological situation that TW owners are faced with. Indeed, one would have to know the yield of the tapped aquifer.

Since it is difficult to obtain data on the aquifer yield, one indicator of the productivity of this aquifer might be the main crop calendar. It may provide a rough idea of environmental constraints, since the intensity of the crop cycle is directly linked to irrigation and more especially to groundwater, which is the only one available from February to September. Nevertheless, crop calendars vary greatly and several elements may be taken into consideration: percentage of land that was dry in the past but which is irrigated today, number of irrigated seasons, type of crop, etc., given that this is combined with the representativeness of these data on a village scale, as well as the various ways in accessing groundwater in the village itself—linked to local geographical disparities in hydrogeological features.

Among irrigated crops on *nanjai* land (or irrigated paddy fields, see Box 7), paddy, sugarcane and sometimes banana, tapioca or groundnut can be found. Non-irrigated crops may also be raised outside the rainy season, such as blackgram, rage, millet and cowpea (*karamani*). Let us take, as an estimation, the number of paddy seasons secured on *nanjai* land as an indication of groundwater availability or of the facility in accessing it (Fig. 3.10). Extreme cases (1 or 3 crops per year) provide us with significant information: where three crops of paddy per year are achieved, groundwater is present in adequate quantities to meet the current demand, with the soil being sufficiently rich and improved to produce this single-crop farming year-round. This can only be observed in some villages in the alluvial area. Conversely, when a single paddy crop is achieved, little groundwater is available: this can be observed systematically in the rocky area (although some well owners do raise two paddy crops) and in some parts of the alluvial area. These parts coincide with those where the presence of open wells was noted in the above analysis of well distribution (and at the time surprised us). Open wells may, therefore, indicate irregular and limited access to water due to the availability of the resource, but the presence of a TW tells us little about water access, a fact that we concluded previously. As for villages, where two paddy crops are achieved, these are found in the alluvial area and the sedimentary area, where TWs are the general rule: the second crop (at least) is cultivated using groundwater. However, with no fieldwork to base our findings on, it is impossible to say whether the number of paddy seasons is limited by water availability or by bore depth which pushes up the cost of accessing the resource—especially in the sedimentary area or the area south of Puducherry where deep groundwater is captured—or by other factors, such as a lack of nutrients in the soil, the readiness to grow another crop, or to respond to projects developed by the Puducherry government which, since 2006 pay farmers for raising crops, other than paddy, in the third season, etc.

In certain cases, the calendar for *punjai* land allows us to make a slightly more accurate analysis. It is more difficult to characterise the calendar for *punjai* land due to its greater diversity, both in terms of the crops raised and the cropping periods (it is therefore impossible to establish a "typical" cropping calendar; few are presented however in Fig. 3.9). As our observations show, sugarcane is systematically irrigated by groundwater throughout the year. Yet this crop does not prove to be a discriminatory and explanatory, element in water access, as it is present in all villages, whatever the environment, and farmers get a low yield in the case of a lack of water. The planted area might be an indicator, though the presence of sugarcane in the cropping calendar is not.

One difference can be observed, however, between villages where all land is irrigated and those where only part of it is, the rainfed crops being millet, blackgram, groundnut (second season is irrigated), black-eye bean or cowpea (*karamani*), gingelly (or sesame), redgram (*tuvarai*), or bringal, over one or two seasons. Furthermore, some crops have a 10-to-12-month cycle, whether irrigated (yam, turmeric) or not (cotton) and plantations last several

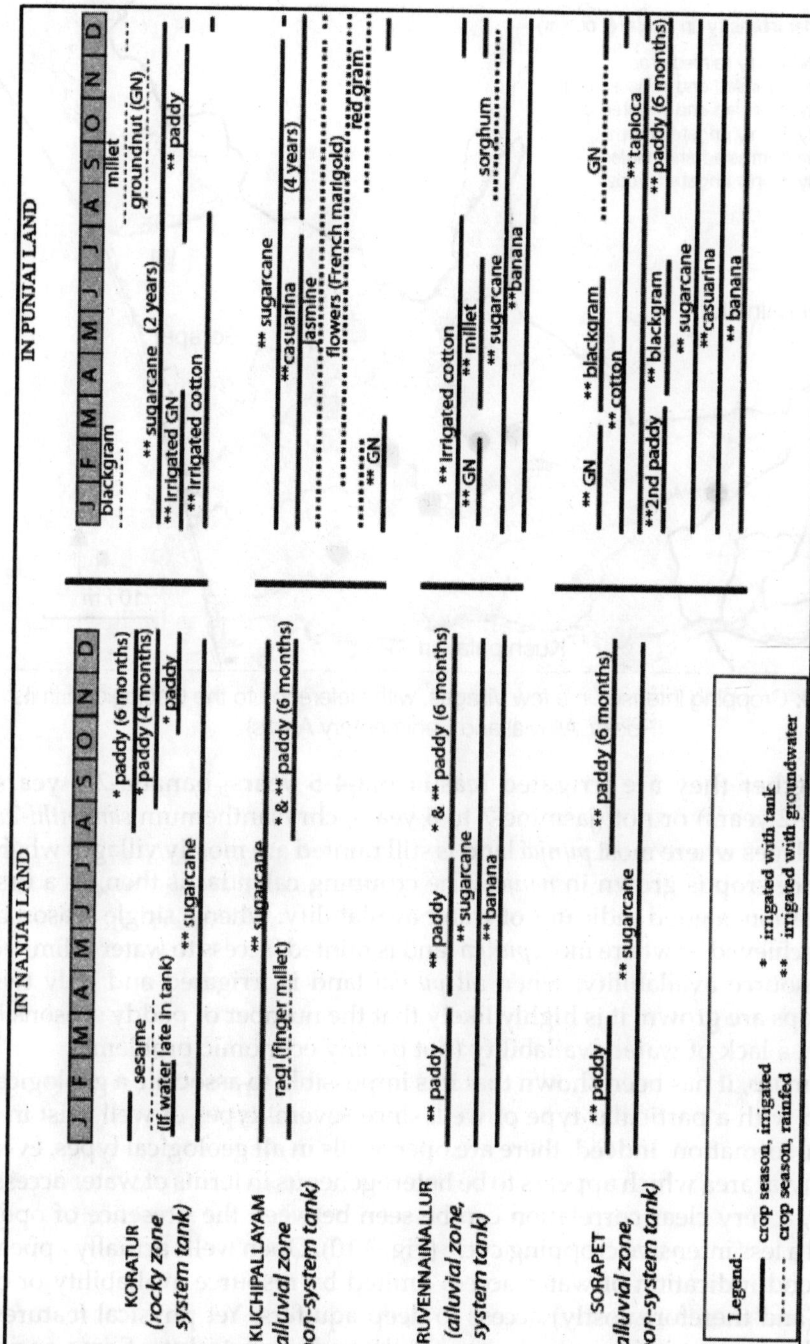

Fig. 3.9: A Selection of Crop Calendars

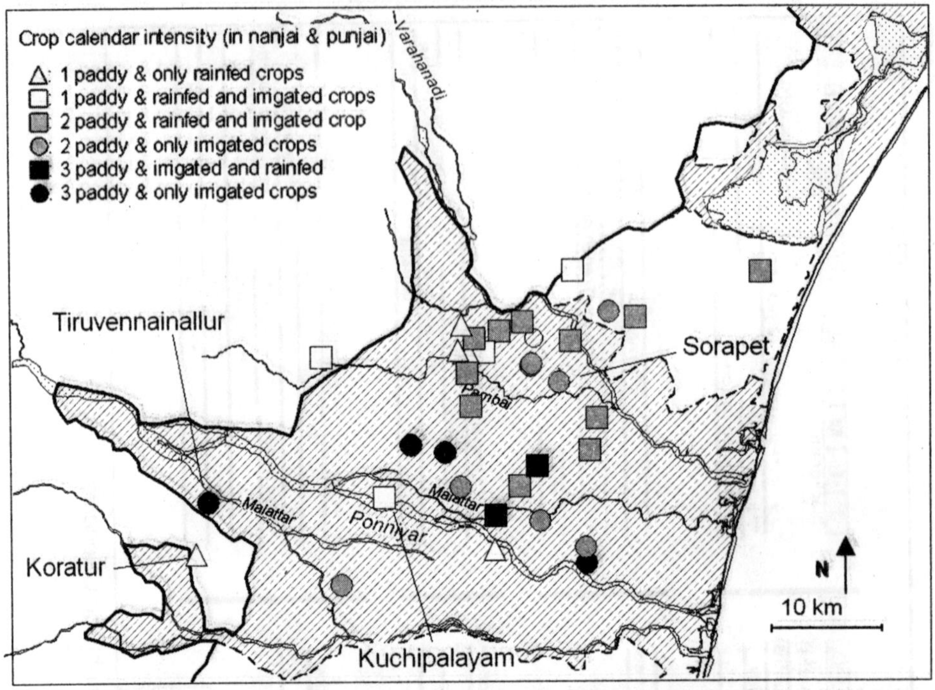

Fig. 3.10: Cropping Intensity in a few Villages, with Reference to the Geological Limits (Rocky, Alluvial and Sedimentary Areas)

years whether they are irrigated (casuarina-4-5 years; banana-2-3 years; sugarcane-2 years) or not (jasmine-7 to 8 years; chrysanthemum *samanthi*-2.5 years). Villages where most *punjai* land is still rainfed are mostly villages where a single rice crop is grown in *nanjai*. The cropping calendar is then, as a first approximation, a good indicator of water availability: when a single season of paddy is achieved or where most *punjai* land is rainfed, access to water is limited due to resource availability; when all *punjai* land is irrigated and only two paddy crops are grown, it is highly likely that the number of paddy seasons is limited by a lack of water availability (not by any economic problems).

To sum up, it has been shown that it is impossible to associate a geological formation with a particular type of well, since several types of well exist in a geological formation. Indeed, there are open wells in all geological types, even in the alluvial area which appears to be heterogeneous in terms of water access. However, a very clear correlation can be seen between the presence of open wells and a less intensive cropping cycle (Fig. 3.10). Open wells actually appear to be a good indication of water access limited by resource availability or of difficult (and therefore costly) access to deep aquifers. Yet physical features, though they are very important, are not the only constraints. Some socio-economic reasons may limit the implementation of TWs or well-*cum*-bores, so that the cropping cycle is less intensive. In areas with open wells, the small amount of land irrigated outside the rainy season belongs to well or borewell owners, showing a striking socio-economic difference in water access. Conversely, borewells are not indicative of any particular hydrogeological situation, and their density or depth does not help to draw any conclusion regarding access to the resource: they may coincide just as much in situations of water abundance as in those of a barely productive aquifer. Generally speaking, they do, nevertheless, allow for more cropping seasons than open wells, which explains why they are associated in people's minds with greater water availability than in the case of open wells. Emphasis must be laid on the fact that open wells tap the upper aquifer which relies on rainfall and surface water for its refill, whereas TWs can tap aquifers with very different hydrogeological features, which prevents us, therefore, from being able to generalise about water access characteristics for TWs. In addition, the implementation of these infrastructures is risky, since, due to the geographical heterogeneity and the complex functioning of aquifers, the availability of this "hidden" water is not guaranteed.

NOTES

1. See for example GoTN (2002: 59), Census of India, 1991.
2. I would like to thank this centre for providing these data and for the warm welcome provided by its staff.
3. The data available are of various categories: types of well whatever their use, different uses (irrigation, domestic, industrial, commercial, others). The number of each type of well for irrigation is not available as such: I, therefore calculated it, based on the hypothesis that domestic wells and industrial wells were tubewells.
4. One element may confound these densities: Casuarina plantations, for example, are listed in woodland and plantations, not as cultivated land. With Casuarina being planted mainly in the plateau region between Tindivanam and Puducherry, the figures for cultivated land are thus lower than arable land in this area and well densities have no doubt been maximized.
5. Well density and effectiveness of the traditional irrigation institution (tii) in Palar area (Janakarajan 1993: A57).

tii effective	tii less effective	tii defunct
2 to 52 wells/100 ha	0 to 32 wells/100 ha	17 to 82 wells/100 ha

6. It would have been interesting to try and confirm Kumar's remarks about the fact that well density does not denote irrigation intensity, by comparing the geographical distribution of these two phenomena. However, the data

we have been using were recorded at different dates (net irrigated area in 1991, cf. Fig. 4.3 and number of wells in 2004, cf. the density Fig. 3.4) and cannot be used for that purpose.

REFERENCES

Cesselin, F. 2001. Irrigation par tank et motopompes à *Tiruvenainallur au Tamil Nadu: compétition ou complémentarité de techniques?* Mémoire du Master of sciences DAT (Développement Agricole Tropical) du CNEARC, Montpellier.

Dubash, N. K. 2002. *Tubewell Capitalism: Groundwater Development and Agrarian Change in Gujarat.* Delhi: Oxford University Press.

Girod, P. 1937. *Programme de grands travaux pour le développement de l'outillage économique et social.* Pondichéry: Imprimerie du government.

Government of Tamil Nadu. 2002. *Groundwater perspectives. A profile of Villupuram district, Tamil Nadu.* State ground and surface water resources data centre, Water Resources Organisation.

Janakarajan, S. 1991. *The extent and impact of ground water irrigation in Tamil Nadu: some macro and micro evidences.* Madras: Madras Institute of Development Studies (Working paper 102).

— —. 1993. In search of tanks: some hidden facts. *Economic and Political Weekly* 28: A53-A60 (Review of Agriculture).

Kumar, M. D. 2007. *Groundwater management in India. Physical, Institutional and Policy Alternatives.* New Delhi: Sage Publications.

Lefebvre, O. 2006. Dynamiques de gestion de l'eau dans un contexte de surexploitation des eaux *souterraines.* La région de Pondichéry et de Villupuram au Tamil Nadu, Inde du Sud, Mémoire présenté en vue de l'obtention du Mastère Spécialisé Développement Agricole Tropical, CNEARC, Montpellier.

Palanisami, K., and K. W. Easter. 2000. *Tank Irrigation in the 21st Century. What Next?* New Delhi: Discovery Publishing House.

Roy, A. D., and T. Shah. 2003. "Socio-ecology of groundwater irrigation in India," *in Intensive Use of Groundwater. Challenges and Opportunities.* Edited by Llamas, M.R. and C. E. Lisse, the Netherlands: Balkema Publishers.

CHAPTER

4

SPATIAL DYNAMICS OF TANK IRRIGATION

Olivia Aubriot

4.1 Tank Use: Tentative Estimation

Tanks are today undergoing a sharp decline in their use in Tamil Nadu and even more so in Puducherry district where nearly 40 per cent of them are no longer used for irrigation purposes, while those that are used supply only an average of 35 per cent of their command area (data from Atkins, 2001, consultant for the TRPP). Admittedly, the data regarding command areas can be discussed and can represent theoretical numbers as the determination of such areas is not necessarily based on the availability of water but can be linked to some political objectives. For example, we do not know if the official command area can realistically be irrigated by the tanks. We have to be cautious with the figures that can hide many situations.

The aim of this sub-section is to analyze the geography of the non-use of tanks and to see if it follows a specific spatial pattern. We will then map tank irrigation, and combine it with some statistical analysis. The data have the specificity on being on a village scale: one set is taken from the 1991 population census and the other has been specifically collected from field enquiries. We will then test the importance of some environmental factors (such as easy access to groundwater, reliability of the tank water supply) to explain the more or less important proportion of tank irrigation in the area studied.

(a) Which Mapping of Tank Irrigation?

Even if irrigation is merely water spread over surface areas and therefore seems simple to be represented on maps, mapping Indian irrigation is not quite so easy, as there is a large diversity in the types of irrigation, and there are different irrigation and cropping seasons. Irrigation data refer to a net or gross irrigated area, or to the importance of irrigation according to the cultivated area (again net and gross areas), or to each type of irrigation. This means that a number of maps are necessary to understand the importance of irrigation in agriculture and to situate an area in a broader context (district, State, Southern India, etc.). Statistics are therefore needed, even if their reliability is often contested: see for example Rawal (2001) who pointed out, for West Bengal, that irrigation statistics are collected by different agencies with various classifications—thus limiting the scope of comparative study—and that some data are missing, such as irrigation intensity, irrigation regularity or areas receiving multiple sources of irrigation.

For mapping, tanks are quite specific as they occupy a lot of space and constitute—whenever they are full—a real "waterscape": on a satellite image[1] or an aerial view, a region with tanks appears as a "spotted" area with water patches; on topographic sheets, the omnipresence of the outlines of these water bodies is striking. Such is the case in our study area. With a GIS tool, we measured the area covered by tanks, using the tank limits[2] represented in the topographic sheets to a scale of 1:50,000. This covers 352 km², corresponding to 8 per cent of the total area (cf. Fig. 4.1). This percentage is considerable compared to the cultivated area, which is estimated as 44 per cent of the total area[3].

Maps of the tank command area (or *ayacut* area or land theoretically irrigable by tank) are scarcely found and even more so of the tank irrigated area. Such depictions are made on a village scale especially by NGOs, but very seldom on a larger scale. One exception is worth noting: with the implementation of water user associations where the members are landowners of the *ayacut* area, Tamil Nadu's PWD has started to prepare maps representing both tanks and their *ayacut* area (Fig. 4.2). On the contrary, in Puducherry, where members of the Tank Associations are not exclusively *ayacutdar* (landowners in the *ayacut*), such maps do not exist: only tanks are represented.

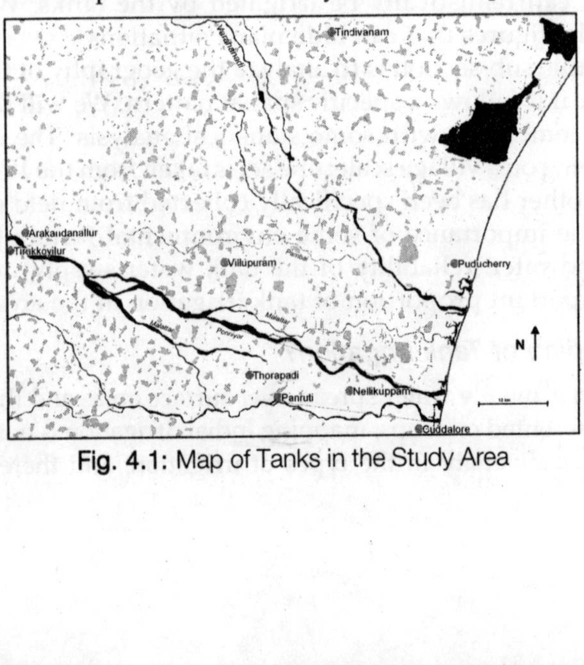

Fig. 4.1: Map of Tanks in the Study Area

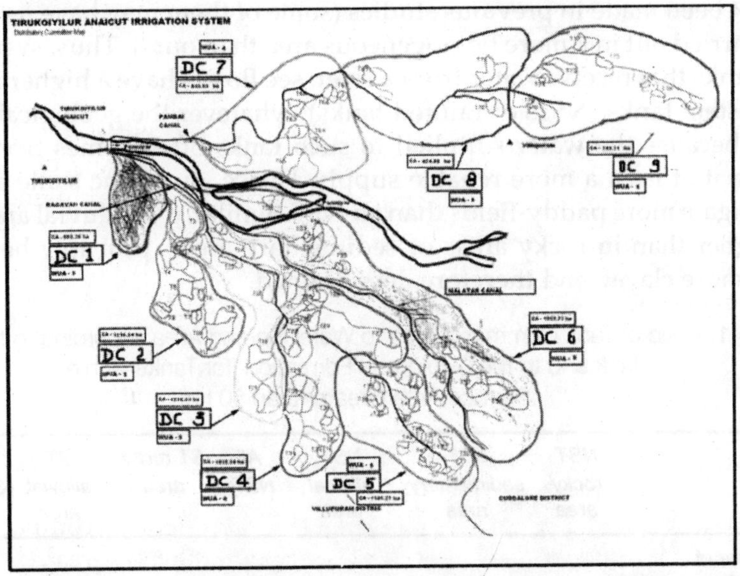

Fig. 4.2: Tirukoylur Tank Irrigation Systems: Canal, Tanks, their Registered Irrigated Area (*Ayacut*) and the Distributory Committees Designed by PWD and in the Process of Being set up (*Source*: PWD)

Drawing tanks does not represent the actual irrigation they ensure for two main reasons. First, the "irrigable area/tank area" ratio varies depending on the tanks. Thus in our study area, for 147 tanks with a spread area of more than 50 ha, the command area ranges from 0.5 ha to 5.3 ha per hectare of tank, with an average ratio of 1.7. For all our calculations, we only took into account these tanks to avoid making too many mistakes when measuring the tank area using the mapping method[4], the command area being the one given by each PWD section. Compared to data given in the literature, this ratio is higher than that given in most of the previous studies: an average of 1.3 in a study dated 1959 carried out by the Planning Commission on 245 tanks in Tamil Nadu (Vaidyanathan, 2001: 15); in Andhra Pradesh, for 45 tanks, Von Oppen and Rao (1987: 13-14) give a ratio of 0.9 for tanks with an *ayacut* of less than 40 ha and 1.5 for tanks with an *ayacut* of more than 400 ha. Yet our ratio is much lower than the one presented by Palanisami *et al.* (2000: 41 and 63) for 67 PWD tanks at various sites in Tamil Nadu: small tanks (less than 100 ha) have a ratio which is close to two, whereas larger tanks have higher ratios which are as much as five. Contrary to results from Van Oppen and Rao, and from Palanisami *et al.*, in our sample, the size of the tank has no influence on the ratio. However, our data clearly show (Table 4.1) that the ratio varies according to the geological support and the type of tank, distinctions that had not been made in previous studies (some of them may have nevertheless been carried out in a more homogeneous area than ours). Thus, system tanks (STs, tanks that receive water from a river, see Box 1) have a higher ratio than non-system tanks (NSTs or rainfed tanks), whatever the geological zone, no doubt because the water supplied to such tanks often comes from a river, making it at least a more reliable supply. Hence, given the same area, a ST will irrigate more paddy-fields than an NST. Similarly, in alluvial areas, ratios are higher than in rocky areas or sedimentary areas, probably because the soil is more clayey and therefore, less drained.

Table 4.1: Ratio of Tank Command Area to Water Spread Area according to the Type of Tank and to the Geological Formation (for Tanks with a Surface Area Greater than 50 ha)

	NST rocky area	*NST sedimentary area*	*NST alluvial area*	*All NSTs*	*ST rocky area*	*ST alluvial area*	*All STs*	*All tanks*
Number of tanks for measurements	70	9	14	92	16	38	55	147
Ratio: *ayacut*/ tank spread area	1.46	1.25	1.62	1.45	1.89	2.02	1.97	1.66

The second reason why mapping tank provides no information on tank irrigation is that the irrigable area is not necessarily equal to the area irrigated by the tank: there are bores and wells everywhere, which compete or even replace tank irrigation. Moreover, the tank irrigated area may vary for several reasons, including the availability of water in the tank. Only field investigations confirm whether water comes from the tank or not, and determine the tank irrigated area.

To our knowledge, no systematic data, and even fewer maps on the degree of tank use—the percentage of the irrigable area actually irrigated—are available, although they would help in understanding the importance of tanks for irrigated agriculture, in particular compared to underground irrigation. Here we propose to look into the issue of tank use on a village scale, and to analyze the importance of these structures for the local dynamics of water management, using two methods: one based on data from the 1991 population census, the other one through field enquiries.

(b) Maps Using Data from the 1991 Census

The 1991 population census has the advantage of defining, for each revenue village, the area for each category of land use and, for irrigated land, the net irrigated area for each type of irrigation (nine types were identified, which we regrouped into four types: tank, canal, well and TW). As explained in Chap. 1, we could use the GIS database prepared by C. Z. Guilmoto, in which all villages of South India (Karnataka, Andhra Pradesh, Kerala, Tamil Nadu and Puducherry) were georeferen ced and associated with data from the 1991 population census. The methodology used to constitute this large-scale spatial database of the South India Fertility Project (SIFP) is described in Guilmoto *et al.* (2004); the first analyses of the geographic distribution of irrigation on a scale of Southern India and the three main irrigation types in Tamil Nadu are presented in Guilmoto (2002). This author also uses the diversity of the census data to make correlations between irrigation and socio-economic data such as population density, the proportion of agricultural labourers, etc. "As expected, irrigation appears to be positively associated with higher demographic density. Irrigated villages support larger populations, most notably because the workforce [is] engaged in agriculture"; in vastly irrigated areas, agriculturists are mainly landless labourers, and statistics show that "when irrigated land has crossed the 50 per cent [of land cultivated] threshold, additions to the agricultural workforce are entirely due to the number of landless labourers and marginal farmers. These observations indirectly confirm that irrigation tends to worsen land inequality." On the other hand, irrigation also increases labourers' standard of living by providing them with job opportunities. Some of the maps from this project are also available on line, in Oliveau (2003).

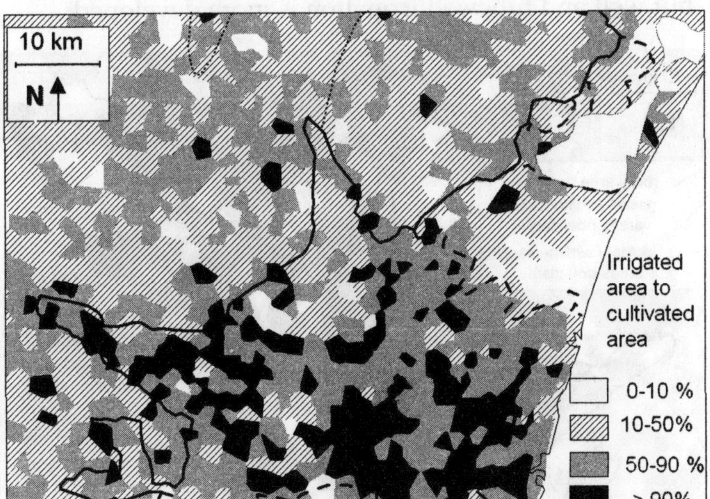

Fig. 4.3: Proportion of Cultivated Land which is Irrigated, according to the 1991 Population Census Data

If we isolate the data related to our study area, and add the limits of geological formations, we can see on Fig. 4.3 that the alluvial area is nearly fully irrigated (except on the western part where the geomorphology reveals that the alluvium deposit is very meagre). Thus, for the following maps that represent the proportion of the different types of irrigation to the total irrigated area, one has to be careful in interpreting them as they do not include the non-irrigated area: see for example, the difference between Fig. 4.4 and Fig. 4.5 representing the area irrigated by surface water, in proportion to the total net irrigated area and to the net cultivated area respectively. Thus, if the area irrigated by surface water appears to be large in Fig. 4.4 for south and south-west of Kalluvelly tank, it is only relative to the total irrigated, not cultivated area. Using data from the census poses another limitation: only the net sown area is accounted for, that means the area irrigated during the monsoon season. Thus tank irrigation is taken into account, yet such data (and then our figures) do not demonstrate the extent of GW irrigation nor irrigation intensity throughout the year (irrigation intensity being the number of cropping seasons permitted by irrigation).

Figures 4.6, 4.7, 4.8 and 4.9 represent the spatial distribution of each type of irrigation in the study area (open well, tubewell, tank and canal respectively) as a percentage of the total irrigated area. The maps show that there is more tank irrigation in the sedimentary and charnockite area than in the granitic gneiss area, since the ratio of irrigated area to cultivated area is more or less the same, but well and tubewell irrigation is more developed.

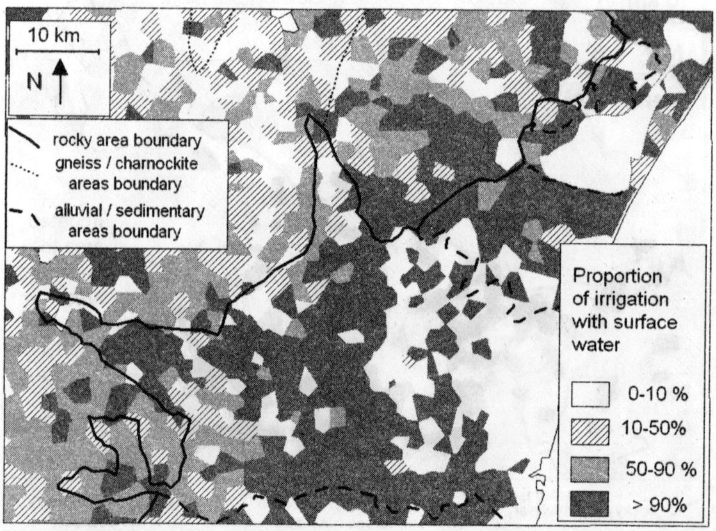

Fig. 4.4: Proportion of the Irrigated Area supplied by Surface Water (Ratio of Net Area irrigated by Surface Water to Total Net Irrigated Area) according to the 1991 Population Census Data, with Reference to the Main Geological Boundaries

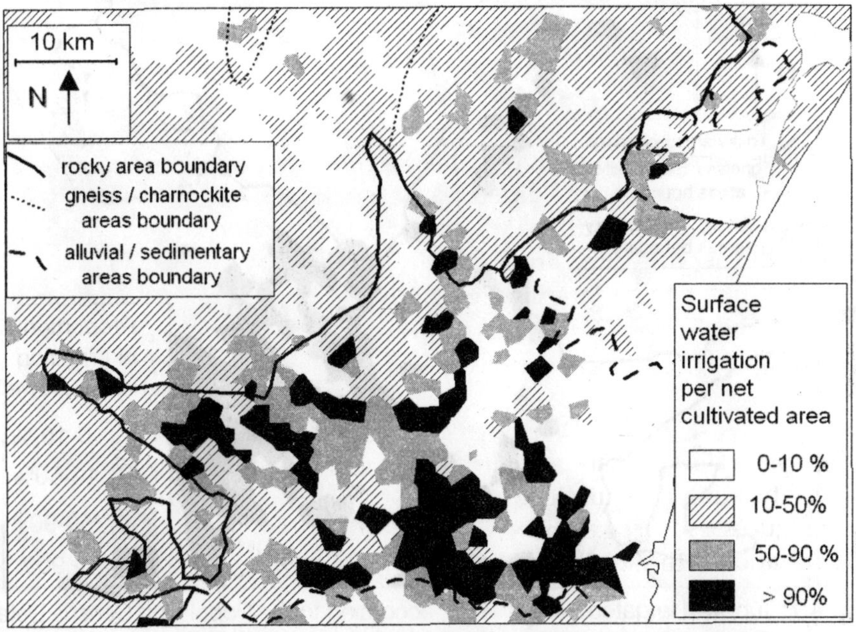

Fig. 4.5: Proportion of Area Irrigated by Surface Water (Ratio of net Irrigated Area to net Cultivated Area) according to the 1991 Population Census Data, with Reference to the Main Geological Boundaries

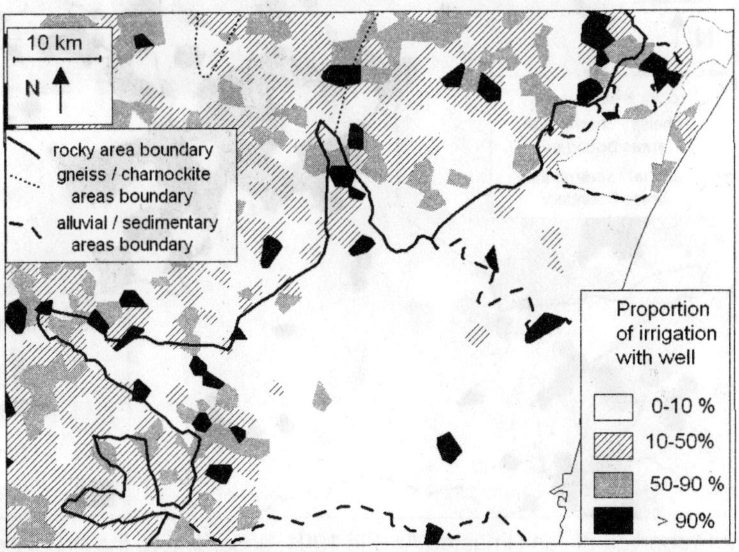

Fig. 4.6: Open Well Irrigated Area in 1991, according to the Population Census Data (Net Irrigated Area to Total Irrigated Area), with Reference to the Main Geological Boundaries

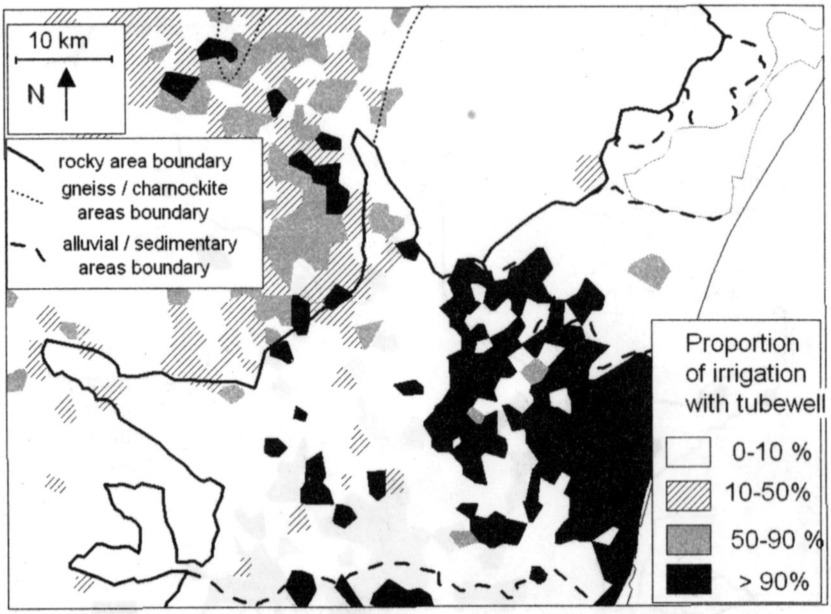

Fig. 4.7: Tubewell Irrigated Area in 1991, according to the Population Census Data (Net Tubewell Irrigated Area to Total Irrigated Area), with Reference to the Main Geological Boundaries

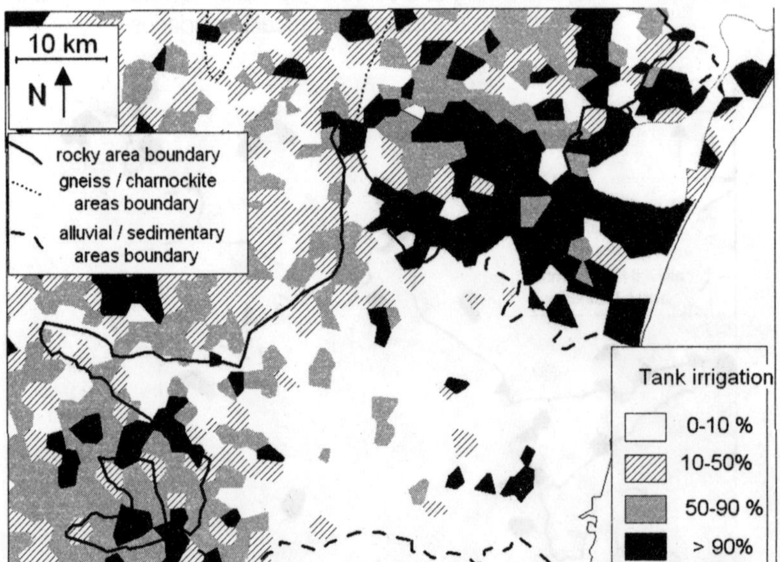

Fig. 4.8: Proportion of Tank Irrigated Area in 1991, according to the Population Census Data, with Reference to the Main Geological Boundaries

If we superimpose on Fig. 4.8, the tank map taken from topographic sheets to a scale of 1:50,000 (or 1 inch for 1 mile for the those dated 1915 and 1930) which give the outline of tanks, as well as the limit of the alluvial area (from the 1:250,000 geology map), we come up with Fig. 4.10. It is clear that many tanks are not used in the alluvial area, while Fig. 4.7 shows that a large part of the alluvial area is entirely irrigated by tubewells. This methodology helps to rapidly locate tanks which are not used, if the census data are accurate. The next method tests this accuracy.

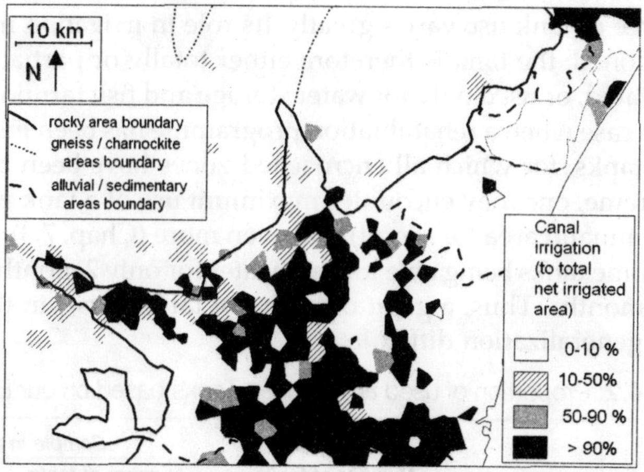

Fig. 4.9: Proportion of Canal Irrigated Area in 1991, according to the Population Census Data, with Reference to the Main Geological Boundaries

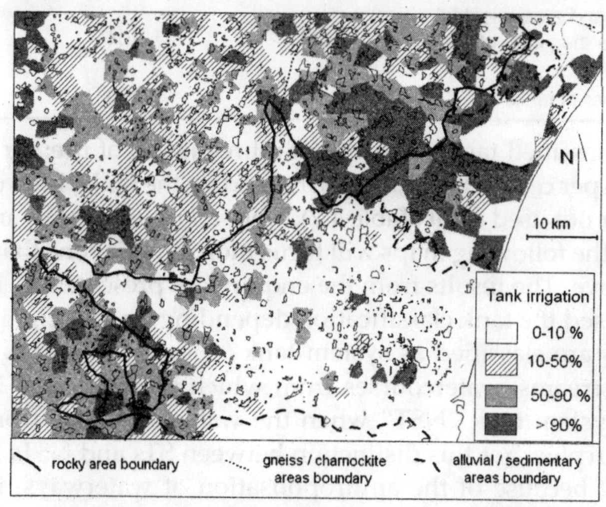

Fig. 4.10: Tank Irrigated Area according to 1991 Census and Tank Spread Area, with Reference to the Main Geological Boundaries

(c) Maps from Field Enquiries

In order to corroborate the above findings, namely that tank irrigation has been abandoned in the alluvial area, we investigated 212 tanks[5]: 170 in the alluvial area (169 to be precise, since one of the tanks has been made into a bus station in Villupuram), two in the sedimentary area and 40 in the rocky area. As we were particularly interested in the impact of TW development in the alluvial area and in finding out if the non-use of tanks is homogeneous from a spatial point of view, we made more enquiries into this area.

The degree of tank use varies greatly. Its role in irrigation may have been totally abandoned: the tank is therefore either totally or partially encroached on and cultivated, or used only for water storage and fish farming. The latter is generally the case when a rehabilitation programme has been instigated, as for Puducherry tanks, for which all encroached zones have been cleared. At the opposite extreme, one may encounter maximum use of a tank which irrigates the whole command area (or *ayacut*) and even more (Chap. 7.3). Tank capacity also varies, some tanks being able to store water for only 2 months while others up to 6 to 8 months. Thus, a great diversity of tanks exists in the study area, making any generalization difficult.

Table 4.2: Proportion of used and unused Tanks based on our Enquiries

	Sample	*Sample in alluvial area*	
	Number of tanks	*Number of tanks*	*%*
Unused tanks for irrigation	53	53	31
Little used (less than 25% of command area irrigated)	13	13	7
Used (more than 25% of area irrigated)	145	104	61
Total number of tanks	211	170	100

First we classified tanks according to three levels of use: for irrigation use (more than 25 per cent of the *ayacut* is irrigated by the tank), little used (less than 25 per cent) or not used at all (Table 4.2). Since all unused tanks are found in the alluvial area, the following maps will focus more on this zone than the maps we presented above. The results from the enquiry are presented in Fig. 4.11.

We also used the tank classification depending on the origin of their water supply: tanks are classified as system tank "ST" when their water supply is from rivers, streams, watercourses from which an inlet canal has been dug; and as non-system tank "NST" when the water supply is from rain and/or upper tank surplus. Yet this distinction between STs and NSTs is not quite so easy to apply because of the anthropolisation of waterways. Indeed, in the landscape around Villupuram and Puducherry, where rivers are not perennial,

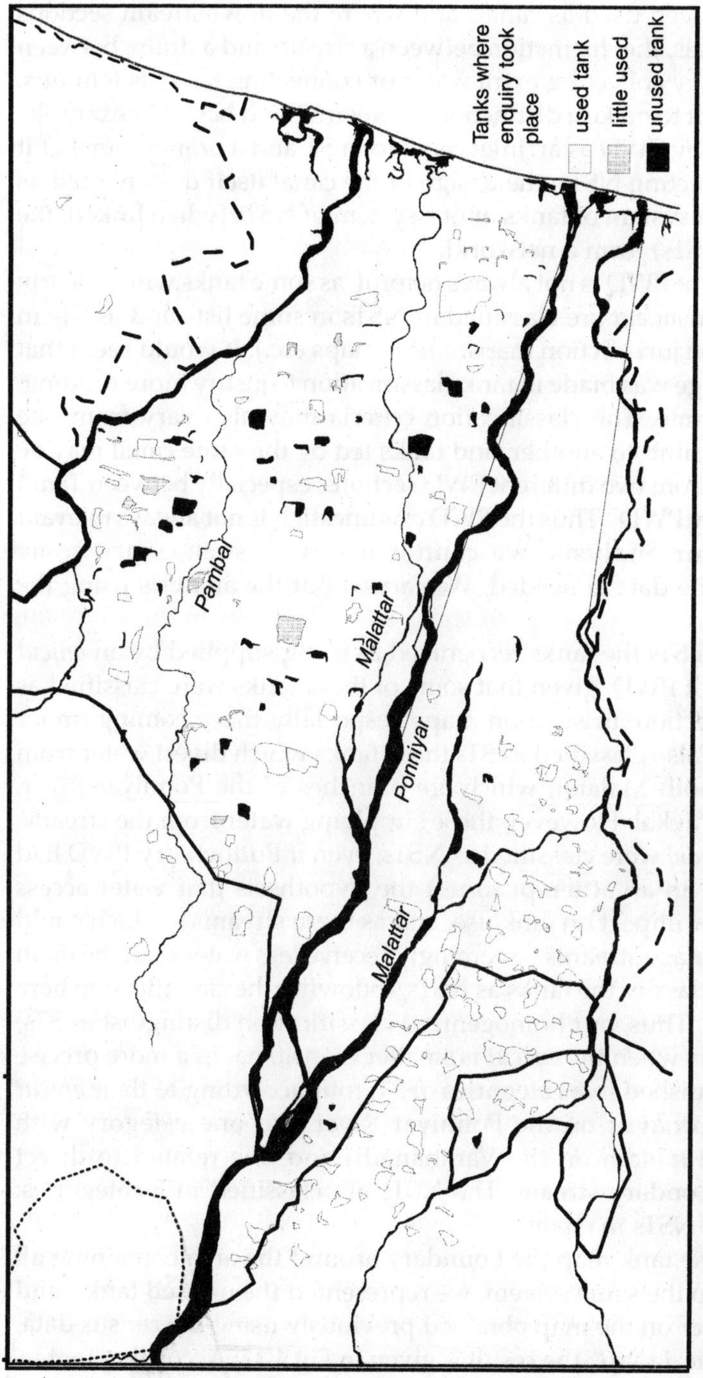

Fig 4.11: Distribution of Used and unused Tanks in the Alluvial Area according to Our Field Survey (only tanks where enquiries took place are represented, about 200 in total)

where streams have been used as canals and where the downstream sections of drains become canals, the distinction between a stream and a drain, between a canal and a waterway collecting extra water or connecting tanks is tenuous. Therefore, it is difficult to make a distinction between ST and NST. For example, a canal is considered by PWD as an inlet canal for a ST and a branch canal of it as a drain canal connecting NSTs: the design of the canal itself does not tell us if the canal is part of a system of tanks or of a system of NSTs [when linked, the non-system-tanks (NSTs) form a network].

The classification of PWD is not always helpful, as some tanks which clearly receive water from an *anicut* are classified as NSTs in some lists and as STs in others (such as section jurisdiction maps, *anicut* maps etc.). It would seem that in some places a change was made in tank classification to justify more funding, at a given point in time. The classification criteria may also vary from one PWD administrative unit[6] to another, and tanks fed by the same canal may be classified differently from two different PWD sections, especially between Tamil Nadu and Puducherry PWD[7]. Thus the PWD classification is not always relevant or coherent. For our analysis, we cannot use it as such, since some homogenization of the data is needed. We carried out the analysis using the following criteria.

We designated as STs the tanks recognized as being supplied by an *anicut* in the "*anicut maps*" of PWD, given that some of these tanks were classified as NSTs on the PWD section jurisdiction maps[8], especially those coming under Ellis *Anicut*. We then also classified as STs those tanks which divert water from North Malatar or South Malatar, which are branches of the Ponniyar River, and from Sellangal Vaykal. However, those supplying water from the streams *Nari* and *Kuduvayar odai* were classified as NSTs, even if Puducherry PWD had classed them as STs. In an atttempt to test the hypothesis that water access reliability has a major impact on tank use, and as these streams—which could be designated as drainage streams—seemingly receive less water than the main ones, we decided to classify the tanks as NSTs, following the classification here of Tamil Nadu PWD. Thus, our homogenized classification distinguishes STs, NSTs, and small NSTs when the *ayacut* is smaller than 40 ha. In a more precise typology, we distinguished five categories of ST, four according to their *anicut* (Tirukoilur, Ellis, Sornavur on the Ponniyar River and one category with Vikravandi and Vidur dam on the Varahanadi) and one related to direct diversion from a secondary stream. The NSTs are classified in 3 categories: isolated, among STs, NSTs in chain.

We marked on the tank map the boundary around the area containing all the STs dependent on the same system, we represented the unused tanks, and superimpose the latter on the map obtained previously using the census data, and on the hydrological map[9]: the result is given in Fig 4.12. According to this, unused tanks are mainly NSTs or downstream STs, while most STs are used. As mapping on this scale conceals the result regarding small tanks, there is a need to analyze these data using a statistical tool, which is the purpose of the next sub-section.

Fig. 4.12 reveals another fact that raises a question about the data: according to our enquiries, there are more used tanks in the western part of the alluvial area than suggested by the census data. How can one explain this difference in results? It is not due to the time difference between the collection of data (15 years between the 1991 census and our enquiries) since GW irrigation developed more during this period—most of the alluvial area is today irrigated, either by tank or groundwater, land previously rainfed now being irrigated by groundwater—and at State level data reveal a decrease in tank use during the same period. Is it due to inaccurate data from our interviews[10], or to some heterogeneity in the census classification since the intensive use of TW irrigation corresponds to Puducherry district (Fig. 4.17) and since the central part of the alluvial area appears as an area irrigated by canals in the census data (Fig. 4.9), though tanks exist? This uncertainty about data homogeneity regarding the tank and canal classification has led us to aggregate tank and canal irrigation into surface water irrigation (Fig. 4.13): contrary to the tank irrigation map, the census let suppose that surface water is used and our enquiries show many tanks are used.

(d) Precision Through Statistical Analysis

Using statistics we analyzed the data from field enquiries—those mapped above. Table 4.2 confirms that unused or little used tanks for irrigation are only to be found in the alluvial area, and that in this area, two-third of tanks are used. When a tank is used, the irrigated area represents on average 60 per cent of the command area, ranging from 10 to 100 per cent. Therefore, tanks in the alluvial area have not been completely abandoned as the 1991 census data may lead us to believe (tank irrigation Fig. 4.12).

If we differentiate the tanks in our sample[11] according to their type (Table 4.3), we see that the proportion of unused tanks is larger among NSTs (44%) than STs (25%). When seeking an eventual difference depending on the ST, some interesting figures emerge: first of all, systems with a larger *anicut* have globally the same percentage of unused STs as systems characterised by a "direct" diversion from the Ponniyar or from a secondary stream. Dependency on a large or small *anicut* has no influence on the non-use of the tanks. Thus, either the type of diversion has no impact, and therefore what is important for farmers is to secure water from the river; or there is an impact, but it has to be measured in another way (such as the percentage of irrigated area). Our data prompt caution regarding the interpretation.

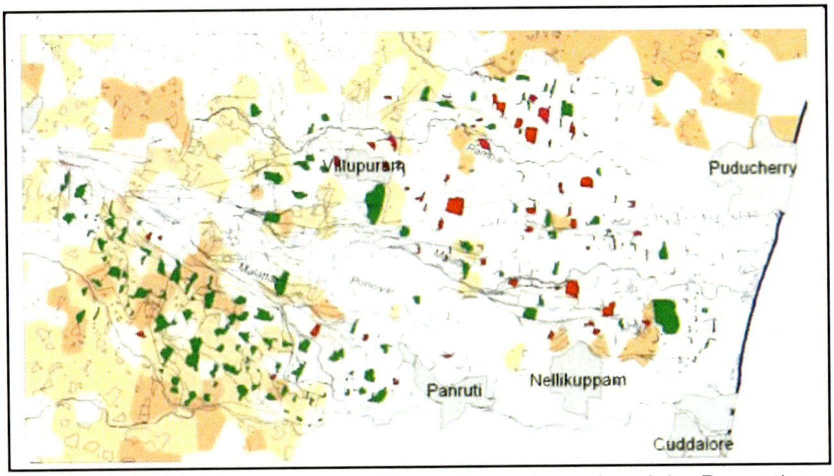

Fig. 4.12: Comparison of the use of Tanks (From Survey) and the Proportion of Tank Irrigated Area (from 1991 Census Data) with Reference to the Command Area of STs

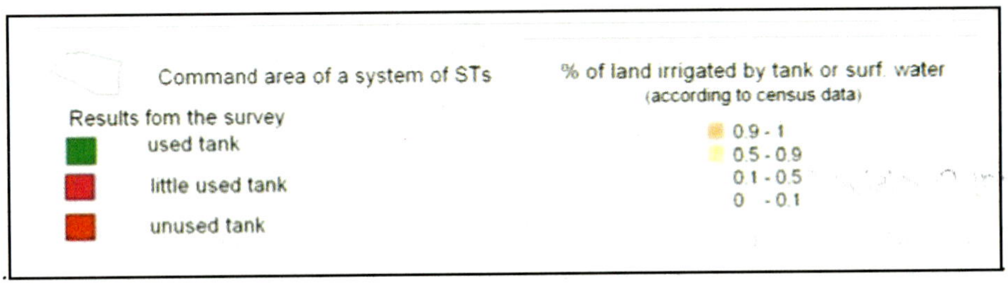

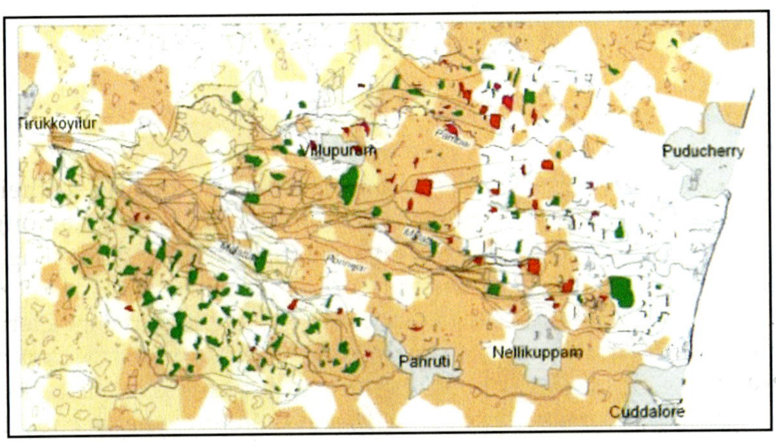

Fig. 4.13: Comparison of the Use of Tanks (from Survey) and the Area Irrigated by Surface Water (from 1991 Census Data) with Reference to the Command Area of STs

Table 4.3: Proportion of Used Tanks according to the Type of Tank, in the Alluvial Area (Based on our Enquiries)

	System tanks		*Non-system tanks*	
	Number of tanks	*% among ST*	*Number of tanks*	*% among NST*
Unused	28	25	24	44
Little used	6	4	7	12
Used	79	71	24	44
Total	113		55	

Table 4.4: Proportion of Unused Tanks according to the Type of *Anicut* (Based on Our Enquiries)

	ST anicut	*ST "direct"*	*ST anicut*							
			ST1 = Tirukoilur anicut		*ST2 = Ellis anicut*		*ST3 = Sornavur anicut*		*ST5 = Vikravandi*	
			Number of tanks	*%*	*Tanks*	*%*	*Tanks*	*%*	*Tanks*	*%*
Unused tanks	24%	25%	6	13	4	26	5	42	5	33
Total number of tanks in sample	87	32	45		15		12		15	

The position of the *anicut* along the Ponniyar is important: the more the *anicut* is situated downstream, the more tanks are unused (Table 4.4). Is this because there is less water downstream? As we did not try to relate data to flow measurement, we cannot categorically answer this question and the hypothesis presented in the following paragraphs is based on interviews with farmers. First of all, let us point out the importance of the Sathanur dam for these systems. Since the construction of the dam (1959), farmers benefit from the regulation of the flow and, at least those located near upstream tanks dependent on Tirukoylur *anicut* note that they can get water for a longer period of time, and especially for a second paddy season. In addition, they know that they can put pressure on the PWD Executive Engineer and ask for water to be released from the dam. Indeed, at the lower part of the Ponniyar basin, the "old system tanks" (dependent on the *anicut* mentioned in Table 4.4) have priority for water allocation over the newly created irrigation area (left and right bank canals) located below the dam, even if they are located much further downstream (the Tirukoylur *anicut* is 50 km downstream from the dam). This led to the absence of any irrigation by the left bank canal during the summers of 1993 and 1994, whereas the "old *ayacut*"—name of the command area for all these ancient system tanks—did receive water [Shaktivadivel *et al.*, 1995 (Vol. 2): 117]. Therefore, the construction of the dam has improved the Tirukoylur *anicut* water supply, while the rules on sharing dam water are less restrictive towards this old *ayacut* than the command area of the new canal (outside our study area).

In Tirukoylur *anicut*, only the tailend tanks are not used, while in Ellis *anicut*, some tanks situated in the middle of the system are not used either. Farmers from these tanks explain their disinterest in tank irrigation because of the lack of a reliable water supply. Tail-end tanks have, therefore, more difficulty in getting water which has been diverted from the river. At Vadanur, the last tank on the Pambai canal, farmers say that the process of fetching water from the upper parts of the canal has always been an important activity for irrigators (*see* Chap. 9). Nevertheless, the lack of water supply in downstream tanks is not the only reason for explaining the non-use of tanks, since in Sornavur *anicut* for example, some upstreams tanks remain unused. This system is, however, quite specific. It brings water to Bahour tank, a famous tank in the history of Puducherry and the second biggest tank in the district, whereby a set of rules gives this tank overall priority for water allocation: located at the end of Bangaru channel; it receives water first and only then do the upstream tanks receive water (there are gates along the channel). Until 1970, date at which the *anicut* was built in concrete, these tanks received water according to a rota system, with each village knowing for how many hours it could divert water into its tank. Since this date, the Puducherry PWD has been in charge of managing water allocation; there is no longer any collective management of Bangaru channel (CNEARC, 2006). Two reasons may explain that some upstream tanks on this Bangaru channel have been abandoned and that most of them are encroached on. The first one is related to a drop in water availability, especially a shortage of water during drier years, since the *anicut* is downstream from other *anicuts* which divert a lot of water. The second may be because of the lack of priority in water allocation, since the Bahour tank and its downstream tanks—fed through the Bahour tank—are used.

Among NSTs, PWD and panchayat tanks show more or less the same percentage of unused tanks and of irrigated area, the latter being very small, less than half the *ayacut* (Table 4.5). The irrigated area is rather small, either because there is an alternative source of irrigation (groundwater), especially for the panchayat tanks in our sample that are located in the alluvial area, or because the tank is encroached on and the storage potentiality of the tank is thus reduced (the groundwater irrigation alternative may also explain encroachment in the alluvial area since cultivation in the encroached area uses GW irrigation), or because the potentiality of the tank is in itself very low because of its dependency on rainfall, or because it is silted. When we examine the state of tank encroachment, we see that there is more encroachment in panchayat tanks, as if local institutions were more lax in controlling this phenomenon. Encroachment may explain the disuse of panchayat tanks: since a reduction in area for small tanks may have more impact on tank disuse than for bigger tanks (Palanisami *et al.*, 2001) and since, when compared to NST-PWD tanks, they suffer more encroachment and are smaller, encroachment may have a bigger impact. Palanisami *et al.* also note that about 30 per cent of the water spread area of tanks is encroached on and smaller tanks have slightly higher encroachment rates (2001: 62).

If we examine tank encroachment between STs and NSTs in our data, we see that the percentage is the same among used tanks, about one quarter of the area: encroachment alone cannot explain the disuse of STs. Besides, a 10 to 20 per cent encroachment may not disrupt tank filling so much since the real water spread area is quite low (see next sub-chapter). However, encroachment is much more frequent for unused STs than for unused NSTs, as if the non-use of STs was more acceptable or perceived as definitive by farmers than that of NSTs. Thus farmers do not hesitate to use the tank area for their own cropping. This may also be linked to the possibility of avoiding the tank water supply more easily in STs than in NSTs, the water of which may damage crops.

Table 4.5: Usage and Encroachment of Different Types of Tank

	NST PWD	*NST panchayat*	*All NST*	*ST*
Unused tanks (%)	39	35	36	20
Ayacut irrigated (%)	43	49	46	69
Encroachment and tank used	19	32	24	24
Encroachment and tank	41	62	48	69

A multivariate analysis of data[12] shows a relationship between encroachment and the disuse of tanks, without determining the effect or the cause: a further 1 per cent of an encroached area is associated with a 0.5 per cent lower probability of being used ($p = 0.0001$). Another interesting result of this analysis is that the water spread area and the command area of the tank are not statistically related in a significant way to the probability of the tank being used, somewhat explaining the fact that PWD and panchayat tanks have more or less the same percentage of unused tanks.

Let us sum up the main results of this statistical analysis. The non-use of tanks is not as homogeneous from a spatial point of view as the map based on 1991 population census data led us believed. Moreover, all the NSTs in the alluvial area have not been abandoned, as might be assumed from the census maps. On examining the tank maps, the fact that only bigger tanks can really be seen on this scale, has minimized the importance of the small tanks which are being used. Nevertheless, we can say that unused tanks are mainly NSTs or STs located downstream from the system because the proportion of unused tanks is larger among NSTs (42%) than STs (25%), and because, for used tanks, the proportion of the command area irrigated by NSTs is lower (46%) than that of STs (69%). As regards NSTs, encroachment along with an unreliable water supply may be the main reasons for explaining the non-use of panchayat tanks.

4.2. Tanks, Their "Shared Water" Areas and Environmental Factors

(a) Filling of Tanks

The results of our interviews on the use of tanks are as follows: STs are used more than NSTs, STs irrigate a larger part of their *ayacut* than NSTs, while unused STs are situated mainly downstream from the system, except for one system privileging the last tank supply, that of Bahour. These results lead us to think that the reliability of surface water access combined with easy access to groundwater (abandoned tanks are all located in the alluvial area) are key factors which explain whether tanks are used or not for irrigation. Let us analyze the water tank filling data.

A study combining satellite images and GIS was carried out in collaboration with the University of Provence (France) by Savin (2007) under the direction of P. Oliva, in order to examine the rate at which tanks fill up[13]. This work was conducted prospectively to determine, first of all, the elements needed for further study and to draw some assumptions from a small number of satellite images, some of which are freely available *via* the Internet. Thus by achieving a precise match between satellite images and 1:50,000 topographical maps—using Mapinfo software—, it was possible to isolate the data concerning the spread area of the tanks and to define—using the Winimage software[14]—the percentage of surface area covered in water (taken from the infrared canal), of damp ground, of ground covered in vegetation and of bare ground. Six dates were used: January 1977, January and August 1991, October 2000, June 2002 and February 2004, with almost the same resolution of less than 30 m, except for 1977, where the resolution was only 79 m (Land sat): the differences in some results between this date and the others are significant enough for us to believe that they are not the result of a difference in resolution[15].

The main results of this analysis are:

- A clear difference is observed between the filling of STs and NSTs: STs receive water more frequently and have a larger flooded area than NSTs (cf Fig. 4.14). This is no doubt due to the concrete and permanent *anicuts* on which a large part of these STs rely, and to the Sathanur dam that regulates the flow of the Ponniyar and guarantees a supply of water to the *anicuts* until the month of February.

- A striking difference exists between the filling of STs and NSTs between the years 1991 and 2000 (cf. same graph), which may be due to a lesser use of NSTs (see above) and therefore to a lack of interest in maintaining feeder canals, whereas in 1977 the proportion of tank surface covered in water was similar for both categories.
- The area of tanks covered in water is relatively low compared to total tank area, even during the monsoon period[16] (Fig. 4.15). From the graph we see that whatever the date, 50 per cent of the tank area is not covered in water. Since trees and shrubs have grown in the tanks concealing part of this water-covered area, the result is underestimated. However, the amount of water stored is not linearly proportional to the surface covered in water. The relation between the area and the volume of each tank differs according to tank topography. According to Bossy (1985: 78), in two tanks in the gneiss region near Gingee (located north-west of the study area), half of the area covered in water represents only one-fourth and two-fifths respectively of the total storable volume. The small covering of water in the tank, added to the fact that the soil of the foreshore has residual moisture and is fertile, explains that farmers are tempted to cultivate the foreshore area of the tank; unfortunately the method applied to analyze land use from the satellite images does not allow us to differentiate between different types of land cover (crops, shrubs or stubble).
- It is difficult from these data to interpret the tank filling process according to the geological environment in which the tank is built (alluvial plain and sedimentary plateau, granite, charnockite) because each date presents different situations for each environment, and the spatial variability of rain makes integrating them in the spatial analysis somewhat complicated.

The six dates here are not enough for a proper diachronic analysis of how tanks fill up, since these dates refer to different periods in the year over several years (one date during the monsoon, three after it, one during the hot season, one before the monsoon), and therefore, make the comparison somewhat inconsistent. However, this study has clearly shown the difference in the watered area between STs and NSTs at these 6 dates and has demonstrated that the tank foreshore is rarely watered. It also raises interesting questions and encourages us, on the one hand, to start a more in-depth study based on the analysis using more dates on which tanks fill up; and, on the other hand, to produce a more accurate typology of land cover inside the tanks (thus based on a local-scale analysis) in order to determine variations in encroachment. Such a study, however, will necessarily have a limited diachronic approach due to the date of the first satellite images available.

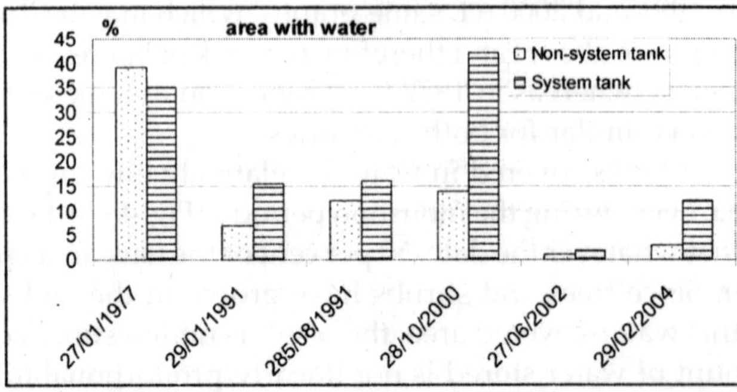

Fig. 4.14: Percentage of Surface Area covered in Water, according to the Type of Tank between 1977 and 2004 (For 611 Non-System Tanks and 148 System Tanks)

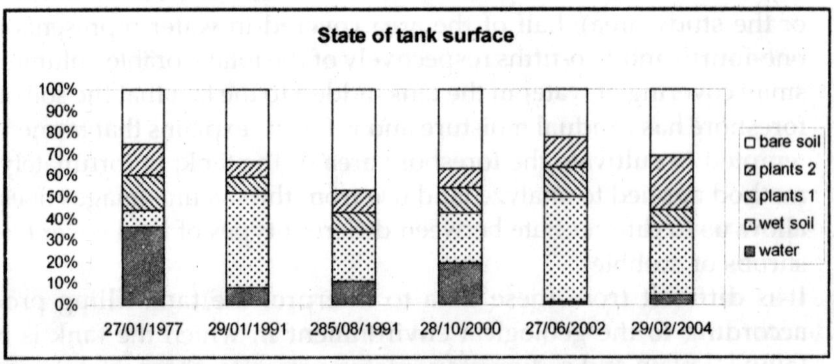

Fig. 4.15: State of Tank Surface (for 1782 Tanks)

(b) Importance of a Secure Supply

All our results highlight the importance of a secure water access: when tank water availability is not guaranteed and ground water is easily accessible (case in the alluvial area), tanks are hardly used and even abandoned. When tank water availability is not guaranteed and ground water not easy to access, the area irrigated by the tank amounts on average to half of the *ayacut*. Once water availability has been guaranteed (the case of most STs), most tanks are used and 70 per cent of the *ayacut* is irrigated. When tanks supply water for less than one crop season (less than 3 months), farmers show less interest in the technique, an issue also raised by Palanisami *et al.* (2003b: 56-64) and in Coste and Ploumpidis (2007).

Thus, environmental factors such as the reliability of water availability in the tank, the duration of water storage, combined with easy access to groundwater, but also the social factors that determine this reliability of the water supply (rules for water allocation along a canal, pressure on engineers to have water delivered from the dam, as described in 4.1.d) are vital in the dynamics of tank use.

Part of this uncertainty concerning the water supply is due to variations in rainfall, especially at some periods of the year where rainfall is crucial for cultivation. As Palanisami and Easter (2000: 57-62) have shown, poor rainfall, especially in September, affects tank irrigated areas: September rainfalls guarantee the first filling of the tank and if they are below expectations, farmers reduce their cultivated area. Other physical factors, such as siltation and encroachment along feeder canals, as well as encroachment on the tank are often mentioned as reasons for a reduction in tank storage capacity and thus for a decline in tank irrigation (Palanisami *et al.*, 2000, Vaidyanathan, 2003, Janakarajan, 1993).

In a way, our results are similar to those of Palanisami *et al.* (2000: 53-55) who noted that "utilization is higher for STs than it is for NSTs" and that "the percentage of command area irrigated in tanks with a command area of less than 40 ha was slightly higher than those with a command area greater than 40 ha", though our approach using GIS is rather original. An interesting point is the comparison between our figures and those found in the literature that highlights significant differences, even though data were collected at a 10-year interval. For example, according to Palanisami *et al.* (*ibid.*), in their study of tanks in various parts of Tamil Nadu, 60 per cent of the NST area and around 80 per cent of the ST area was irrigated in the mid 1990s (these figures are about 46 per cent and 69 per cent in our data collected in 2006-2007). The tendency towards a decline in NSTs rather than STs is the same in Palanisami's work as in our analysis for the Puducherry hinterland, although there is a more dramatic fall in tank use in our sample. Is this due to a bias linked to the methodology applied to measure this tank use? We are not of this opinion. Is it due to encroachment, since encroachers often damage the sluice and surplus weir in order to lower the water level in the tank? This cannot be the main reason, especially as there are fewer cases of encroachment in our sample of tanks (it concerns a quarter of the tank area according to interviews, *versus* 30 per cent in Palamisami's study, 2000: 62). Is it therefore due to a phenomenon which affects our study area in particular, since it is mainly situated in alluvial areas, as opposed to Tamil Nadu which is mainly characterised by rocky areas? Or to a rapidly increasing rate of abandoning tanks over the past few years, especially because of the 2001-2003 droughts that affected the region and were a major factor in the non-use of tanks (*see* Fig. 2.10 in Chap. 2). The last two reasons may for the most part explain the difference.

We can interpret this difference between STs and NSTs as an uncertainty regarding tank irrigation which is no longer accepted once another alternative exists (e.g. groundwater use). However, this area of uncertainty is one of the characteristics of tank irrigation, especially for NSTs that rely on rainfall (representing 90 per cent of tanks on a Tamil Nadu scale, cf. Box 1). The production system linked to tank irrigation "is inherently unstable, changeable, fluid and subject to calamity" (Mosse, 2003: 7). Can we characterise the tank crisis by this rejection of tank vulnerability? And how can one explain that this area of uncertainty is no longer accepted? Is it merely due to the increase in groundwater irrigation and to present-day agricultural production objectives? Might we therefore interpret the decline in tank irrigation as a consequence of a technique ill-adapted to changes in society (demographic pressure, intensive agricultural production, multiple small landowners, etc.)? Or, if one accepts Mosse's argument (p. 7) that in this part of South India "local and regional institutions not only respond to the inherent uncertainty of water resources, but also actually contribute to or amplify this uncertainty", and that the book "the rule of water shows how water control systems come to inscribe new forms of power and new articulations of State and society", can we interpret this rejection of tank uncertainty as a change in the balance of power, and especially as a way of transferring the balance of power from tank to groundwater management? We certainly tend to think so. Does this then imply that the tank crisis reveals a more general crisis in the relationship between State and society?

(c) What Use of Tanks?

A clear difference in tank utilization has thus been observed in our study area, a "tank intensive region"[17]: on the one hand, between the alluvial area and the rocky area, with farmers in the latter being more dependant on surface water for irrigation because of the lack of groundwater availability and, on the other hand, between STs and NSTs in the alluvial area, with the latter being used to a lesser extent. In tanks not used for irrigation, their spread-area has been converted in various ways according to their location: encroached on individually for cropping thus preventing collective use, converted to bus stations in cities, used mainly for fish farming, or left unused. Palanisami *et al.* (2003b) identify several factors that might influence the neglect of tanks, as far as irrigation is concerned: among the physical and environmental factors, storage capacity and rainfall variability are two majors ones (small rainfed tanks storing water for less than 3 months in areas with a deficit and a very variable rainfall, are more prone to be abandoned). The main hypothesis offered here to explain the difference in the use of tanks for irrigation is the reliability of the water supply depending on a combination of both groundwater and surface water characteristics.

Does this imply that the tank water supply needs to be increased, especially in areas dependent on surface water? Prior to asking such a question, one should ponder the functions of tanks. Tanks are multiple-use infrastructures with multi-functions (*see* Chap. 11), but they have been of interest to researchers and policy-makers mainly for collective irrigation. As there has been a decline in this type of use, the tanks have been said to be "in disuse". Moreover, the relevance of the tank technique as a means of storage and irrigation in a very densely populated region has been questioned, since the tank uses a lot of space and loses a lot of water through evaporation (Von Oppen *et al.*, 1980).

However, since their function of groundwater recharge has been emphasized quite recently (see Palanisami *et al.*, 2003b, 2006)—whereas in the book by Palanisami and Easter (2000) this idea of percolation tank was lacking, since tanks were mainly analyzed with the aim of increasing their surface irrigation performance—unused tanks are now called "percolation tanks" by researchers and policy makers. The acceptation of the non-use of tanks for irrigation—often described as a decline in use—must be linked to the recent policy of recharging groundwater through different infrastructures. These researchers mentioned the benefit of such a conversion: it leads to "improving cropping intensity and crop diversification and it ultimately provided more farm employment and farm income. Thus tank conversion will lead to better cropping and crop diversification when compared to non-converted tank commands" (2003b:.35-36). Since tanks are seen as GW recharge infrastructures—hence of use for GW irrigation—their water storage function has therefore been recognized. The question then is whether to distinguish between an irrigation and a percolation tank, in other words to know which tank fulfils the irrigation function better, and which one should be consigned to the percolation tank function. Palanisami *et al.* (2003b: 56-62) studied the environmental and social factors influencing the transformation of tanks into percolation ponds. Among these we might mention the small size of the tank, hence low storage capacity, the fact that most marginal farmers are not well owners, and the number of times the tank refills. Moreover, tanks for which farmers take no share either in tank maintenance or in managing tank irrigation affairs, should be converted into percolation tanks.

Our results related to the small covering of water may prompt us to question the way to use the upstream part of the tank which is fertile, humid and rarely covered in water. If the use of space and of soil moisture in this part has to be effective, it is not necessarily by proscribing agriculture in the spread area that tanks would be the most productive (since the foreshore is often left uncovered by water). However, this is not discussed in rehabilitation programmes which forbid any encroachment, whether the tank is used for irrigation or percolation purposes. And since full use has to be made of their storage capacity, development projects lead to the eviction of encroachers and the desiltation of tanks.

In defence of these projects, it is true that the use of the tank spread area for agriculture by encroachment is often detrimental to tank performance, because encroachers very often damage the tank structure (such as the gate and the surplus weir) in order to lower the water level in the tank and thus avoid damage to their crops in the foreshore area. However, growing crops in the tank is not in fact an illegal practice, since the Government rules that cultivation is forbidden if water stands for more than 21 days in the foreshore area of the tank (Palanisami and Easter, 2000: 62). The form of misuse described above explains why encroachment has often been singled out as an illegal practice, even if it has been more or less accepted by villagers for various reasons: generally speaking, encroachers are either politically influent persons, so that nobody is likely to object to their practice, or groundwater sellers whom water buyers need or, on the contrary, they are landless people for whom such cropping is an unique opportunity. Water sellers, of course, earn a substantial amount on the water they sell to encroachers. Encroachment thus implies complex social relations. It has even become accepted by local administrators, who may have also taken an interest in such a practice for themselves as farmers, or have loss any interest in this collective structure (Prabhakar, 2007). The acceptation of tank encroachment, which has often been described as a lack of interest by local elites in tanks, can be explained by the fact that tanks are no longer the seat of local power, which has been transferred to well owners. However, in some cases (*see* Chap. 9), WUA members use their newly acquired power to curry support from district administrative officers and evict encroachers, and thus counterbalance the power of those who appropriate land illegally[18]. In some specific tanks, such as Koratur where the tank is directly linked to well irrigation (*see* Chap. 7), tanks, through WUA, may again become the seat of local power.

(d) Implications for Tank Management

From the discussions above and due to the multiple uses of tanks, tank performance may come in various forms: performance related to surface irrigation (tank performance usually relates to this and is defined as the actual area irrigated compared to the registered one), performance related to water storage, to GW recharge, to the production of collective resources (fish, grass, trees), and possibly to use of the foreshore. Since irrigation and percolation functions have been recognized as being of prime importance, the main objective now is to boost tank storage capacity. Thus, increasing the tank water supply is also vital, since it would make more water available and would improve the tank's function of recharging groundwater or its use for irrigation, should—as we believe—the reliability of the water supply prove to be an important factor. Greater water availability in tanks must thus be encouraged. This could be done by cleaning the inlet channels that bring runoff water to the tank and by evicting encroachers on such canals. The runoff reaching tanks has indeed been reduced in many places since "feeder [= inlet] canals have become filled with silt and abandoned. Encroachment and extension of villages (...) have also reduced their length and capacity" (Palanisami *et al.*, 2000: 38). Cleaning and enlarging inlet channels should increase the possibility of harvesting runoff water. The NRGEA programme, thanks to which a number of NST inlet channels in our study area were cleaned in 2007-08, should have a visible impact on the tank filling process. The hypothesis whereby the state of the inlet channel is essential in tank supply could be verified. Let us not forget that the drop in water availability in the tank is the result of the combined effects of encroachment, siltation, rainfall variation and mismanagement (Palanisami and Easter, 2000: 73). Although nothing can be done to change the unreliability of rainfall, this is not the case for the other factors.

In tanks still used for irrigation, cleaning the inlet channel should be combined with the implementation of water management rules that would put the water to better use (as suggested by Palanisami). The sluices in particular could be closed in the evening since most farmers do not irrigate their land at night. Without taking part in the technical debate about the effectiveness of tanks for recharging GW (*see*, for example, Kumar *et al.*, 2008; answer by Rao *et al.*, 2008), we can say, just like Rao *et al.* (2008: 63) that one of the problems of rehabilitation programmes is that they lend too much importance to water resource augmentation without promoting a better use, and without developing any demand side management.

Indeed, our field enquiries show that such programmes provide little incentive for improved use of underground water resources and, on the contrary, may emphasize the strong social demand for accessing this resource. Some farmers in the alluvial area do recognize that they waste GW, even if they know that its level is dropping. Due to the irregular supply and cuts in electricity, they simply switch on the motor and come back a few hours later to turn it off, whilst in the meantime any surplus water has flowed out of the field. This waste of GW is linked to the farmers' facility in handling this resource, and they confess ironically that they are interested in the tank rehabilitation programme precisely because of it aims at recharging GW !

In the case of STs, a secure water supply explains that these tanks are still largely used, even in the alluvial area where the densities of wells and bore-wells are nevertheless high. The PWD thus has a role to play in managing the distribution of water within a network of STs and especially in privileging the distribution of water in rocky areas where villagers depend more on surface water than in alluvial areas. Yet such control no longer exists in the cases studied, and no rota system is used in allocating water between tanks[19] (contrary to what Sivasubramaniyam, 1997, describes for system tanks in the Palar). Thus, on the Rhaghavan Channel, which depends on the Tirukoylur *anicut*, the tanks upstream from the canal are obviously (based on satellite images) better supplied with water than those located downstream. Engineers do, of course, regularly patrol the canal and report on whether water arrives in the tank. Yet no gate supplying the tanks' feeder channel exists, no *laskar* (PWD agent) wishes to risk preventing a farmer or group of farmers from damaging any gates and from diverting as much water as possible to their tank. "We can't say anything to farmers; they do what they like", one of these *laskars* complains. However, some tanks located in the rocky area and in the downstream sector of the network would need this water more than the tanks situated in the alluvial area with good access to groundwater. However, the PWD's own administration does not necessarily apply a consistent form of management within the system, since several PWD sections are responsible for tanks relying on the same feeder canal, and there is little coordination between these sections.

Interestingly enough, villagers view this lack of control over water distribution—illustrated by the tanks' unequal access to water—in terms of social phenomena. Thus the tanks in two villages, situated 3.5 km apart, are not filled at quite the same time: they tap water from the Raghavan channel at a distance of 4 km from each other, with the most upstream village diverting more water and earlier than the other. An old man from the second village explained that since the first village uses more water, villagers from the second village no longer wish to marry anyone from the village that takes all the water for itself. Here sharing water even has repercussions on sharing women ! This line of thought is in keeping with Adicéam's analysis (1966: 318): he analysed in these tank systems a system of sharing water diverted from the river. Water distribution was formerly governed by a strict set of rules, yet today it is very often based on the principle of "first come, first served". It is to be hoped that the new organisation into "distributory committees" which federates the representatives of several WUAs for the same ST network (cf. Fig. 4.2) ends up compensating for this weakness, and contributes to managing water distribution on a scale of several tanks, if not of the whole system. These committees, in fact clusters of WUAs created by decentralization and participatory management policies, should play the role of arbitrator in water sharing, if they receive enough encouragement and if they dispose of the necessary means (financial, as well as legal means, access to hydrological data, information dissemination facilities, etc.). Yet this is unlikely to happen if we compare such tank systems to large irrigation systems, all of which face the same kind of upstream/downstream competition, and especially to that of Andhra Pradesh where WUAs have been in existence for longer (*see* Gulati, 2005; Reddy and Reddy, 2002, 2005; Mollinga, 2002).

This aforementioned organisation appears to be inconsistent with management on a river basin scale that is scheduled to be implemented very gradually.[20] Indeed, ST networks make up waterscapes that may challenge the hydrological limits of the watershed [by contrast, non-system tanks (NSTs) recuperate runoff water and are thus fully integrated in watersheds]. Canals, which divert river water in order to carry it through the countryside, may exceed the hydrological limits in this flat landscape. Therefore, from Fig. 4.16, it is obvious that water from the Ponniyar is thus diverted to the Varahanadi basin by the Pambai canal: the notion of watershed is not appropriate here in managing water in system tank networks, since water flows from one river basin to another. It would be preferable to define water management institutions according to the "area of shared water"[21], not to the watershed.

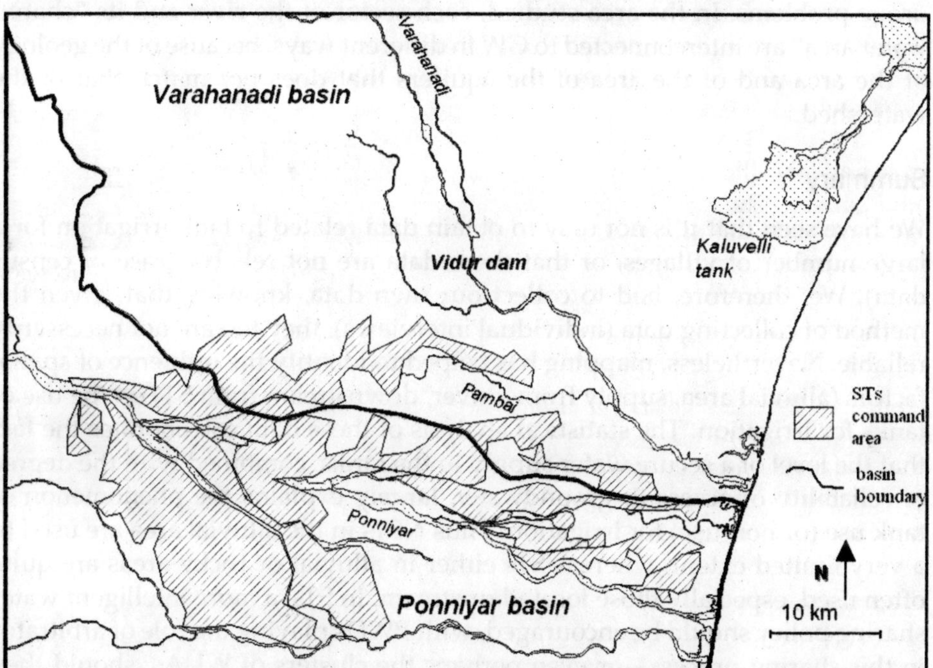

Fig. 4.16: Basin limits and the ST "shared water" area

This point of view is linked to that of French agro-geographers working on social water management. The French ISIIMM group suggests talking of "overflow catchment" (Ruf *et al.*, 2008: 15) or "overflow basin" (pp. 7; 91), which defines the area of hydraulic and hydrological influence that depends on infrastructures and networks using the basin water. Due to various territorial systems and to the areas of influence of the different interlocked and interdependent systems, "the notion of basins centered on the water offer must be completed by new notions of basins focusing on water demands and on the organization of sustainable development policies leading to spatial interpretation that enables managing the economic system and building up institutions and conventions for environmental protection and promotion of social fairness. The construction of innovative policies should be based on the representation of the hydrological and hydraulic complexity: deversing basins, drainage basins, reversing basins are common spaces that must be taken into account together with the watersheds"(*ibid*: 40). The river basin is not necessarily the relevant unit for understanding the management of water at a more local scale. This has been shown in some examples such as in Moench *et al.* (1999), since each sector of the river (and of the watershed) faces different usage problems. In the area studied, each sector of the river and its "shared water area" are interconnected to GW in different ways, because of the geology of the area and of the area of the aquifers that does not match that of the watershed.

Summary

We have seen that it is not easy to obtain data related to tank irrigation for a large number of villages, or that these data are not reliable (case of census data). We, therefore, had to collect our own data, knowing that, given the method of collecting data (individual interviews), they too are not necessarily reliable. Nevertheless, mapping has helped to identify the influence of spatial factors (alluvial area, supply from a river, downstream from ST) on the use of tanks for irrigation. The statistical analysis of these data corroborates the fact that the level of a secure water supply to the tanks, combined with the degree of reliability of access to groundwater, largely explains the phenomenon of tank use (or non-use) for irrigation. Thus NSTs in the alluvial area are used to a very limited extent, whereas STs either in alluvial or rocky areas are quite often used, especially those located upstream. In STs, a more intelligent water sharing policy should be encouraged, with PWDs playing the role of arbitrator in this sharing process—or even perhaps the clusters of WUAs, should they have the incentives and the means to do so, which seems unlikely when compared to other experiences of the decentralisation process. We have, moreover, shown that in STs the watershed is not an appropriate geographical unit to manage surface water, since some canals may divert water to another watershed: we therefore propose the notion of "water shared area" which encompasses the area that shares the same surface water.

Our results are similar to those of Palanisami *et al.* (2000, 2001 and 2003b) regarding the difference in use for irrigation between, firstly, tanks in rocky areas and alluvial areas and secondly, in the latter between STs and NSTs, though our approach using cartography to analyse tank water coverage (which is quite low) is rather original. The figures for our sample show a more dramatic fall in tank use for irrigation, probably due to the fact that in our sample a large part is located in the alluvial area. This, combined with the droughts of the years 2001 to 2003 has left people to rely more on groundwater. Such non-use of tanks for irrigation may question the relevance of these infrastructures in a densely populated country since tanks take up a lot of space. However, though this issue was the subject of much discussion in the 1980s, this was no longer the case in the early 1990s when their water storage and GW recharge functions were brought to the fore. Tank performance should, therefore, be defined by taking into account the multiple uses and functions of tanks.

Indeed, tanks have been of interest to researchers and policy makers, mainly due to their irrigation function. However, their GW recharge function has recently highlighted percolation tanks rather than unused tank. Since this function is the focus of water management policies, it is interesting to see that some elites have again started to show some interest in tanks (*see* Chap. 9 for more detail). However, the lack of demand side management in rehabilitation programmes, as well as the idea that the water supply is increased through recharge, does not avoid wastage of GW and its overexploitation. On the contrary, it may increase the strong demand for GW: surface water is therefore seen as a resource that supplies groundwater; that is a collective property resource (surface water) which is canalized and stored for the benefit of an privately appropriated resource (GW). The relationship between surface water and groundwater has thus been highlighted, but it does not lead to any resource management or to demand side management, because GW and surface water are not managed using an inclusive approach.

Such management does indeed heighten the tank crisis. There are two main hypotheses, neither exclusive of the other, to explain this situation: the rejection of tank vulnerability is linked either to the fact that the technique is ill-adapted to present-day changes in society, especially to demographic pressure and agricultural production objectives, or to a change in the balance of power, i.e. a transfer of the balance of power from tank to groundwater management.

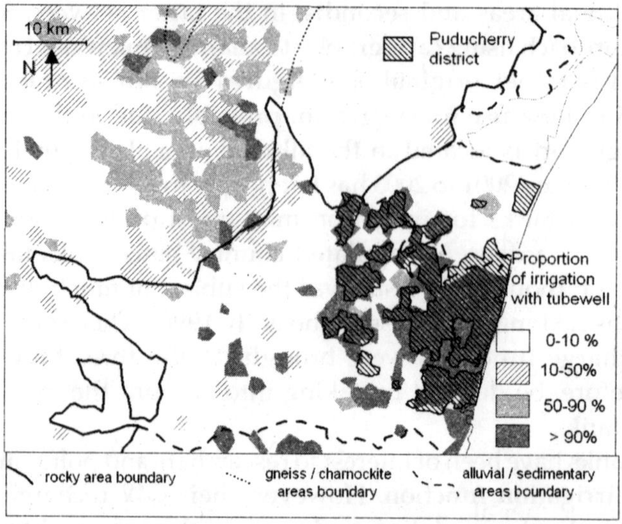

Fig 4.17: Tubewell Irrigated Area in 1991, according to the Population Census Data, with Reference to Puducherry District Boundaries

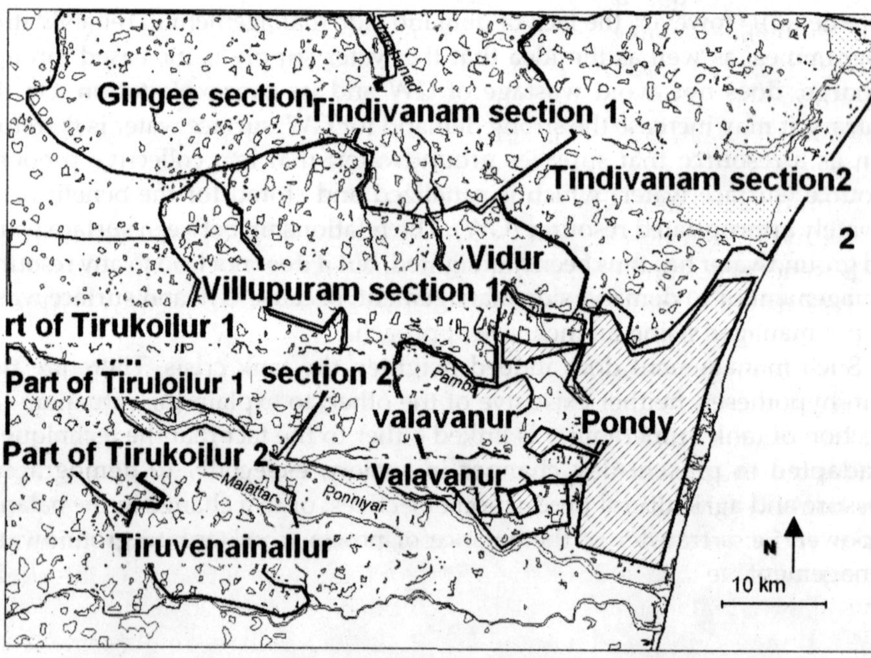

Fig. 4.18: Command Area of the Main PWD Sections or the Study Area

NOTES

1. See the image on page 248 in Agarwal *et al.*, duplicated as the front cover in Vaidyanathan (2001).
2. The tank limits were drawn partly by Siva from IFP and Méas (2007) – thanks to both of them.
3. Forty-four per cent is the proportion of cultivated area in Puducherry and Villupuram districts, 58 per cent in Cuddalore, as taken from the 2004-05 Season and Crop Reports of Pondicherry and Tamil Nadu.
4. The area was automatically calculated on the map, using the GIS software. But a half-millimetre line is equal to 25 metres and a 1 mm difference in the drawing could introduce a 50 per cent error for an area of 50 ha, and even more for smaller areas. We therefore calculated ratios only for tanks larger than 50 ha; otherwise our results might have been totally misleading.
5. All field enquiries made with the help of D. Chakravarthy (whom I would like to thank) either alone or as a translator for O. Lefebvre or myself.
6. PWD is organised into divisions, subdivisions and sections. In the study area, sections are entrusted with the management of tanks, 30 to 85 according to sections. See Fig. 4.18.
7. In Puducherry tanks are the responsibility of PWD whatever their size. Thus, from the 84 tanks, 47 have an *ayacut* of less than 40 ha, and would be managed by the *panchayat* if situate in Tamil Nadu.
8. We were able to obtain the jurisdiction maps from each PWD section and the *anicut* map from the Villupuram office: I would like to thank here the institution and its engineers for their help.
9. The hydrological layer was drawn by G. Savin from the Université de Provence according to topographic sheets.
10. The data come from interviews with 1 or 2 farmers for each tank. This is, therefore, a very rough approximation.
11. In our sample there are more STs, which is not representative of the NST and ST distribution in Tamil Nadu, where STs only represent 10 per cent (Vaidyanathan, 2001: 15).
12. This multivariate analysis was carried out by Victoria Reyes Garcia, UAB, whom I would like to thank.
13. For various reasons, including the cost of colour photocopying, it has not been possible to include all the results of this study here. Part of this study used the GIS data related to tanks compiled by Méas (2007).
14. Winimage '97 was designed by P. Oliva, Centre de Treatment de l'Information Géographique [geomatic center], Université de Provence, Aix-en-Provence, France.
15. Resolution of 28.50 m in 1991 and 2000 with Landsat and of 23.5m in 2002 and 2004 with the IRS satellite.

16. The rain analysis was carried out using monthly data. It is useful to know if the months observed were dry or not according to the average taken over a 100-year period.
17. To use the distinction made by Palanisami *et al.* (2003: 1-5), who have shown a difference in water use dynamics between a "tank intensive region" and "less tank intensive region" in Tamil Nadu where the percentage of defunct tanks (= tanks not used for irrigation) is higher. The reason they give is the predominance of well irrigation resulting in the disintegration of irrigation institutions.
18. This case of Koratur is quite unique in the sense that farmers owning a well in the tank command area use this water to fill their well and transport water to their "dry fields": they are interested in the tank mainly because it supplies their wells! See the rules along the Bangaru Channel presented above.
20. The river basin organization is known worldwide. As F. Molle (2006) reminds us, the river basin is quite a recent concept dating from the 18th century that satisfied "this search for adequate administrative and political boundaries" (p. 1). It "was used by political establishments to symbolize and strengthen their power" (p. 23), the 'natural' aspect of its definition being used as a justification of its legitimacy. At the beginning, in the 19th century it was an engineering concept for the development of infrastructures, but also a political concept to centralize and control water. Then the organizations changed into coordinating agencies and had to be restructured in order to accommodate local scales, processes and stakeholders. The way the concept has become operational varies greatly between countries (p. 24).
21. Which is, to a certain extent, done by splitting them into distributory committees for a same ST system.

REFERENCES

Adicéam, E. 1966. *La géographie de l'irrigation au Tamil Nadu*. Paris: Ecole Française d'Extrême Orient.

Agarwal, A., and S. Narain. Editors. 1997. *Dying wisdom. Rise, fall and potential of India's traditional harvesting systems*. New Delhi: Centre for Science and Environment (State of India's Environment, 4, a citizens' report).

Bossy, A. 1985. *Evaluation des ressources en eau et de leur gestion. Cas d'un village en zone tropicale sèche (Tamil Nadu-Inde)*. Ecole Nationale du Génie rural des Eaux et Forêts (GREF), Diplôme d'Etudes Approfondies.

CNEARC. 2006. *Student report from a collective training course in four villages in Puducherrry hinterland, about water management*. Institut Français de Pondichéry and Centre National des Régions Chaudes.

Coste, N., and N. Ploumpidis. 2007. "Bittersweet water. Irrigation practices and modern challenges in South India". Trilingual. DVD (Eaux douces, Eaux ameres—Thaneer: inippum kacappum). Co-authors: K. Anupama, O. Aubriot, Y. Gunnell, Puducherry: French Institute of Pondicherry (Publications Hors Serie 5).

Guilmoto, C. Z., S. Oliveau, V. Charles, R. Delage, and S. Vella. Editors. 2004. *Mapping out Social Change in South India. A Geographic Information System and its applications.* Pondicherry: French Insitute of Pondicherry.

Gulati, A., M.D. Ruth, and K. V. Raju. 2005. *Institutional reforms in Indian Irrigation.* New Delhi: Sage Publications and IFPRI.

Janakarajan, S. 1993. In search of tanks: some hidden facts. *Economic and political Weekly* 28: A53-A60 (Review of Agriculture).

Kumar, M. D., A. Patel, R. Ravindranath, and O. P. Singh. 2008. Chasing a mirage: water harvesting and artificial recharge in naturally water-scarce regions. *Economic and Political Weekly* 44:61-71.

Meas, W. 2007. *Les "tanks" dans Varriere-pays de Pondichery (Inde): distribution spatiale et correlation avec les donnees demographiques et socio-economiques (cenus of India), integration SIG.* Memoire de Master 1 "Sciences geographiques", Université de Provence, Centre de treatment de l'information geographique.

Moench, M., E. Caspari, and A. Dixit. 1999. *Rethinking the mosaic.* Kathmandu: Nepal Water Conservation Foundation. http://www.i-s-e-t-.org/rethinking%20the%20.mosaic.pdf edition.

Molle, F. 2006. *Planning and managing water resources at the river-basin level: Emergence and evolution of a concept.* International Water Management Institute (IWMI) Comprehensive Assessment Research Report 16.

Mollinga, P. P. 2002. "Power in motion; a critical assessment of canal irrigation reform in India," in *Users in water management.* Edited by R. Hooja, G. Pangare, and K. V. Raju, pp. 265-281. New Delhi: Rawat Publications.

Mosse, D. 2003. *The Rule of Water. Statecraft, Ecology and Collective Action in South India.* New Delhi: Oxford University Press.

Palanisami, K., and K. W. Easter. 2000. *Tank Irrigation in the 21st Century. What Next?* New Delhi: Discovery Publishing House.

Palanisami, K., P. Paramasivam, B. Chandrasekaran, and S. Karunanithi. 2003 *Participatory Irrigation Management in Tamil Nadu.* Coimbatore: Water Technology Centre, Tamil Nadu Agricultural University.

Palanisami, K., P. Paramasivam, C. Karthikeyan, and A. Rajagopal. 2001. *Sustainability of Tank Irrigation Systems in South India.* Coimbatore: Water Technology Centre, Tamil Nadu Agricultural University.

Palanisami, K., A. Raviraj, S. Thirumurhti, S. Gurunathan, G. Dhanalakshmi, and R. Shanti. 2006. *Augmenting Groundwater Resources by Artificial Recharge. A case study of Kodangipalayam village in Coimbatore district, Tamil Nadu.*

Coimbatore: Centre for Agricultural and Rural Development Studies (CARDS), Tamil Nadu Agricultural University.

Palanisami, K., D. Suresh Kumar, V. Vediappan, and L. P. Swaminathan. 2003 b. *Conversion of tanks into percolation ponds.* Coimbatore: Water Technology Centre, Tamil Nadu Agricultural University.

Prabhakar, P. I. 2007. "The illegal practice of cultivation in irrigation tanks: an approach to legal pluralism through actors' narratives," *in Land Law and Natural Resources Management Comparative Perspectives (Land: Lessons and Stakes).* Edited by C. Eberhart. Puducherry: Institut Français de Pondicherry.

Rao, K. V., P. K. Mishra, G. R. Korwar, B. Venkateswarlu, V. U. M. Rao, and M. Osman. 2008. Myths about small water harvesting systems. *Economic and Political Weekly* 44: 60-63.

Rawal, V. 2001. Irrigation Statistics in West Bengal-1. *Economic and Political Weekly* 36 (27): 2537-44.

Reddy, R. V., and P. P. Reddy. 2005. How Participatory is Participatory Irrigation Management? Water Users' Associations in Andhra Pradesh. *Economic and Political Weekly* 15: 5587-5595.

Reddy, V. R., and P. P. Reddy. 2002. Water Institutions: is formalisation the answer? (a study of water user associations in Andhra Pradesh). *Indian Journal of Agric. Economy* 57: 519-534.

Ruf, T., and J. Riaux. 2008. "General synthesis, ISIIMM Project (Institutional and Social Innovations in Irrigation Mediterranean Management)," in *Euro-mediterranean regional programme.* http://www.agropolis.fr/pdf/isiimm/Synthese Eng.pdf.

Savin, G. 2007. Les modes d'irrigation par tank dans l'arriere pays de Pondichery: organisation et evolution entre 1977 et 2007. Approche par teledetection et SIG., Memoire de Master 1 "Sciences geographiques", Universite de Provence, Centre de traitement de l'information geographique.

Vaidyanathan, A. Editor. 2001. *Tanks of South India.* Delhi: Centre for Science and Environment.

Vaidyanathan, A., K. Sivasubramaniyam, and S. Mariasusai. 2003. *Conditions and characteristics of well irrigation under Palar basin, Tamil Nadu.* Madras Institute of Development Studies, Working Paper No. 179.

Von Oppen, M., and K. V. S. Rao. Editors. 1987. *Tank irrigation in semi-arid tropical India. Economic evaluation and alternatives for improvement. Research Bulletin no. 10.* Patancheru, Andhra Pradesh: ICRISAT.

Von Oppen, M., and K. V. Subba Rao. 1980. *Tank irrigation in semi-arid tropical India: historical development and spatial distribution.* International Crops Research Institute for the Semi Arid Tropics (ICRISAT).

Chapter

5

Watershed and Groundwater Management in Coastal Region of Tamil Nadu: Problems and Remediation Approaches: A Case Study of Kaluvelly Watershed and Puducherry Sedimentary Basin

Pierre Jorcin and *Judith D'Souza*

5.1. The Resources: A Threatened Potential, Assessment and Perspectives

(a) Groundwater Level Depletion

The Kaluvelly (or Kazhuveli) watershed is located on the coastal area of southern India, in Tamil Nadu State and adjacent to Puducherry territory (Fig. 5.1). This chapter will present the situation in the watershed, focusing on the part lying within Puducherry sedimentary basin. The region has a semi-arid climate with no perennial rivers. Water resources depend on rainfed tanks filled only during the monsoon peak season. Therefore, the pressure on groundwater is very high, as it represents the most important potential of the region, mainly for agriculture.

In the state of Tamil Nadu, with the government subsidies on electricity for agricultural usage in the last 30 years as part of the Green Revolution initiatives, the area has seen unprecedented growth of deep bore wells which have ultimately resulted in an overexploitation of the ground water. Indiscriminate use of chemicals to aid agricultural growth has increased manifold

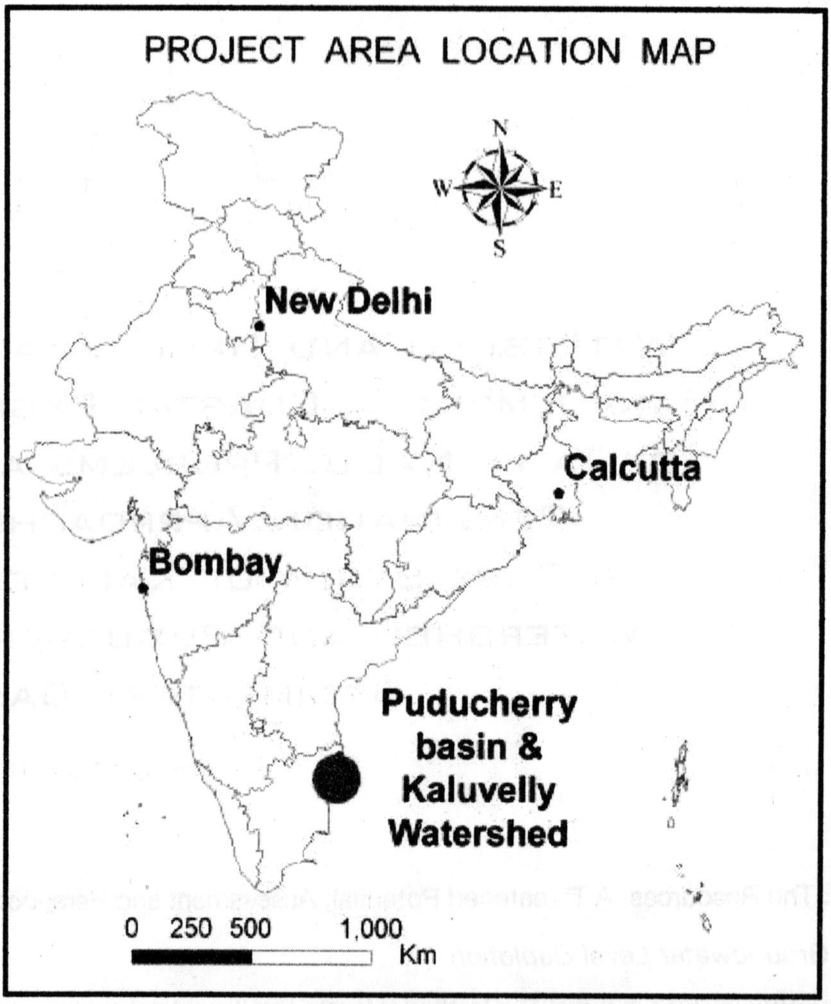

Fig. 5.1: Project Area Location Map

which has also lead to the contamination of the groundwater. With no control boundaries and bodies created to take in a holistic perspective of the critical issues, there has been rampant exploitation and over-utilization of resources in the area.

The lack of regulation from the government on groundwater exploitation and surface water, combined with the low awareness of the population leads to a very severe depletion and deterioration of fresh water resources, those intensifying the already negative effects that the population and the environment is facing.

In the Puducherry coastal sedimentary basin, groundwater resources exist in a multilayer aquifer system with alternatives layers of sandstone, limestone and clay (*see* Figs. 5.5, 5.6 and 5.7). However, the fairly good quality and capacity of the aquifers in the region are now being threatened by the ever increasing deep borewells. The hydrogeological studies conducted by Harvest and the UMR Sysiphe of the University of Paris VI, clearly highlight a severe deficit in the yearly water balance (Violette *et al.*, 2004). The groundwater extraction has been evaluated through a census of all bore wells and their parameters covering an area of 250 sq. km within our study area. The average bore well depth in the area is about 150 metres, ranging from 60 to 280 metres depth. The extraction is mainly done from the deeper sandstone aquifer, but it is important to mention that very few wells are cased and water is drawn from all the layers of the aquifers. From the survey done in 2006, we can assume that approximately 180 million cubic metre of groundwater is being extracted annually from 6000 borewells within the operational area of 250 km^2 (Fig. 5.2). Most of the wells are equipped with a 25 HP pump, providing a water discharge of 25,000 litres per hour. An average motor pump running 6 hours a day will extract about 150,000 litres a day. The heavy extraction of groundwater accompanied with limited distribution of rainfall has lowered the level of groundwater in many parts of the area. The pressure on this resource is very high and its potential is seriously threatened by overexploitation. This situation has led to a severe drawdown of the water table. The water table everywhere is below the mean sea level in the whole study area. In the deeper and main aquifer, Vanur, the water table drawn down is about 63 metres from 1977 to 2007. Existing groundwater level is at 84.75 metres below ground level in September 2007, equal to 57.35 metres below mean sea level. The yearly groundwater extraction was estimated to be 4 times more than the natural recharge.

Reasons for Groundwater Depletion

- Water drawn from borewells beyond the capacity of the recharge potential. The natural groundwater recharge cannot compensate the water extracted every year from the coastal sedimentary aquifers for the cultivation of water intensive crops like paddy, sugarcane, casuarina plantations and other crops.
- Inadequate care on surface water structures like tanks and ponds.
- Lack of knowledge about the crop water requirements prevailing among the farmers. The frequency of irrigation is usually 4 times more than the actual requirement.
- The free electricity for the farmers induces them to draw as much water as possible from the borewells.

Fig. 5.2: Irrigation and Borewell Density in the Area Studied

In this coastal aquifer, due to hydrogeological parameters such as the aquifers structure and dynamics, porosity and transmissivity, the water table depletion increases dramatically the risk of sea water intrusion inland. It is today threatening to affect the entire aquifer system. Therefore, retention of water in the water bodies like lakes and ponds for recharge of groundwater over a long period of time is a necessity.

Moreover, due to factors such as mineralization process and other groundwater pollution, the aquifer is getting more and more saline (Gassama *et al.*, 2002, 2003). Thus even potable water for the people is getting seriously affected. In this context, scientific assessments through in depth hydrogeological studies are a most important to understand the aquifer system, its dynamics and evolution trends.

(b) Watershed and Drainage Parameters

The surface water drainage is the network of the irrigation tanks, where water flows through the feeder channels that link one to another surplus outlet channels (Fig. 5.3). The drainage system has a very complex structure, highly modified by the human intervention to divert and control the flow. From a natural watershed it has evolved to a complex network system including numerous water diversions to regulate the flow to optimal use of water. Water flows from the tanks through a direct catchment area and a combined catchment area including the upstream tanks and their catchment (Paulik, 2001). As indicated earlier, the surface water drainage has no perennial river. The geographic shape of the watershed induces heavy runoff at the outlet of the watershed due to the concentration of flows, with runoff water from all sub-watersheds reaching the outlet during the same period of time. This runoff happens during rainfall events of the monsoon season.

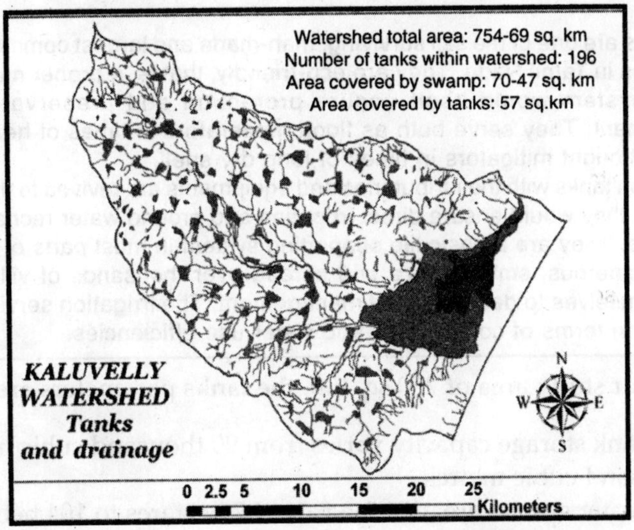

Fig. 5.3: The Kaluvelly Watershed and its Tank Network System

Elevation Parameters

- Maximum elevation = 53 m
- Minimum elevation = 2.5 m (measurement at the watershed outlet)
- Average elevation = 22.5 m

Apart from a small portion on the plateau of Auroville, the entire watershed area is flat lands, with an average slope of 0.7 per cent, more favourable for infiltration and siltation.

Box 5: The Importance of Tank System

- Irrigation tanks are earthen bunded reservoirs usually constructed to capture monsoon runoff. They are found in all soil types except sand, located in all socio-ecological, agro climatic and rainfall areas of Tamil Nadu. They are in existence for several centuries serving the water needs of communities.
- As one of the oldest man-made ecosystems, the tank system consists of water bodies, tank structures, feeder canals and supply channels, wells, wetlands, semi-dry tank fed lands, soils and plants, animals and birds, aquatic plants and fishes.

(Contd.)

Box 5 (*Contd.*)

- As an agricultural system tank fed agriculture is distinct in cropping practices, varieties and water management. As an engineering system it is historically one of the oldest in irrigation engineering design. As a management system it had institutions built around them.
- As a social system the tank serves and benefits various sections of the village community such as farmers, fisher folk, artisans, animal hearers and especially women.
- The tanks are one of the still surviving, man-made and largest common property resources in Tamil Nadu. They are eco-friendly, that is, a proper management of the system would itself ensure protection and preservation of the environment. They serve both as flood moderators in times of heavy rainfall and as drought mitigators in times of long dry spell.
- If irrigation tanks with their tributaries and equipments are revived to their original capacity, they would ensure direct irrigation and ground water recharge in rain fed areas. They are a basic life supporting system in most parts of the state.
- Being numerous, small in size and spread over thousands of villages, they lend themselves to decentralized management. The irrigation services are far superior in terms of conveyance and water use efficiencies.

Within our study area of 250 sq. km, the tanks parameters are as follows:

- The tank storage capacity varies from 90 thousand cubic metres to 610 thousand cubic metres.
- The water spread areas range from 4.7 hectares to 104 hectares.
- The average water storage time in a tank is about three to four months, corresponding to cultivation season.

Lots of tanks have a limited capacity and need to be rehabilitated. This is due to the natural siltation process of the storage lake and its feeder channels (Photos 5.1 and 5.2), as well as open breaches happening in the bunds and channels. Encroachment by farmers on the catchment areas and channels are also reducing the storage capacity. The hydrological parameters of the subwatershed have been strongly modified by the human intervention in order to optimize the water management for irrigation purpose. From a natural watershed it has evolved to a complex network system including numerous water diversions to regulate the flow and respond to the needs of filling the tank and letting excess water flow.

Through time this system has slowly degenerated and affected by various natural process combined with a severe lack of maintenance leading to various problems taking place:

- An important siltation due to the topography, the change of land cover and agricultural practices, resulting in changes of flow.

- Higher risks of flood as a consequence of changes in land use, deforestation, change in the regulation of flows, degradation of the channel network system.

Photo 5.1: Drainage Degradation, Channel with Very High Siltation Resulting in Reduced Flow

Urban Development and its Impact on the Surface Water Drainage

In the surroundings of Puducherry, as well as in Kalapet area, along the East Coast Road, a very fast and unplanned urbanization is taking place. The larger density of infrastructure will have a direct impact on the surface water flow and drainage. The groundwater recharge will also be affected by a heavy change in the runoff pattern. This area is subject to rapid soil erosion due to its topography.

Photo 5.2: Natural Degradation of the Tank due to Heavy Siltation and Vegetation Growth in the Water Spread Area

Photo 5.3: Shared Water Supply and Uncontrolled Drainage in Villages

Photo 5.4: Kaluperampakkam Women Collecting Water from a Hole near the Pond—this Village is situated 15 km from Puducherry

Nevertheless, the complex surface water management system in the area is not maintained properly in the area due to radical changes in the social tissue and strong political shift brought over time. It can be indicated that this decline has been escalating in the last 50 years though it may have had its beginnings earlier than the last century. As a result, the rich and positive impact it generated for the environment, the society and the economy for centuries is now reverted and a general degradation of these intricate components is observed. The present situation is critical. Surface water is rushing to sea bringing with it masses of fertile top soil creating heavy lost in soil fertility. Erosion is developing fast. Groundwater recharge is largely reduced because of the shorten time surface water is present. Swift increase of groundwater extraction is monitored, though creating a fast depletion of the groundwater table, increasing cost for extraction, fast salinity increase and in some place seawater intrusion. A heavy reduction of forests, flora and other ecological niches is monitored, with related vanishing of endemic species.

Drainage in the Settlement Areas and Pollution Risks

Bad water supply structures leading to water stagnation, un-controlled waste water and runoff create pollution problems for the village. There is no proper drainage for flood control as well as for waste water in the rural area of the sub-watershed. Common sights of stagnant water mixed with household waste mark the habitable areas within the village. Underground water supply pipes are being cut for better access to water creating water contamination problems. The piped system for distribution of potable water is marked with leaking connections and perforations. A study revealed that almost 49 per cent of total water of the overhead supply tank was lost through the faulty pipe lines. The lack of drainage near these illegal water supply posts add to the contamination.

The problems existing in the settlement areas need to be included in a study for an integrated water management at the regional scale. Issues like the water supply structures, the drainage in the village streets and the solid waste disposal are integral to understanding the social angle of the village in relation to water. Water management has become a serious issue for a sustainable development of agricultural activities and related livelihood security. Therefore, water harvesting programmes come as priority actions among the stakeholders involved in water management and farming.

(c) Water Quality in the Kaluvelly Tank and Immediate Surrounding Area

Even though Kaluvelly tank contain fresh water, the tank water gets contaminated more often, by the invasion of sea water during the periods of high tides, due to the damaged condition of the control structure arrangements. On examination it was found that due to the presence of marine sediments, the quality of water is affected, Electro Conductivity (EC)* values of surface water samples collected on site in 2005 are in the range of 66300 ms/cm to 70600 ms/cm. The TDS value are ranging from 46250 to 50890 mg/I. Aquaculture and prawn culture and saltpans have been established in the immediate surrounding, inducing heavy pumping of sea water into land and even more water quality degradation.

The people in the area are fully dependent on the groundwater for their drinking and agricultural needs. In some places the people have no other option, except to drink the salty water due to non-availability of good water source from nearby area. The people are also complaining that their hair becomes

* Electrical conductivity (EC) estimates the amount of Total Dissolved Salts (TDS), or the total amount of dissolved ions in the water.

reddish and sticky after bathing with this water and the vessels containing the water is becoming reddish as if coated with red paint.

The crop yield is also reported to be very low. In some places it was seen that the seeds sown in the nursery area, a majority of them do not germinate. This forced the farmers to change their cultivation patterns.

Moreover, with the Tsunami of December 2004, the situation of the coastal area has seen exacerbated deterioration. There was already a serious risk of contamination by fecal matter, salinity, and industrial pollution with the lack of proper sanitation infrastructure and hygiene knowledge. The Tsunami further increased the levels of contamination, both fecal and saline, in the water tables. Salinity was measured 10 times higher than normal range in most of the village located along the sea shore from Kottakupam to Konimedu, north of Puducherry city (Fig. 5.4). This situation remained for about 6 months. However, the post-Tsunami studies and monitoring have shown that only the superficial layers of the aquifers were affected, whereas the deeper aquifers have been preserved from such degradation (Violette *et al.*, 2005).

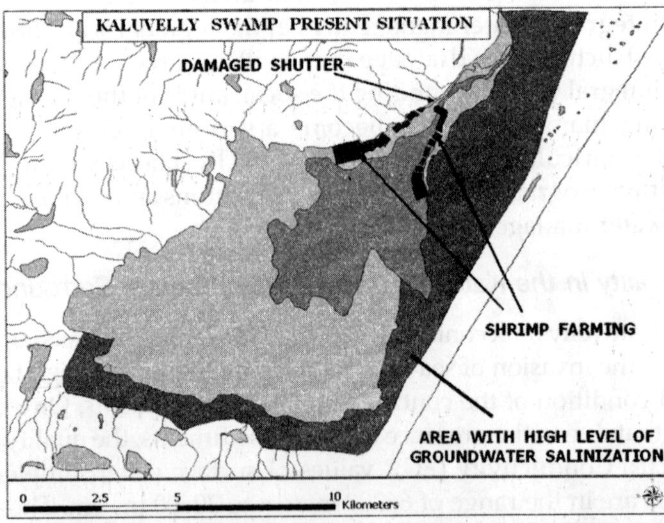

Fig. 5.4: The Salinization of Groundwater Around Kaluvelly Swamp

5.2 Related Issues Faced Among Rural Populace in Borewell Irrigation

Due to the lack of proper maintenance of the tanks over the years, they slowly fell into a state of disrepair and farmers had to resort only to the monsoons. Due to the heavy costs involved in digging a borewell, many small farmers leave their lands fallow or depended on the rains and the tanks or even purchased water from borewell owners. Seeing the erratic rate of rainfall, good agricultural land was left fallow or sold.

With land slowly becoming infertile, the rural areas have seen increased daily migration of the men to industrialized and urban areas for work. In many instances, families break up and migrate permanently to urban areas thus deteriorating the social fabric even further. A very common phenomenon now seen in rural areas, with the advancement of education opportunities, is of the young educated youth leaving their traditional homes and work to work in extreme situations outside their village especially in urban setting. With this, we see the drain of potentials and of talent, into a wider environment where competition is fierce along with depressing levels of lifestyle.

Many small and marginal farmers, due to the decrease in soil fertility and heavy loans for agricultural inputs, have sold off their lands and migrated to urban areas in search of low paying jobs, thus acerbating social problems in those areas. There is a very high level of indebtedness of the agricultural sector to private moneylenders for livelihood reasons coupled with medicals and food. More and more large farmers are converting their tracts of land towards cash crop cultivation which is water intensive and there is less food grain cultivation. With this trend, as cash crops don't need much maintenance, many daily wage labourers now have to find other means of livelihoods and turn to migration.

With the current spate of distress migration even for the landless to urban settings in search of employment, the age old social fabric gets disrupted. The urban areas get even more clogged up leading to unhealthy living conditions for all.

The women are left to take care of the lands leading to a critical shift in the gender relations of the area. With the men folk having migrated yearly to industrialized areas, women take on the burden of the household and the farm lands. Women and girl children are the main home makers. They manage and maintain the family and provide care. It is seen in most rural areas, basic necessities like potable water and sanitation facilities are grossly neglected. Women and girl children have to face unnecessary indignities in having to perform their ablutions either early in the morning or late at night thus giving rise to gynaecological problems at an early age.

This being one of the key problems that this area has to face and the consequential contamination of groundwater through various interrelated factors has brought about the need to develop sanitation infrastructure that addresses the situation.

Fetching water being one of the main responsibilities, they have to endue long queues and heavy burdens. If there is a water crisis, it is the women and the girl children who have to go long distances in search of water and bring back enough for the needs of the entire family.

The present situation is a complex socio-economic condition leading to circumstances that can spiral out of hand if not given proper direction. So, in essence, when we address the issue of bore well irrigation, we are addressing a whole range of issues surrounding it which in turn affect the social fabric of rural life.

Presently what is needed is an integrated approach to water management. The catchment is the smallest complete hydrological unit of analysis and management. It should also be understood that a catchment or watershed approach implies that water should be managed alongside the management of co-dependent natural resources, namely soil, forests, air and biota. A lot of it is in effect of man's play on the environment. And hence this has to be addressed alongside peripheral issues that contribute in the long run to the depletion of the groundwater.

5.3. Remediation Initiatives of the NGO 'Harvest'

Water management has become a serious issue for a sustainable development of agriculture activities and related livelihood security. Therefore, water harvesting programmes come as priority actions among the stakeholders involved in water management and farming. Large scale programmes have been implemented in the basin with the support of government agencies, local bodies, NGOs and research institutions.

(a) Research Programmes and Hydrogeological Studies

Groundwater Resource Monitoring

Thematic mapping of the groundwater has been modelled by a long term monitoring programme. An interactive database, updated regularly, allowed the creation of a direct link from the field situation to the planning stage for decision making.

The groundwater table can be measured and studied efficiently using GIS. To know the evolution of the water table and its spatial variations, the water level can be mapped for each aquifer. A selection of observation borewells chosen for their particular location and depth is monitored on a monthly basis. The difficulty is to find existing wells answering these conditions, or drill observation wells for this specific purpose. The location of these borewells is taken using Global Positioning System (GPS) technology. A high accuracy differential GPS calibrated on referential benchmarks is used to measure precisely the altitude of the borewell. This elevation data is essential as it will be the base to trace contour maps of the water level in reference to the mean sea level. The measurements taken on the field in reference to the ground level will be entered in the GIS and converted to mean sea level values. This will allow to give reports according to scientific standards, taking in account the absolute water level, free from variations of the elevation on the ground.

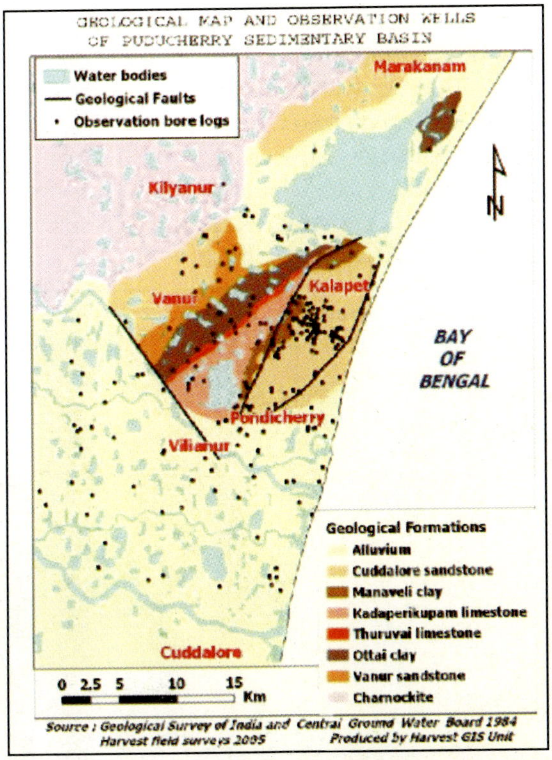

Fig. 5.5: Geological Formations of the Puducherry Sedimentary Basin with Location of Observation Bores

The groundwater quality is being monitored monthly to study the current situation and the evolution trend. The salinity is measured through Electrical Conductivity (EC) values. The GIS allows analysing the extent and spatial variations of the salinization of the aquifers, and locating the most degraded areas. Through such a regular monitoring programme, any specific event can be highlighted, such as the impact of the Tsunami on the sandy dune aquifer. For planning the exploitation of the resource on sustainable bases, such an information system is essential, as it gives strong support to build knowledge based on scientific assessments, and prepares for decision making.

The groundwater extraction had to be estimated in order to evaluate the water demand, the resource potential and its evolution. There are different ways to estimate groundwater extraction, such as estimation of the irrigated land from land use maps. For this project, the evaluation has been done based on data collected from the source of extraction, the borewells. An intensive field survey has been carried out, where each and every borewell has been identified and locate by a handle GPS. Engineering details on the borewell are integrated in a database, such as the total depth, the pump type and pump power, the casing size and depth etc. The groundwater extraction database covers an area of 250 sq. km and includes more than 6000 borewells. Quantitative data have been collected to estimate the average water extraction per day for all the different seasons in a year.

Based on the processed data in GIS, a 3D geological model of the aquifer system has been completed. The results of the surface contour mapping of the different formations have provided the base to build the 3D model The 2D

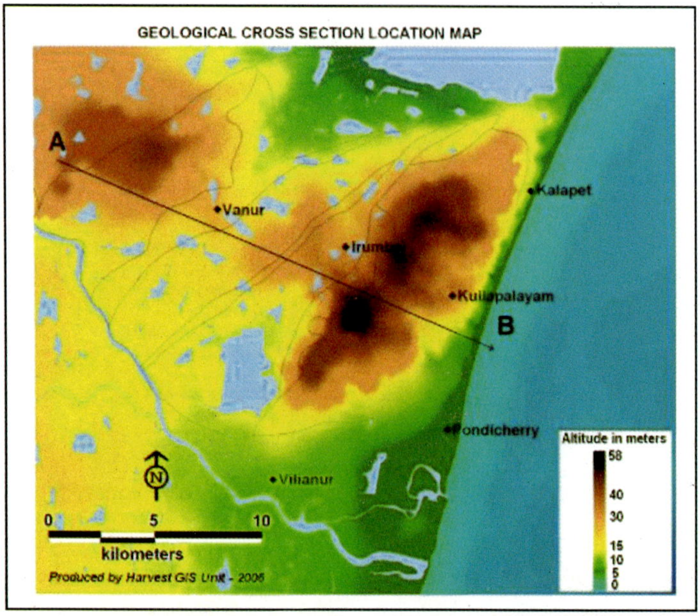

Fig. 5.6: Geological Cross-section Location Map

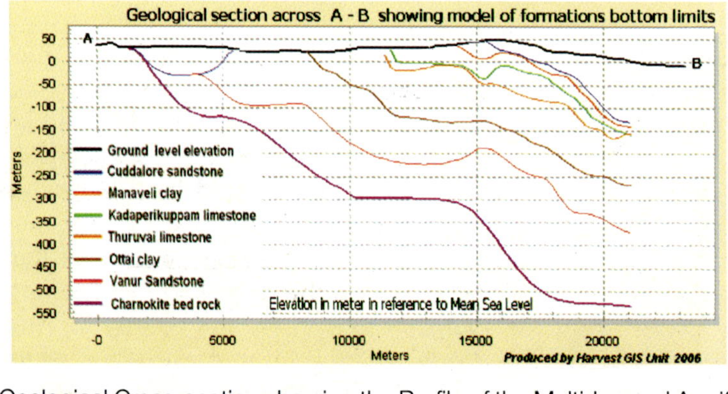

Fig. 5.7: Geological Cross-section showing the Profile of the Multi-layered Aquifer System

layers (Figs. 5.5 and 5.7) created in the GIS have been converted to 3D layers. The result is 3D layers of the formations showing their geometry, their shape and structure. According to the 3D model, the volume of the aquifers could be estimated. Then, the water table depth was integrated in the calculation to get a precise figure of the aquifer volume.

An estimation of the aquifer capacity has then been generated based on porosity parameters: the porosity parameters have been taken from the available literature and experts reports. Pumping tests are required to redefine this index on a more valuable scientific basis.

Drainage Modelling

All the hydrological parameters of the watershed can be studied using GIS, by analysing the runoff patterns, the extents of the catchment areas, the drainage network system, etc. To create a hydrological information system, all the different hydrographic features have to be localised and measured precisely. The drainage network and its characteristics has been precisely mapped using GPS technology. Topographic surveys have been carried out for most of the important storage tanks of the region. An inventory of the irrigation structures has been realized. Quantitative data as well as qualitative data have been entered, giving reports on the soil conditions, the state of siltation, the vegetation cover, etc. From this mapping work, the parameters of each tank and feeder channel are well known, and this helps to estimate storage, runoff and flow in the catchment areas. During monsoon season, an intensive data collection programme is carried out. The flow is measured in the channels at specific points of the drainage network. The water level in the tanks is measured using calibrated scales installed on the tank sluices. The data are entered in the GIS to provide a base for further and advanced hydrological modelling.

A watershed delineation processing can be realized based on a digital elevation model to define the limits and extents of all the different catchment areas. This work is done by spatial analyses using a model based on the slope directions. Using the slope direction values in a raster model, the flow accumulation areas and the drainage pathways can be identified. These kinds of model have strong limitations as they can only be built on high accuracy data, and this involves intensive topographic surveys on the ground. The results are better for high lands than flat lands. On flat lands, the results of the computer processing need to be adjusted based on ground truthing.

Land Cover Mapping for Groundwater Recharge Studies

For hydrogeological studies and watershed management it is important to know more about the groundwater recharge potential for a particular place. This is defined by studying the runoff parameters and the soil infiltration parameters. The land cover characterizes the soil infiltration capacities.

The land cover has been mapped for the whole study area from satellite imagery. Satellite imagery offers precise data giving information on the actual features of the land. The reference satellite image used is a LISS III and Panchromatic merged image, from IRS 1C satellite. The source data is a multispectral image with 3 bands: green, red and near infra-red. This allows to identify the vegetation cover through vegetation index and image classification methods. The satellite image has been precisely georeferenced using GPS and integrated in our GIS. Identified benchmarks on the image have been surveyed using GPS to register the image in the existing database, with geographic coordinates. The satellite image has been processed and a classification has been done to create a land use map. This work has been carried out by image interpretation and processing, based on ground truthing. The methodology used is based on land features identification and sampling from field surveys. Different types of land features have been identified on the image and classified by type of land use. The vegetation cover was first classified according to the type of land use, forest area or cultivated area. The cultivated areas have then been subdivided according to the cropping patterns in the study area. The classification includes different type of cultivated crops and plantations, the areas under irrigation, the forest extents, the non-cultivated areas and barren land etc. Later on, other available satellite images from other sources and other period of times will be processed using the same classification method to get a comparative time evolution data set.

Advanced Hydrological Research Studies and Modelling

Whereas the GIS has limitations in terms of hydrological modelling, it still plays a major role by building a data structure directly usable for advanced studies and simulations. The GIS will provide processed and calibrated data to run hydrological models such as MODFLOW, MODSUR. All parameters needed for a water balance can be taken from the high accuracy GIS database.

A recent concluded doctorate study on field observations and different model outputs show that the transient run (monthly time step) done on the period 1950-2005 (Vincent *et al.*, 2006):

- an upward drainage from the Ramanathapuram aquifer to the Vanur aquifer is modelled with its volume increasing in time, which confirms the geochemical study of D'Ozouville *et al.* (2006).
- Important water fluxes are exchanged between the Alluvium aquifer, the Cuddalore and Vanur aquifers, as well as between the latter two. Direct recharge by rainfall of the Alluvium and Cuddalore aquifers is then distributed between all three aquifers, namely, Alluvium, Cuddalore and Vanur.
- Sea water intrusion in the Vanur aquifer at the coast level is being modelled since 1990.

A marine intrusion is suspected by the model NEWSAM model (ENSMP), which had not been detected by the previous geochemical study (Violette *et al.*, 2004). An estimation of a potential offshore freshwater storage show that such a storage could have protected the Vanur aquifer against a saline intrusion for several years. However, this storage is estimated very sketchily, thus we cannot determine the time period when this protection will cease. Nevertheless, once the salted limit reaches the coast level, the marine intrusion will increase rapidly (within a few months).

(b) Physical Remediation

Large Scale Rain Water Harvesting

Harvest has conducted the rehabilitation of 80 irrigation tanks plus other structures (Fig. 5.8), 37 water user associations and 17 drinking water user associations has been created with multiple training programmes, awareness campaign through street theatre. Large scale rain water harvesting can be done in the irrigation tanks. These tanks provide good storage capacity and seepage into the ground thus aiding the groundwater tables. However, to make such structures sustainable and manageable, it is important to involve the dependent population. Tank rehabilitation is a way to involve village communities to water management:

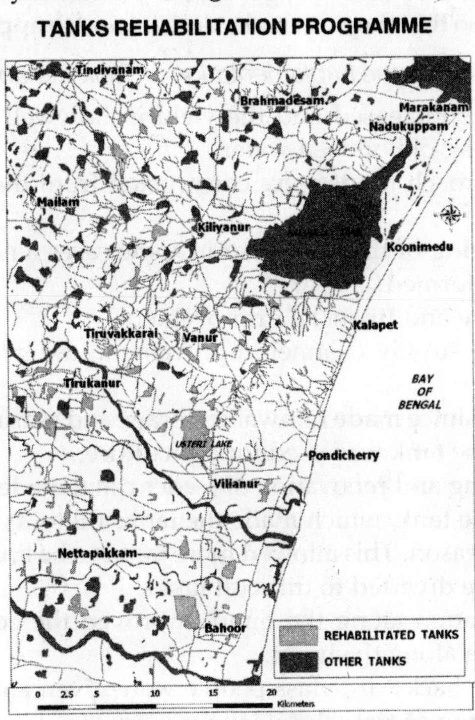

Fig. 5.8: Map of the Tank Rehabilitation Progress

It is a crucial part in the overall water management plan at a watershed level. It involves desilting the tank, deepening it to increase its storage capacity (so there is a longer duration of water supply to the wetland), strengthening the bunds (walls of the tank), repairs of the sluices and weirs and cleaning drainage channels to improve in the distribution of water to the irrigated area.

The project implementation is done completely with the participation of the village community. All strata of society are involved in the decision making of the implementation of the rehabilitation of the tank and not just those who utilize the water. As the tank is an ecosystem in its own right, and that it benefits not just the ones who use the water for irrigation and agriculture, it is a source of income to others. Hence, it becomes all the more important that all stakeholders are involved in crucial decision making regarding the tank.

The water users are imparted with training and information regarding the technical aspects of the tank. A volunteer is selected to measure the water levels of the tank. He is also responsible to monitor the weather and note it down on an information board located in the village. This board also records the temperature (minimum and maximum), rainfall (through a rain gauge situated within the village) and the relative humidity levels. All this information is easily accessible by everyone in the village. This is a means to build the knowledge base of the village so that they can make decisions with appropriate knowledge.

Achievements seen after the Implementation of Tank Rehabilitation Programmes

- Water table increased and the yield in borewell increased. Reduced the pumping of borewell water.
- Many approach roads were constructed from the silt and mud from the tank.
- Strengthening of the bund and in some cases a notable road over the bund was formed.
- Fish culture and financial improvement.
- Due to the supply channel improvement, the irrigation practice was achieved.
- Improved sluice made rainwater storage and rearing of fish in the tank.
- Silt from the tank was used for agriculture.
- The cleaning and renovation of feeder canal made the water flow easy towards the tank, which made the tanks fill to its full capacity during the rainy season. This allowed more tanks to be linked and water could be easily be diverted to difficult tanks.
- Tree plantation along the bund improved the ecology, and reduced soil erosion along the bund.
- Financially backward class people were given part-time employment for plantation of trees, local labour work etc.

- Through the project the farmers were trained on Organic farming, Tree plantation, Fish culture, Cropping pattern methodology, Horticulture, etc., which improved their standard of living, attitude and skills.
- Foreshore bund was created achieving the stoppage of encroachments and water inundation to the field.
- Saving of electricity was also achieved in some cases as the pumps were stopped to pump water from the groundwater table as the storage capacity in the tank was increased and the farmers used the water for irrigation.
- Tank water is more fertile than groundwater and thus the productivity in the field is found to increase for a minimum of 10 per cent.
- Due to participatory approach water distribution schedule is prepared which facilitated equal distribution of water till the tail-end farmers.

Augmentation of Groundwater through Artificial Recharge

Due to the insistence of deep borewells and its rapid development in the area, we are seeing an alarming decreasing of water tables. With the increasing number of borewells and its extraction rate it becomes imperative that we utilize artificial means to recharge groundwater. To depend on natural methods of refilling an aquifer would take 15,000 years.

There are a number of structures that aid in harvesting rainwater and help in percolation. Check dams, both earthen and stone, help in preventing both water and soil from flowing away. It aids in preventing soil erosion and conserving water. We have made many check dams in areas which have high percolation and infiltration rates. These have aided in raising the water tables to a considerable amount.

The objective is to maintain a proper hydraulic balance between the extraction and the recharge. It will help to stabilize the system and preserve the water resource potential and its quality. This can be achieved by augmenting the recharge potential by artificial adequate methods. To recharge the aquifer, it is, therefore, needed to focus on two main objectives: artificial methods can be applied to improve the natural recharge of the Vanur Aquifer; recharge structures can be installed at selected areas for direct recharge of the aquifer, these can consist of large percolation ponds, recharge wells and field bundings.

- Concentrate and maximize the flow reaching the suitable recharge zone on the outcropping areas of Vanur sandstone aquifer.
- Improve and facilitate the groundwater recharge through appropriate structures such as percolation ponds and recharge wells.

Flow Diversions

The heavy runoff happening during monsoon and reaching Kaluvelley swamp can be diverted and concentrated to the recharge zone of the aquifer. A large amount of excess water is presently flowing to the sea and this volume of water can be saved and used for direct recharge of the Vanur Aquifer. This can be achieved through various means including check dams, field bundings and diversion channels. A large buffer zone could be created up stream of the Kaluvelley swamp to harvest the excess water flow reaching the tank.

Direct and Effective use of GIS in Projects Implementation

Tank rehabilitation and groundwater recharge projects are defined with the support of GIS analyses. First, the GIS gives status report of the current situation of the water resources, regarding their potential and quality. It is, therefore, easier to analyze and face the problems, and propose a solution at a proper scale. GIS will help to define where an intervention is needed for a sustainable watershed management. The studies conducted will allow choosing suitable areas for groundwater recharge. The knowledge of the geology and aquifer system will allow defining precisely the type of structures to put in place on the ground. The viability of the programme can be assessed with the GIS by a simulation of the runoff and estimate of the potential groundwater recharge. In this case, the GIS acts as a practical tools to setup engineering solutions.

(c) Building Capacities of Communities to Take on Their Own Development

Community Management of Resources

Ownership of the community of a project is very important for sustainability. This is the key to any successful project. Development should be seen more as a change from the bottom up than from top down.

Through the various projects related to water resource management, Harvest looks at institutional building to ensure sustainability of the activities. Committees are formed based on the activities and in certain cases also registered to ensure autonomy. The wetland farmers who utilize the tank for irrigation are formed into the Water User's Association. They are instrumental in maintaining the tank and the structures and ensuring the sharing of water in an equitable manner. Alternative and appropriate farming techniques like organic farming are imparted to the farmers through trainings and exposure visits to successful farms.

The development process should be managed as a natural organic process rather than according to plans, goals, objectives, targets and schedules, implying that goals and targets may change and, therefore, their timing should be tentative and flexible to make room for adaptation to local conditions.

The development process is supported by local institutions with village panchayats, primary cooperatives, religious, youth, community-based users and self-help groups playing a lead role. It is more important to make sure that the development process is rooted in a strong local institution than ensuring that local institutions have a grasp of all the finer technical points. It is comparatively easy to arrange technical services from outside than to bring about social involvement and willing popular participation in the development process. Strong local institutions are necessary as support posts quite independently of whatever technical skills and other background they may have.

What is important in people oriented projects is that their capacities are built up for them in order to take on their own development. It is not just imparting training to people but also recognising and giving importance to the existing traditional knowledge of the people.

Awareness Generation

We strongly believe that through awareness generation and creating enabling environment for change will people change. Awareness generation, campaigns and educational projects in schools and villages play an important part of our work and cuts through all our project village meetings, displays at the weekly market in the main villages, dramas, staff trainings for other NGOs, and exposure visits for the beneficiaries take place regularly.

Our integrated water management social activities envisage participatory approach of implementation of the activities. This involves involvement of the beneficiary community in all stages of the project.

Weather and Environment Monitoring

We train village volunteers to monitor the weather and water levels. They are following imparted training and refresher courses. A weather board is installed in the village where the information regarding temperature, humidity levels, rainfall etc. are recorded for everyone's benefit. In this way, the farmers can access the information free of cost and utilize it for their work.

A network of rain gauges and weather stations have been installed in the area so that the data can be collected, analyzed and disseminated. The dissemination is done either through meetings or through reports. Soil sampling of the agricultural land is done in association with the government soil testing laboratory. Distribution of soil cards are done through a meeting with the laboratory technicians and the farmers. At these meetings, the farmers raise their queries regarding the quality of the soil. In this context, we link the farmers with the services provided by the government which are not known to the farmers themselves. This enables the farmers to access information and services provided by the government even when Harvest withdraws from the village.

5.4. The Need for an Integrated Approach

In this part of India, the remediation approach has mainly been focused on large scale rain water harvesting systems and surface water management. Among all, the tank rehabilitation programmes are the first to take place in Puducherry territory and Tamil Nadu State, where needs are still not fulfilled in many districts. It is still a priority to set up appropriate community water management to optimize the resource management at their best.

While many solutions exist to target certain specific surface problems, the interaction of various human activities with nature is not properly addressed and integrated in the planning process. The lack of coherence and integration of the multiple components and inverse consequences on the human populace has lead to a situation where even large scale rehabilitation programmes seem to fail.

The Puducherry region has a rich history regarding development of activities around water. The complexity of the actual problems calls for in-depth answers, based on a solid comprehension of the water mechanisms and the socio-economic dynamism of our area, articulated around a systematic transfer of knowledge and of the methods suitable for sustainable and integrated development. Auroville Water Harvest plays an important role in this regard by setting up not only scientifically grounded solutions, also environmentally, socially, organizationally and economically viable, with the local population and main actors.

Becoming aware of the importance of these structures for the life of our area, and its interlinking with society and its rapid pace of development, large efforts are being carried out by Harvest for the last twelve years not only to rehabilitate these structures, but also to recreate the necessary institutions in the villages with the participation of all the beneficiaries. Harvest today is the most recognized actor in this field, as a project carrier but as much as partner and adviser to the government and international agencies.

The area of operation of Harvest lies in a complete sub-watershed by itself. The series of tanks in the sub-watershed area has either already been partly or not at all rehabilitated. At the same time, this area is under constant investigation as far hydrology and hydrogeology are concerned, giving a very solid scientific base on which solutions can be promoted.

The process of improving the groundwater level and the desalinization of the aquifer is a long term and time consuming process. It requires a larger vision on in-depth studies, heavy construction work and population awareness. While Auroville Water Harvest is persistently working on improving the water situation of the area, the perverse trend of evolution, chiefly links to over extraction of groundwater, is not yet changed. By studying carefully the groundwater level fluctuation and the soil characteristics, the conclusion was arrived at that a particular area is both highly affected and is also one key area for the penetration of seawater in the region.

Working closely with rural communities on water issues, Harvest realized that it was important not only to work with just one focus in an entire resource management plan. Water can be looked at from different angles, as it touches different aspects of life: agriculture, irrigation, health, access etc. In the overall scheme of things, it is not only important to address just the irrigation practices of the farmers but also to provide alternatives and solutions to them. In turn we also need to address market economics, lifestyles changes and attitudinal changes. It was necessary to involve all stakeholders and address important issues in relation to water.

In addressing all interlinking issues of water consumption for agriculture, an integrated approach may be used so that there is maximum benefit accrued to the population and over a wide range of issues. Information and education of the population has played a key role in social changes all over the world. Thus it becomes imperative to provide people with information so that they have informed choices and make an informed decision.

Considering the general public health and environmental situation in India it is indispensable to integrate solid waste management component in an integral strategy for public health and hygiene. If not properly managed, solid waste interferes negatively with roadside drainage and wastewater flow, leading to mosquito breeding and spread of other disease vectors (flies, cockroaches, rats).

While most of the villages are today equipped with a water supply network, the water quality and quantity is far to reach appropriate standard. Due to very limited supply hours (usually 2 hours per day), the piping networks are heavily punctured by villagers as they expect to get longer supply by that means. This results in heavy pollution of domestic water, as punctured pipes are laid in dirty stagnating water. To complement this main supply, villagers develop hand pumps. Because of the lack of sanitation facilities and solid waste management, these hand pumps extract polluted water (fecal contamination), though creating direct contact with the people and the pollutant. Children, elders and young women are the most affected part of the population, and is a constant threat to human health. Lack of basic hygiene precaution to handle water is also source of contamination.

Auroville Water Harvest has developed a very good understanding of ground realities and relationship with the communities. Having experienced the problems first hand and worked on consolidating efforts to build consensus and cohabitation in the area, the people are slowly realizing that any work on the environment has to be in collaboration with all those residing in the area and with perfect unification. Though work has been on in pocket areas, it has become imperative to take the next step forward and build up a network to look at resource management in the entire bioregion. As people are finding their way in organizing themselves for the task, appropriate trainings are necessary to make them carry out their work efficiently and effectively. In order to make the network competent and able to sustain such an endeavour, initial landholding and capacity building is much required.

In the meantime, the need for an integrated approach is growing and is now giving orientation to future programmes targeting sustainable development. Irrigation is part of global water resource management and watershed management programmes need to integrate other side issues and environmental components like water pollution control, reforestation and soil conservation, reducing pesticides and chemical fertilizers, appropriate agriculture techniques reducing the water needs, change in cropping pattern for less water demanding crops according to market possibilities. There is a need to address issues like reducing the water usage by laws, rules and regulation, and cost of water. Due to foreseen water scarcity in the future, we need to explore water recycling technologies and waste water treatments for a safe re-use of treated waste water for irrigation.

Disposal of untreated wastewater has been a major cause of water pollution in India. Strategies for control of water pollution have focused mainly on implementation of expensive and energy-intensive conventional treatment technologies. The limited successes of such strategies can be attributed to the high capital investment requirement, on-going operation costs and non-availability of skilled operators.

The adverse effects of the Green Revolution on society is still being felt in the rural areas. One of the primary and most visual is that of the contribution of women to the agricultural sector. The feminization of agriculture has taken place largely due to the male population migration to urban areas for work after the fallout of their lands. Due to this female headed households have also seen a precedence.

Legitimizing the role of women as care givers and providing them with a space to explore potentials to be the change agents in bringing about a social change in the communities they live in, the interventions need to be linked to community asset creation.

Alternative agriculture is needed to conserve the natural resources and to improve the livelihood standards of the farmers and resource people in the area by integrating the efficient management of the natural resources available in the villages with the participation of the people.

REFERENCES

D'Ozouville N., Violette S., Gassama N., Dia A. and N. Jendrzejewski. 2006. "Origin and modelling of water salinization in a coastal aquifer of the Bay of Bengal: The Kaluvelley Watershed, Tamil Nadu, India". *Bulletin de la societe Geologique de France* 177 (November 2006): 333-345.

Gassama N., Violette S., D'Ozouville N. and A. Dia. 2002. "Multiple origin of water salinization in a coastal aquifer, South India, geochemical point of view". *Geochimica and Cosmochimica Acta 66* (15A): 265.

Gassama N., Violette S., D'Ozouville N., Dia A. and N. Jendrzejewski. 2003. "Multiple origin of water salinization in a coastal aquifer, Bay of Bengal". *Hydrology of Mediterranean and Semi-arid Regions,* IAHS Publ. No. 278: 471-476.

Paulik M. 2001. *Diagnostic des structures drainantes d 'un des sous bassin-versants du bassin versant de Kazhuveli, Inde,* Master thesis, Engineering School of Environment and Hydrology, ENSA, Montpellier, France.

Vincent A., Violette S., Marsily G. de, Boulicot G., Jorcin P., Krisnamurti R. S. and K. Sivasubramanian. 2006. "Infiltration quantification at the watershed scale using hydrological models, in a semi-arid context (Kazhuveli-Puducherry coastal sedimentary basin, Puducherry territories and Tamil Nadu State, India)". EUG Annual Meeting, Vienna, April 2-7 2006, *comm.*

PART 2

A Frantic Quest for Groundwater

Parts 2 and 3 are based on more detailed case studies focusing on seven villages whose main characteristics are described in Box 6. Part 2 concentrates on some aspects of groundwater irrigation, while Part 3 refers more specifically to tank management, its institutions and its multiple uses. However, in Part 2 we focus on groundwater (GW), not as a research subject in itself, but through its relationship to surface water (SW). That is why we try to understand factors which in existing water management add to separating GW and SW, and accentuate non-integrated water management, and even prevent any form of integrated management from being implemented, giving way to a frantic quest for GW despite dwindling GW levels.

Groundwater has become the first source of irrigation in India, in Tamil Nadu and in our study area. Social scientists, especially economists and socio-economists, have studied different aspects of this GW development, such as: its importance for agricultural intensification and consequently for its capacity to mitigate rural poverty because of food security and employment opportunities, especially in small holder farming systems (Marius-Gnanou, 1991; Shah, 1993, 2007; Saleth, 1998; Vaidyanathan, 1999; Narayanamoorthy *et al.*, 2005); the policies to be implemented in order to regulate directly or indirectly its use (Shah, 1993; Moench, 1992a) with some recurrent discussions about the relationship between electricity price setting and the efficiency of GW use (Narendranath *et al.*, 2005; Shaheen *et al.*, 2007); the issue of GW overexploitation and its impact in terms of inequity of access to water (Bhatia, 1992; Janakarajan and Moench, 2002; Kumar, 2007), with some specific in-depth work about informal water markets in Gujarat (Dubash, 2002; Prakash, 2005; Kumar, 2007[1]), the last two putting forward the social setbacks of such markets and the need for new institutional arrangements for sharing GW, also the subject of Mukherji's work (2007).

Here we will particularly focus on local practices and on how farmers perceive the water crisis and how they react when faced with drought situations (which was the case from 2000 to 2003 due to a shortage of rain during the monsoon season). We consider as acquired the legal and administrative aspects as well as the development policies, which limit integrated management and have been presented in Chap. 1. How is one to understand that farmers make no attempt to protect this underground resource and still drill deeper and deeper? What is the logic determining these practices? Why is the lack of rain, the "dry sky", always accused of being the main cause for the water crisis? We will see how, under existing water management conditions, programmes to recharge the water table—with surface water—do not help to transcend this groundwater/surface water dichotomy, but on the contrary sometimes even strengthen competition between irrigation techniques rather than complementing each other.

As a digression, let us examine an ethical issue regarding groundwater that I encountered during my fieldwork. Farmers always require more water and demand more privileges[2] for accessing this precious resource. Yet for me, curbing water consumption is vital if we are to achieve sustainable use. This disparity between the farmers' demands and my perception of the problem left me with a feeling of betrayal of these people who had given me their time and had willingly answered my questions. I did genuinely listen to their arguments. Here I will attempt to explain their logic. There was I making a critical analysis of their practices—as being non-sustainable and marred with social inequities—while all the time they were asking me to support their requests! I do not want to give the impression that I hold farmers—whom some people qualify as greedy and selfish—the only ones accountable for the overexploitation of groundwater, since political incentives and the lack of regulation are the leaders' responsibility. I have also submitted my point of view to farmers: we will see their reactions.

Box 6: Village Case Studies: Main Characteristics

There are seven village studies that were more thoroughly analyzed during the IFP programme and to which we will often refer in Parts 2 and 3. Three of the villages are situated in the rocky area, in Villupuram district. The others are in the alluvial part, two in Puducherry, one in Villupuram and one made up of a mosaic of "English" and "French" plots as indicated on the village map, which refers to the colonial period.

Two villages have non-system tanks (NST), one in the rocky area and one in the alluvial area, and all the other five rely on the same river, the Ponniyar. Four of them rely on Tirukoilur *anicut*, three are supplied by the Raghavan channel (two in the rocky area) and the fourth is the end-tail one along the Pambai channel. The fifth village among the system tanks studied is supplied by the Bangaru channel starting at the Sornavur anicut. See Fig. 1.

Villages in the rocky area have the following irrigation characteristics: irrigation mainly by tank but also by open wells, the latter having gained some importance over the last 30 years, especially on *punjai* land where they used to be very rare; 1 or 2 crops of paddy according to the availability of water in the tank; large area of rainfed agriculture (*see* Box 7 for the definition of *punjai* and *nanjai*).

Villages in the alluvial area are nearly fully irrigated (at least 80 per cent, cf. Table 1 below), mainly by groundwater. They produce two to three crops of paddy on *nanjai* land.

In most of the villages studied (Table 2), the Scheduled Caste (S.C.) population amounts to more than one-fourth of the village population—as is the case in this region (see the population map in Oliveau, 2003)—except for one village where there are none and another one where they are numerically more important (67%). The proportion of cultivators in the latter is thus very low and

(*Contd.*)

Box 6 (*Contd.*)

that of agricultural workers very high, as most S.C. people are still landless in the study area, as in Tamil Nadu (NCRWC, 2002: 4-5).

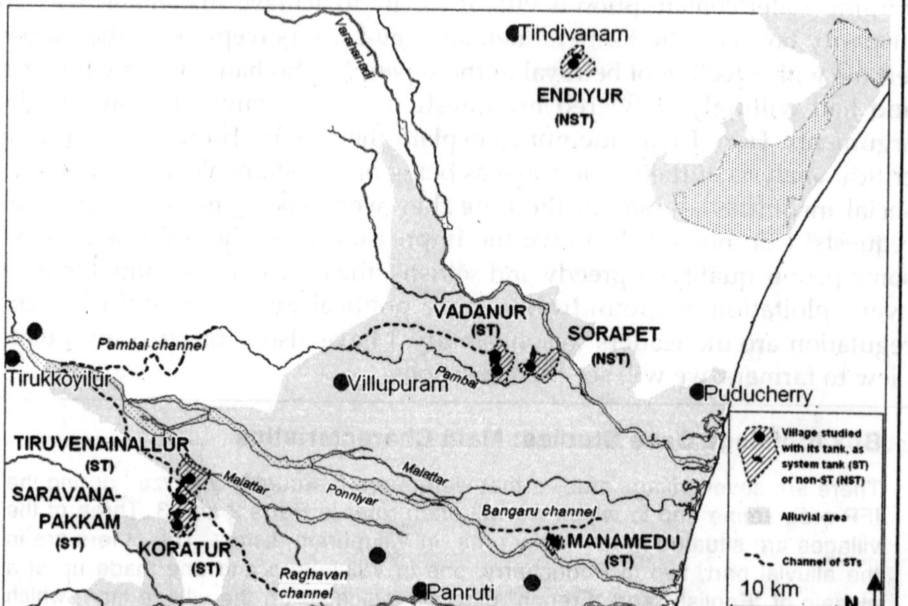

Fig. B 6.1: Localisation of the Village Case Studies

Reasons for Choosing these Case Studies

Endiyur (END) has two non-STs, in the rocky area where a WUA was set up by a NGO during a tank rehabilitation programme in Kaluvelley watershed. Studied in Chap. 11, it will also be compared to the neighbouring village that refused to play any part in the rehabilitation project.

Koratur (KOR), Saravanapakkam (SRVP) and Tiruvennainallur (TVN) are neighbouring villages (SRVP is formed, however, by four scattered hamlets), whose tanks are supplied by the same channel (Raghavan channel). While the first two are in the rocky area, TVN is situated near the river Malattar. They form an interesting transect where it is possible to compare the ready access to groundwater with irrigation practices. Moreover, part of SRVP's land adjacent to TVN can benefit from GW from TVN, whereas KOR had to find another solution to extend its irrigated area.

Tiruvennainallur (TVN) is like a burg, a market village (and was reclassified from special village panchayat to town panchayat in 2006), with two large temples to Shiva and Vishnu. Inscriptions about the purchase of land by King Uttama

(*Contd.*)

Box 6 (*Contd.*)

Chola for the village community and their gift to the temple attest to the existence of the village, its land and the temple during the 2nd part of the tenth century (Balasubramayam, 1971: 200). Local history recounts how King Sadaiappa Nallal reigned during the eighth century over TVN territory and how the poet Kambar contributed to writing the Tamil version of the *Ramayana* and died in the twelfth century in TVN where a temple has been dedicated to his memory. The Shiva temple has great local influence[3], with neighbouring villages still taking part in organizing the 10-day temple festival. This village was chosen for the study for two main reasons: even though there was a large number of wells and tubewells when we visited the village for the first time in 1999, the local institution managing the tank was quite active (this site is hence a counter-example of the principle whereby traditional irrigation institutions are less active when there is high well density); secondly, there are two areas with different access to groundwater, which proved to be an interesting point when analysing water management within a village. Since 2004, Tiruvennainallur and Saravanapakkam share the same water user association (WUA).

Vadanur (VAD) has land in Tamil Nadu and Puducherry. It is a village panchayat administered by the two States and made up of two hamlets, Vadanur and PS Palayam (PSP). These two hamlets share the same tank, called the Vadanur tank. This village caught our attention when its Tank Association (TA) published an advertisement for a public fish auction of the tank fishes in the local newspaper in April 2005. Placing an advertisement for such an auction was not common at the time, and showed the weight of the project and how active the Tank Association was. We then learnt that it was one of the pilot tanks within the Tank Rehabilitation Project of Pondicherry (TRPP) launched in 1999. From a water management point of view, this example is also of interest since the tank has not been used for the last 30 years.

Sorapet (SOR) is located three kilometres east of Vadanur. It was chosen as a case study because, even if located in the so-called alluvial area, only two crops of paddy are grown on *nanjai* land. This village was selected in order to analyze its capacity to adapt to water scarcity and to understand why the tank is still not used.

Manamedu (MAN) tank is part of the system whose channel (Bangaru) supplies water to Bahur, one of the oldest and most famous tanks in Puducherry, still in use. But the MAN tank is not used for irrigation, even if there has never been any encroachment on it and even though a rehabilitation programme (TRPP) has been set up there. This case is of great interest when analysing conflicts of interest between farmers and those raising fish in the tank, mainly an S.C. group, since this caste represents the majority in the village (67%).

Table B 6.1: Characteristics of Irrigation and Agriculture in Village Case Studies

	END	*KOR*	*SRVP*	*TVN*	*VAD*	*SOR*	*MAN*
Village name	Endiyur	Koratur	Saravanapakkam	Tiruvennainallur	Vadanur	Sorapet	Manamedu
District	Villupuram	Villupuram	Villupuram	Villupuram	Villupram/Puducherry	Puducherry	Puducherry
Geology	rocky (charnockite)	rocky (gneiss)	rocky (gneiss)	alluvial (A)	alluvial (B)	alluvial (B)	alluvial (C)
Tank water supply	rain + upstream runoff	Ponniyar	Ponniyar	Ponniyar	Ponniyar	rain + surplus of upstream tanks	Ponniyar
System tank	Non ST	Raghavan channel (Tirukoilur anicut)	Raghavan channel (Tirukoilur anicut)	Raghavan channel (Tirukoilur anicut)	Pambai channel (Tirukoilur anicut)	Non ST	Bangaru channel (Sornavur anicut)
WUA	through NGO (Palmyra) Kalvelley watershed programme	through PWD	through PWD	through PWD	through TRPP (NGO + Pondy's PWD)	through TRPP (NGO + Pondy's PWD)	through TRPP (NGO + Pondy's PWD)
Wells	open wells	open wells	open wells	tubewells and wells-cum-bore	tubewells and wells-cum-bore	tubewells and wells-cum-bore	tubewells
Depth of wells (in m)	15-20	10 to 20	15-20	15 to 25	15 to 150	15 to 60	30 to 160 (250 m. for gvnt TW)
Irrigated area (1991 census)	74%	35%	40%	80%	61%	82%	86%
Crop in *nanjai*	1 paddy	1 or 2 paddy	1 or 2 paddy	3 paddy	3 paddy	2 paddy	3 paddy
Crop *in punjai*	1 or 2 crops	1 or 2 crops	1 season	3 crop seasons	3 seasons	3 seasons	3 seasons

Table B 6.2: Characteristics of Population in Village Case Studies (from 2001 Census)

	END	*KOR*	*SRVP*	*TVN*	*VAD*	*SOR*	*MAN*
Total population	3033	2134	3286	8582	4960	3579	1968
% of *S.C.*	0%	32%	27 %	26%	27%	29 %	67%
% of men cultivators (working on their own or leased land)	45%	55%	21%	12%	14%	19%	5%
% of agricultural labourers	49%	53%	56%	46%	66%	58%	82%

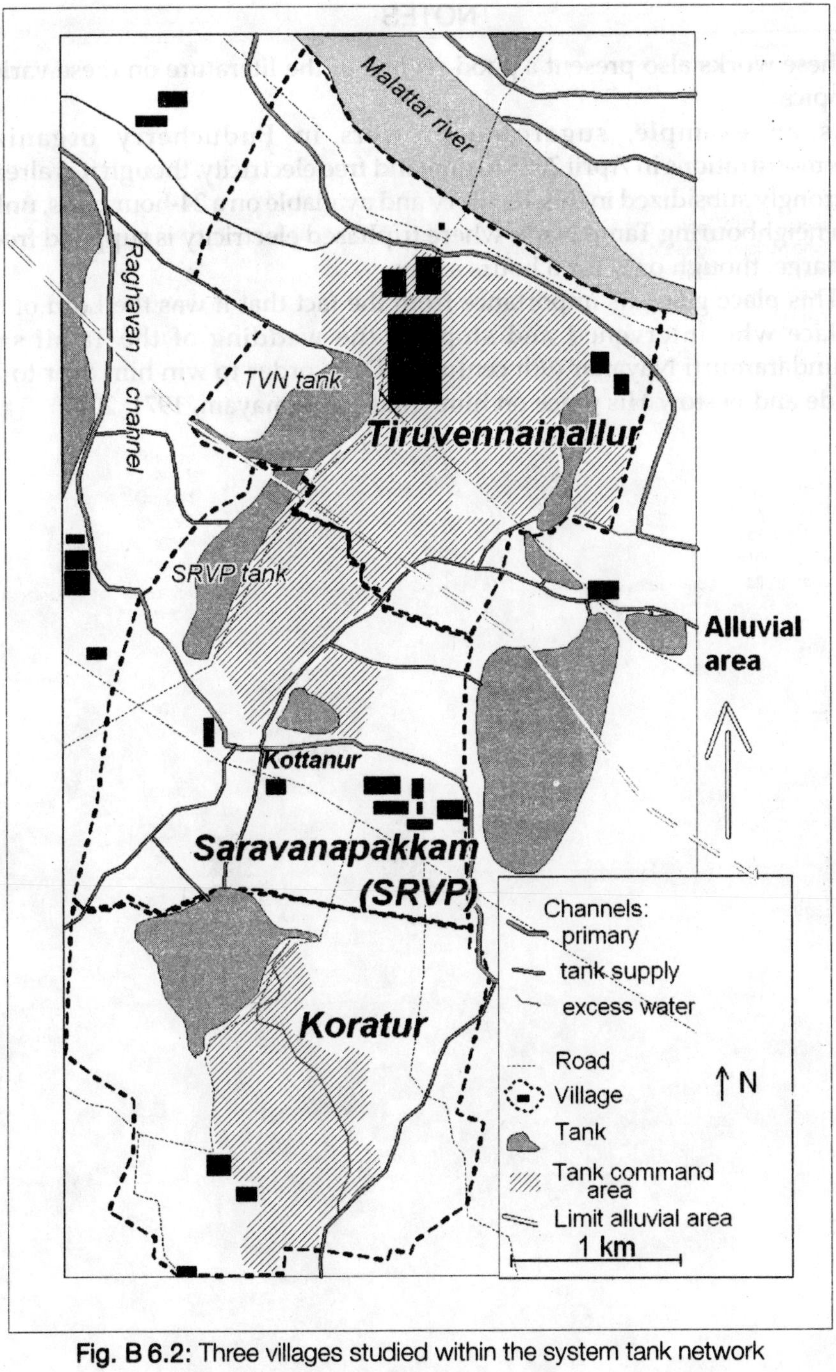

Fig. B 6.2: Three villages studied within the system tank network

NOTES

1. These works also present a good review of the literature on these various topics.
2. As an example, sugarcane growers in Puducherry organized demonstrations in April 2008 to demand free electricity, though it is already strongly subsidized in this Territory and available on a 24-hour basis, unlike in neighbouring Tamil Nadu where triphased electricity is supplied free of charge, though only for 6 hours a day.
3. "This place gains its importance from the fact that it was the Lord of this place who intervened and stopped the wedding of the Tamil saint Sundaramurti Nayanar (9th century AD) in order to win him over to His side and bestow His grace on him" (Balasubramayam, 1971: 200).

CHAPTER

6

No Pumping Limit?

Olivia Aubriot

The alluvial part of the study area was already qualified in 1997 as being "overexploited" and most of the other parts of the study area were found to be at a "semi-critical" stage (GoTN, 2002). Factors to explain this overexploitation of groundwater include the bore-owners' open access to GW, free energy for pumping (in the case of electrical pumps), very limited regulations governing the exploitation of the resource, along with the technical, agronomical and social advantages already cited (*see* Chap. 1.1). In order to collect data for our study of non-integrated water management, two characteristics of this excessive use of water will be analysed here: the deepening of wells and bores, and the sale of water (through informal water markets).

6.1. Increase in Well and Bore Depth: Limited only by Environmental Constraints

The opinion, widely prevalent in literature and among people, that farmers respond to the constant drop in the groundwater level by deepening their bores and wells has to be moderated given the geological and hydrogeological context. The level of water tables is indeed declining and many wells are deepened and bores drilled deep in order to face this depletion: striking examples can be found in North Gujarat where some tubewells were already 240 metres deep in 1996 (Dubash, 2002: 45), with an average depth of 31 m in 1971 increasing to 150 m in 1996 (*ibid.*: 5); studies also observe this tendency in Tamil Nadu: an average open well depth was 12.5 m in 1960 and 20 m after 1985 in Nooyyal basin, while bores were made at 30 m and 80 m respectively (Janakarajan and Moench, 2002), well-*cum*-bore depth increased by 90 m between 1985 and 1995 in some areas of Tamil Nadu (Ramasami *et al.*, 1999: 61). This process of for ever increasing the depth of wells also exists in our study area—and will be described using the example of PSP-Vadanur—though it only applies under specific geological and hydrological situations. One can also observe cases where the physical drilling limit has been reached—this will be analysed using the case of Sorapet in the alluvial area and Koratur in the rocky area: such cases are of interest to us in our attempt to understand farmers' reactions to the water crisis and to anticipate scenarios on a larger scale. Let us focus on the history of well and tubewell installations in these three villages.

(a) A "Typical" Increase in Bore Depth—Case of Vadanur

The historical process of deepening bores and wells in PS Palayam-Vadanur (PSP-VAD) villages was explained by a group of farmers as follows (*see* Fig. 6.1 for a diagram[1]): prior to 1945, a dozen wells fitted with a bucket that was lifted manually or by animal force (*etham*) were located in the paddy fields and only two on rainfed land. Then a dozen wells, equipped with diesel motors were set up on dry land. With the electrification of the village in 1960, villagers set up more wells and invested in electrical pumps. With the use of motors and their greater pumping capacity compared to human- or animal-generated energy for water extraction, farmers had to deepen their wells progressively in order to increase water availability in the well and to address the already growing decline in GW level. In the 1970s, when the water level was below the depth of the wells (about twenty metres) which was the maximum for the latest motors' full pumping capacity, villagers suffered a water shortage for a few years. No farming could be done. There were then about 20 well-*cum*-bores in the 200 acres of paddy fields and 50 in the 550 acres of dry fields, that means a similar density of bores, 1 per 10 acres (= 4 ha) of land.

In the 1980s, when submersible pumps arrived in rural areas, some farmers were able to tap deeper aquifers, by using these powerful motors and by drilling new bores or deepening old ones (up to 40 or even 110 m). Since then, the issue has taken a different angle, going from a water shortage affecting everyone to a situation of a lack of water in which users of deep tubewells are not affected at all. During the summer of 1985, when water was only available at 50 m, those who had not deepened their bore (10 out of 20) could not use their pump during the dry season and had to wait for the water level to increase with the rain. Since 1990, farmers have not tapped into the upper aquifer water (up to 18 m deep) and use only the second or third aquifer, which are about 25-50 m and 90-110 m deep respectively, according to the sites. And today, farmers drill bore at a depth of more than 160 m.

Here, we see how farmers explain why the upper aquifers are progressively left unused, since they no longer provide water during the dry season; hence the farmers tapping into deep aquifers. It must be mentioned that there is a misinterpretation in Fig. 6.1: the depth of the tubewell does not necessarily indicate the depth of the tapped aquifer, since the motor can be placed in a higher position, as we will see below:

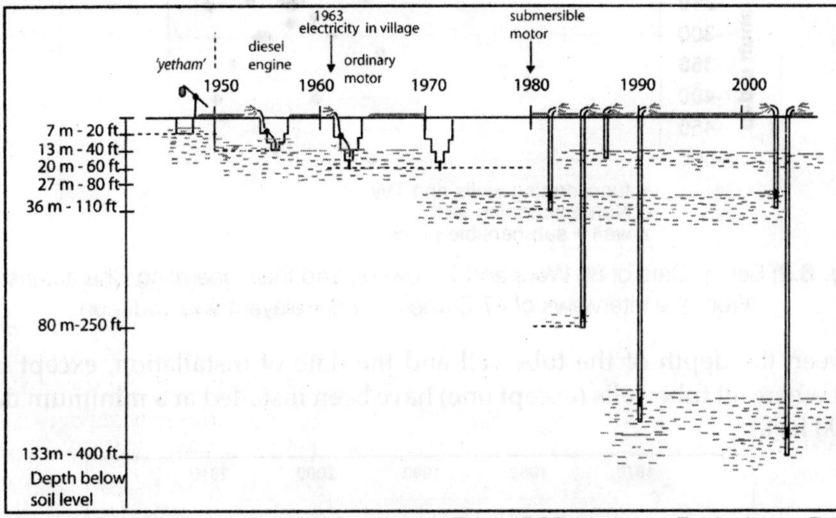

Fig. 6.1: Change in the Depth of wells and TW in PSP-Vadanur, Puducherry. Data Generated by a Group Discussion with five Tubewell Owners, I. Prabhakar and O. Aubriot.

Enquiries made among 47 landowners about the history of their 80 wells confirm the diagram in Fig. 6.1, which was made before these enquiries were analysed: the fore-mentioned group of five tubewell owners therefore has an accurate perception of the history of their village wells—we will see that people who do not own a bore have a very different perception—and that this method of enquiry through group discussions is effective in painting a broad description of historical facts. Nevertheless, only individual enquiries can precisely make the whole process understandable.

In our sample, most wells have been either deepened and changed into well-*cum*-bore, or abandoned and replaced by tubewells, as shown in Fig. 6.2. Only three wells from the sample still work with centrifugal motors which are referred to as "ordinary motors" by farmers. And since 1991, such centrifugal motors are no longer set up. More bores are thus equipped with sinkable motors in this village (which is not the case in the neighbouring village which will be presented hereafter). Note on the figure that very deep wells have been set up for more than twenty years and that no apparent correlation exists

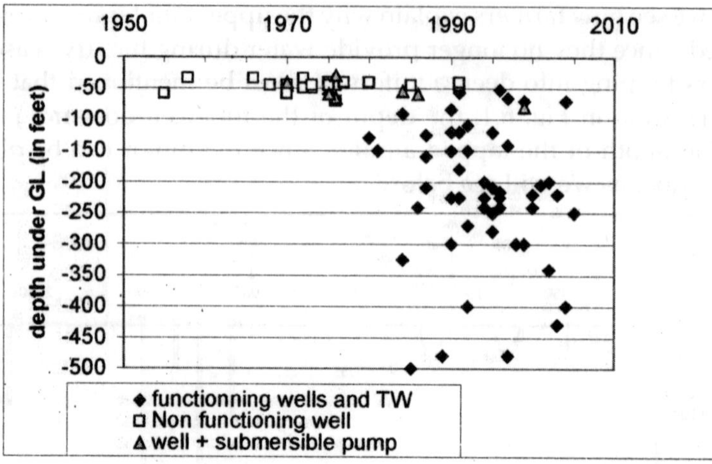

Fig. 6.2: Set up Date of 80 Wells and Tubewells, and their Operating Characteristics (From the Interviews of 47 Owners in PS-Palayam and Vadanur)

between the depth of the tubewell and the date of installation, except since 2000 where all tubewells (except one) have been installed at a minimum depth of 200 feet.

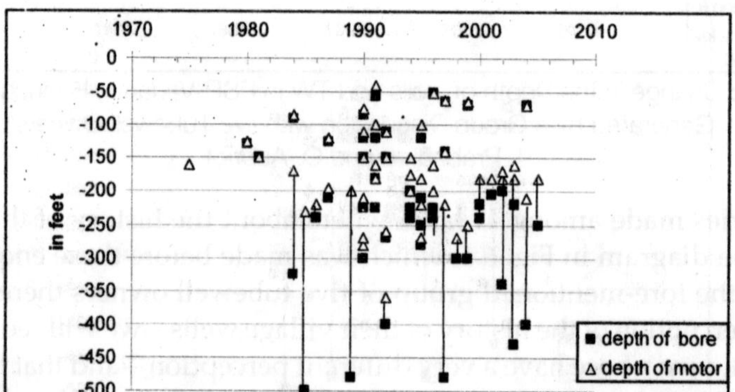

Fig. 6.3: Differentiation Between Bore Depth and Motor Depth, in PS-Palayam and Vadanur

By differentiating bore depth and motor depth (Fig. 6.3), it can be seen that bores of up to 200 feet (61 m) deep are equipped with a motor at more or less the same depth, while beyond this, bore depth does not indicate the depth of the aquifer which is tapped. One may also note that the implementation of a bore far deeper than the necessary length to tap the aquifer is not a recent phenomenon and has been in existence since 1984. This difference in depth between the tubewell and the motor can be explained in various ways: farmers express their fear of seeing the water level decrease and then anticipate this phenomenon to avoid it deepening any further; we may add that there is also a question of prestige in having a deep bore. Indeed it first of all shows an ability to invest in such an infrastructure; also, bore depth is a way for its owner of distinguishing himself from others ("I am the first to have sunk a tubewell. It is 510 feet deep. No one has spent as much money as I have" explains one of them in the documentary by Coste and Ploumpidis, 2007) and to gain access to a resource which is perceived as "endless" and therefore reliable by most people (see Chap. 7.2). Today the owner of a deep TW could not bear the idea that his bore might not provide water and that his neighbours would ask embarrassing questions about a potential dysfunction (*ibid.*). Added to this is the idea that the owner of a TW, which is deeper than today's requirements, is therefore protected from any water level decrease; indeed if this drop in water level—which is currently observed for the upper aquifers—carries on, a deep bore can cope with the situation because submersible pumps can easily be positioned deeper if necessary. Finally, as a consequence of the previous point, such a bore avoids further deepening of the well if the water level decreases, which is more costly than if the bore is initially drilled deeper.

The fact that we have not encountered a case where these oversized bores have had an actual technical utility (that is, where motor depth had to be modified to take advantage of bore depth) indicates that social- and prestige-related aspects are central to the initial decision. Indeed, during a discussion the first mention of a bore concerns its depth, rather than motor depth or the aquifer depth. Fear and anticipation of a water level decrease are indeed important factors, which have been conditioning recent choices not to drill tubewells less than 60 m deep, but they alone do not explain the oversizing of tubewell, which has been observed for the last ten years. In my opinion, this is a technical argument to justify a practice which has more social undertones. Indeed, this practice only concerns one-fifth of our sample: only the wealthiest can afford this and thus mark themselves from the ever-growing number of bore owners; this would also explain that this practice has been developing for the past ten years, as a means of increasing one's social distinction.

If one analyses bore depth taking into account the categories of farmers (based on their land holding), the earlier idea is proven and provides a clearer image of the drilling dynamics. Fig. 6.4 shows the results of this analysis, in which we left aside bores which have been installed in the shallow aquifer (which amount to five in our sample). Bore owners have been sorted into four categories: marginal farmers (below 2.5 acres = 1 ha), small farmers (from 2.5

to 5 acres), big farmers from 5 to 10 acres and very big farmers (more than 10 acres). For very big land owners, one notes that only they own oversized wells (more than 300 feet). But since they also own shallower bores, these extremes do not help to distinguish a particular tendency in their practices. On the other hand, marginal farmers drill to a maximum of 120 feet (37 m), with economic constraints explaining this limit. And if, for small farmers, a very obvious increase in depth has emerged since the nineties, and mid-eighties for big and very big farmers, this is not the case for marginal farmers who cannot afford it. Since 1994, they are the only ones to drill to a maximum of 120 feet deep, all other farmers drilling to a minimum of 200 feet (61 m). If we consider that the necessary depth to tap a productive aquifer is given by the minimum depth of small farmers' bores, one can see that this depth started to increase in the early 1990s and that this minimum is the same for both big and very big farmers. Therefore, the increase in minimum bore depth really indicates the search for a more reliable aquifer (the shallow aquifer being avoided in new installations), the other depths depending on the farmers' financial resources. Nevertheless, the motor is not positioned too deep as this would impede its ability to pump. These results are consistent with the discourse of five bore owners from Vadanur, who explained that those who have not deepened their bores depend on an aquifer whose level has decreased to such an extent that it can no longer always provide water in summer.

Local hydrogeological features might include different levels of aquifers and explain the differences observed between levels of one aquifer in various places. Nonetheless, since 1994 there has been a definite trend to install all new bores at more than 200 feet (except for marginal farmers). In the same way, while it is difficult to explain the maximum limits for bores, for which there may be multiple causes (local environmental conditions, the owner's willingness to drill deeper than the actual level of the aquifer, the necessity to tap into a deeper aquifer...), the fact that the maximum depth for big land owners is systematically greater than for the small farmers cannot be explained by local environmental conditions. This clearly reflects a social phenomenon: the need for big land owners to differentiate themselves from others, to avoid a drop in the water level, and to access a deeper aquifer believed to be more productive. They also set an example for other farmers, who will in turn progressively install their bores at the same depth.

If we differentiate between setting up the tubewell on *punjai* and *nanjai* land, we obtain Figs. 6.5 and 6.6, where we see that depth first increased on *nanjai* land, then on *punjai* land. We do not really know the reason for this (and we did not ask how farmers perceived this), but it may be because of the greater amounts of water needed for paddy cultivation (on *nanjai*) and therefore of the yield of the aquifer which may have decreased earlier than on *punjai* land.

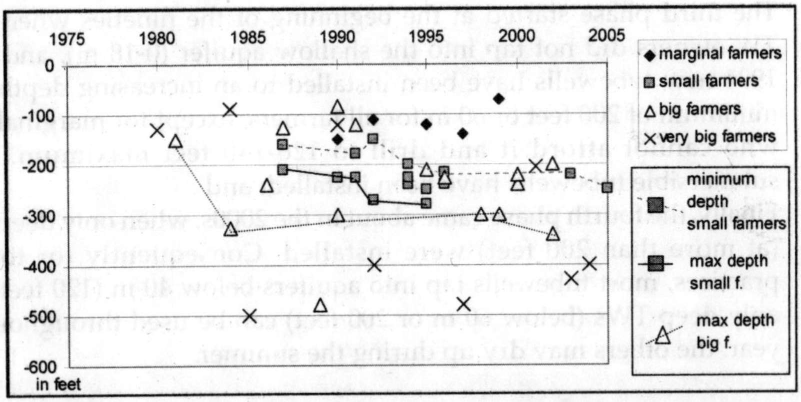

Fig. 6.4: Change in Tubewell Depth according to Land Holding and Date of Bore Installation, on *Punjai* and *Nanjai* Land in the PSP-Vadanur Alluvial Area. (Marginal Farmers < 2.5 acres; Small Farmers 2.5 to 5 acres; Big Farmers 5 to 10 acres; Very Big Farmers = More than 10 acres—5 Tubewells Tapping the Shallow Aquifer (up to 80 feet) have been left out)

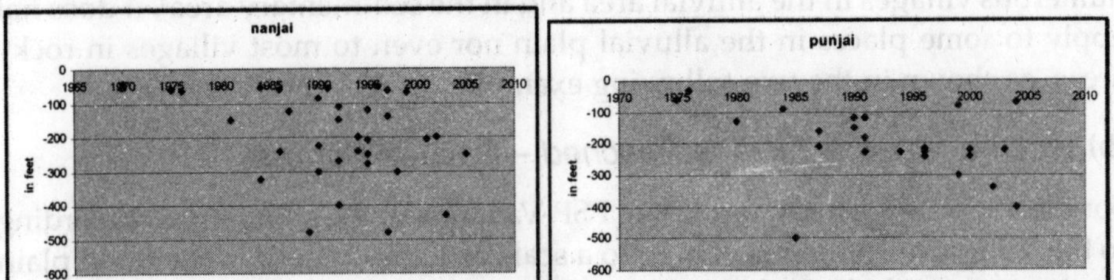

Figs. 6.5 and 6.6: Change in Tubewell Depth on *Nanjai* and *Punjai* Land Respectively

We gather from this case study that the history of GW use is characterised by four main phases, where the first two can be generalised for the whole study area:

- The massive use of electrical pumpsets dating from the sixties;
- Then a second technical revolution took place in the eighties with the arrival on the market[2] of powerful submersible pumps. Big farmers start to drill below 300 feet. Since then, in villages such as Vadanur where there is easy access to GW, there has apparently been an abundance of water and a certain self-sufficiency on the part of agriculture with regard to rain due to the use of submersible motors and deep tubewells. Indeed, since the late eighties the village tank has not been used and all farmers have relied on groundwater for irrigated crops (and even for all their agriculture, since the whole area has been irrigated);

- The third phase started at the beginning of the nineties when most TW-owners did not tap into the shallow aquifer (0-18 m), and since 1994 new tubewells have been installed to an increasing depth, to a minimum of 200 feet or 60 m for all farmers, except for marginal ones who cannot afford it and drill to 120-140 feet maximum. Only submersible tubewells have been installed; and
- Finally, the fourth phase came about in the 2000s, when only deep TWs (at more than 200 feet) were installed. Consequently for today's practices, most tubewells tap into aquifers below 40 m (120 feet) and only deep TWs (below 60 m or 200 feet) can be used throughout the year: the others may dry up during the summer.

Moreover, a clear distinction regarding maximum TW depth arises between categories of farmers: certain economic aspects for marginal and small farmers and certain social aspects for big and very big farmers explain their choice. Though this process of drilling more or less deep tubewells, with a clear trend towards an increase in depth since the beginning of the 1990s, applies to numerous villages in the alluvial area and in the sedimentary area[3], it does not apply to some places in the alluvial plain nor even to most villages in rocky areas, as shown in the two following examples.

(b) When a Physical Limit is Reached – Case of Sorapet

Sorapet (SOR) is located 4 km east of PSP-Vadanur, in the alluvial area according to the Villupuram geological map to a scale of 1:250,000 and in the flood plain on the geomorphological map (but because of the inaccuracy of the latter—1:1,500,000—the village may be situated in the pediment area). The date of well-*cum*-bore and tubewell installation is given in Fig. 6.7. When we compare it to the graphs related to PSP-Vadanur case, we can identify the following main differences between SOR and the previous example:

(i) Some wells dug in the 1930s or 1960s are still in use, whereas the oldest infrastructure still in use in PSP-Vadanur dates from 1970 and most of them only date from the mid-1980s,
(ii) Some bores of 50, 150 or even 250 feet deep are in dysfunction, and
(iii) Since 2000 there has been a strong tendency to drill shallow tubewells (max. 100 feet or 30 m).

Thus, the idea that due to a drop in the water table, farmers drill deeper bores does not apply here. Let us first explain this point. In this village of SOR, below 45 to 60 m, the water in Ramanathapuram sandstone contains hydrogen sulphide[4] which colours it red and makes it smell bad. Furthermore, due to a

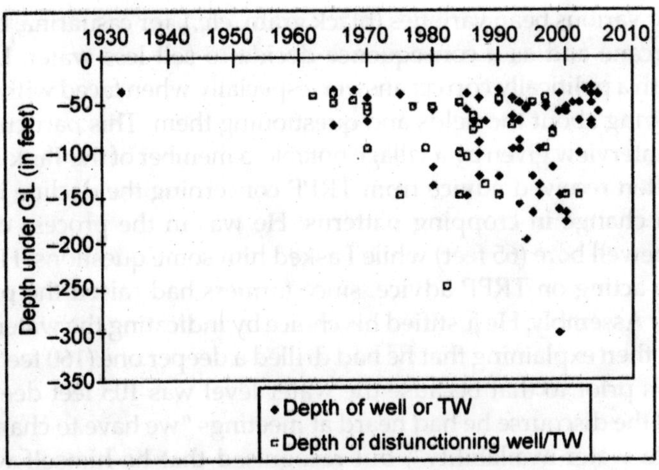

Fig. 6.7: Change in the Depth of Wells and Tubewells according to the date they were set up in Sorapet

rocky formation, farmers cannot drill below this physical limit (or they obtain poor quality water and need a very costly drill). Drillers and engineers from the Agriculture Department have advised farmers not to drill below this level. Yet the idea of drilling deeper is so strongly ingrained that some do try. Some farmers irrigate their land using malodorous coloured water tapped at 400 feet which, others say, spoils the crop. One farmer stopped using his 350 feet-deep bore and one drew no water from his deep TW.

It is interesting to analyze the explanations bore owners give to justify the choice of depth for their bores and the reason why bores have been shallower over the last few years. Some—perhaps with no overall perspective of the phenomenon—admit that their choice is guided above all by the advice of the driller depending on sub-soil texture and by economic considerations (manual drilling to 30 m is less expensive than drilling deeper with machinery). One farmer, influenced by a particularly unsuccessful attempt of a close relative, explained the logic behind his choice of drilling less deep: one of his uncles invested in a 45 m (150 feet) deep tubewell, which had never given any water and he had to sell several plots of land to pay for the drilling. Since then, other villagers have been afraid to invest with no guaranteed result. He himself still buys water and is trying to set up a TW. He is also envisaging changing the cropping pattern in order to use less water: the number of people who recognise the need to use less greedy crops is still too little, and it is difficult to know whether they came to this conclusion on their own or thanks to TRPP policy and recommendations. However, we did see some farmers, those encountering some problems with their bores during the summer, diversifying their crops more.

They opt for various bean varieties (black gram, etc.), for casuarina, but they still plant sugarcane and as a consequence decide to sell less water. Lastly, some came up with a politically correct answer, especially when faced with a European woman moving about the fields and questioning them. This particularly brings to mind an interview given by a village notable, a member of the Tank Association, who had often received advice from TRPP concerning the decline in the water table and a change in cropping patterns. He was in the process of making a shallow tubewell bore (65 feet) while I asked him some questions. He explained that he was acting on TRPP advice, since farmers had raised the problem at a Puducherry Assembly. He justified his choice by indicating the water level (30 to 40 feet) and then explaining that he had drilled a deeper one (160 feet) in another area 15 days prior to that because the water level was 105 feet deep there. He clearly held the discourse he had heard at meetings "we have to change the crop according to water availability", but recognised that he himself would grow mainly sugarcane using the water from the shallow tubewell. It seems that only farmers who have actually encountered difficulties with the amount of well water have started to change their cropping pattern. Let us remember, however, that government funding to grow less greedy plants during the summer has also helped bring about this change.

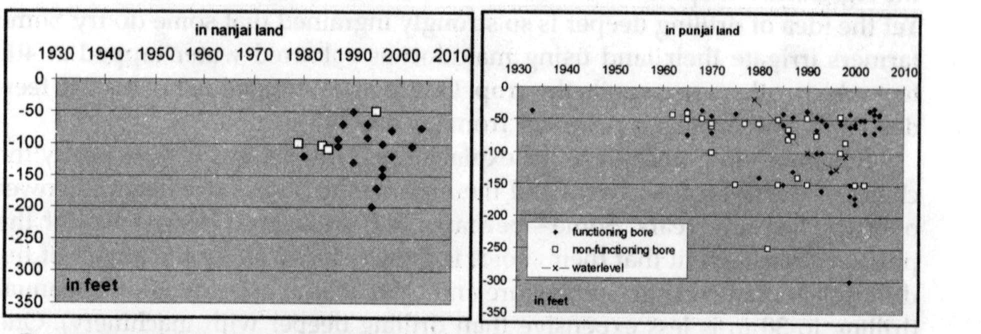

Figs. 6.8 and 6.9: Depth of well/tubewells on *Nanjai* and *Punjai* Land respectively in Sorapet (Water Level recorded by the Department of Agriculture, in GoP, 1999)

The second difference between SOR and VAD is the malfunctioning of some bores above 80 feet, especially on Punjai land (cf. Figs. 6.8 and 6.9). In SOR, owners of wells and bores know that drilling is a risky business: many are those who have seen their wells or bores dry up or become unproductive, as in the example mentioned above. These defects may be due to various situations: either the outflow is unsatisfying, with periods of more than 15 minutes with no water, or it dries up during the dry season, or the pump stops working due to a badly installed bore, with a blockage in the pipes pumping the water; the well may be the property of brothers, which makes deepening it harder to manage from a social point of view, especially if some parties reject the idea (cf. examples in Janakarajan *et al.*, 2002) and therefore, the well is left undeepened and abandoned. Many farmers have had to make several attempts. Thus, of a sample of 40 farmers, together owning 46 pumps still in use, 16 of these have replaced infrastructures that were defective (cf. Fig. 6.10). This means that in all 74 wells had to be drilled: in this village, 38 per cent of the infrastructures were defective or required reinvestment. Since most well owners have small properties (Fig. 6.11), reinvesting in a well or TW is problematical.

One last difference with Vadanur village is that in Sorapet some old wells are still being used even though they have a poor output, since underground water, however deep it is, is not perceived as a resource with an unlimited flow. It became quite common during the dry season to see tubes spurting out water intermittently with a force diminishing to nothing for one or two seconds before once again pouring out some 10-12 l/s.

The story of wells in this village is thus marked by doubts as to the success of a bore, of tubewells with a fluctuating flow and of the management of investment risks. This has discouraged the sinking of deep borewells since the year 2000, maximum depth now being 30 m. Some people have started to change their cropping pattern in order to adapt to water availability, but they still only number a few. Interestingly, these recent problems of TW supply have not changed practices regarding tank water, which has not been used for irrigation for 30 years.

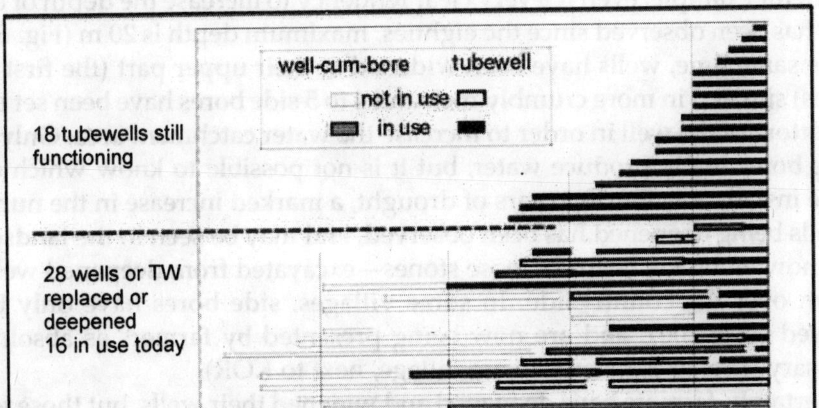

Fig. 6.10: Date of Well/Tubewell Installation on *Punjai* Land In Sorapet, according to the Type of Well and its State of Use

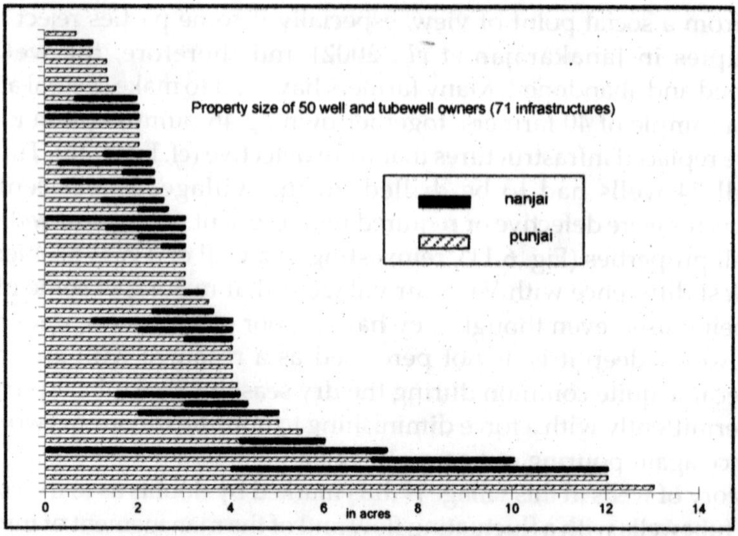

Fig. 6.11: Size of Property of 50 well owners in Sorapet

(c) Open Wells in the Rocky Area—The Case of Koratur

In rocky areas, open wells cannot be deepened beyond certain limits. At Koratur (KOR), for example, even if a very clear tendency to increase the depth of open wells has been observed since the eighties, maximum depth is 20 m (Fig. 6.12). At the same time, wells have been widened at their upper part (the first 7-10 metres) situated in more crumbly rock and 4 to 5 side bores have been set up at the bottom of the well in order to increase the water catchment area. Only 1 or 2 side bores really produce water, but it is not possible to know which ones before installation. During years of drought, a marked increase in the number of wells being deepened has been observed, and may be seen in the landscape from now on by the heaps of loose stones—excavated from deepened wells—strewn over the countryside. In some villages, side bores have only been installed since 2001, and are now being presented by farmers as absolutely necessary (case in Saravanapakam village, next to KOR).

Certainly, farmers have deepened and widened their wells, but those wells essentially collect an upper aquifer, i.e. groundwater, which is greatly dependent on rainfall for its replenishment. So that during the summer, such wells produce very little water. They provide less water nowadays than before due to the increase in water use. Indeed, a well that would have been used 20 years ago for 6 months at a time, now discharges the same amount in only four months. This reduction is a sign of the overexploitation of the upper aquifer and of its depletion (even if the duration of well production is not, strictly speaking, one of the five criteria[5] mentioned by Ballukraya *et al.*, 2002). This overexploitation is quite widespread in this area, the respective blocks (Tiruvennainallur and Kandamangalam) already categorised in 1992 as "overexploited" as regards GW exploitation (GoTN, groundwater perspective, 2002).

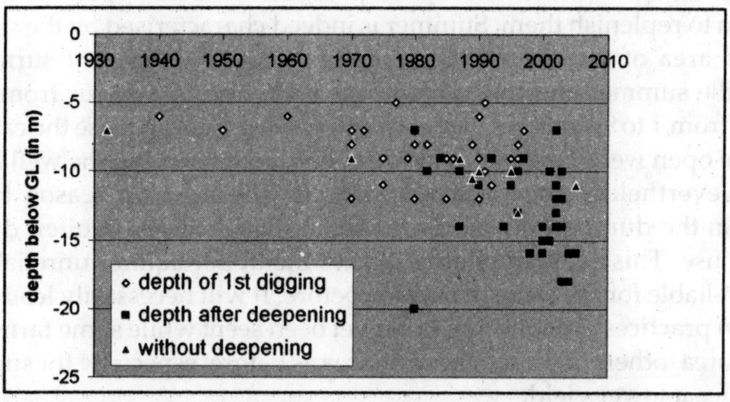

Fig. 6.12: Deepening of 32 Open wells in Koratur out of a Sample of 40 wells

During the summer, crops may suffer from drought: farmers often sow a larger area than what can be properly irrigated during the summer. In years of drought, especially from 2001 to 2003, even well owners had problems and could only farm on a reduced scale, while many could not even cultivate their land and had to practise temporary migration to look for work. In general, migrations cannot be merely explained by the absence of irrigation alone. They introduce much more complex phenomena involving the distance to town, the standard of education, and personal aspirations to free oneself of local social constraints, or else to flee agricultural work, a totally devalued activity in the eyes of many young people. It is interesting, all the same, to note that the practice of "seasonal nomadism of the rural manual worker" is an ancient tradition and has already been analysed by Adicéam (1966: 236) as being linked to dry and relatively humid periods which prompt the displacement of hamlets.

No borewells have been installed in KOR village. Some rare specimens exist in the region (see also figures in Chap. 3 resulting from the well census) dug to a depth of 60 to 150 m but not guaranteeing any success, since they tap water in the fractured rock.

(d) Summary

The decline in the water level of the shallow aquifer during the summer months and its rise during the rainy season is a perfectly normal thing that has always taken place. But the speed at which the level is sinking today, the lowering of the mean level of the water table right after the rainy season (versus a few months later some decades ago) are signs of an increase in the use of this aquifer since the mechanization of pumping.

In the case of the rocky areas, farmers have no choice and depend mainly on this upper aquifer and on the rain, either for their rainfed agriculture or their irrigated agriculture, with the majority of tanks and open wells relying on the rain to replenish them. Summer is indeed characterised by the shrinking command area of the open wells: each one covers a different surface area between the summer and the rainy season—varying in Koratur from 0.1 to 3 acres and from 1 to 9 acres per well respectively. Farmers increase the catchment area of the open wells by installing side bores and deepening the well by a few metres. Nevertheless, they suffer during the second crop season from the decrease in the duration of water availability in the upper aquifer, due to its increased use. This overexploitation makes the upper aquifer unreliable or at least not reliable for the same periods as before. It will necessarily lead to some changes in practices even if this has not yet been seen: while some farmers sow a smaller area, others still sow the same area as before, especially for sugarcane, but produce a lower yield.

In most cases of alluvial areas, peasants are increasingly abandoning this upper aquifer to dig deeper, when geological, hydrogeological and economic conditions permit. Indeed, they have been observing a drop in the different water tables used for tubewells, so they try to dig deeper. Yet according to local hydrogeologic characteristics, the response to this decline differs: when deep bores are possible and productive, farmers progressively install deep bores, abandoning the shallower water tables; when the lower aquifers become unproductive and thus uncertain (case seen in Sorapet in the alluvial area), farmers endure bitter experiences and then limit the risk of such an investment by drilling less deep. In such cases, the image of farmers drilling deeper and deeper since 2000 is not applicable.

As for the sedimentary area, it is a sort of cross between the alluvial and the rocky areas, in so far as access to the productive aquifer is possible but very expensive and reserved for an elite, who do not hesitate to pump ever more deeply (TW ranging from 60 to 280 m). According to farmers, open wells have hardly been used since the drought of 2001-2003 due to a surge in the installation of tubewells: a point which corroborates the analysis in Fig. 1.7 (Chap. 1) related to the scale of Tamil Nadu, where only surfaces irrigated by tubewells were increased during the years of the drought.

The effect of the drop in these various water tables is felt not only in the use of water for agricultural purposes, but also for domestic purposes. In the alluvial area, the wells which used to provide potable water are no longer in use because, new techniques, such as hand-pumps and taps supplied by the overhead tank, are easier to use but also because they are dry for a few months of the year. Thus, at Tiruvennainallur, water does not seem to be a problem at first, since the irrigation bores have not been excessively deepened (maximum 30 m). On the contrary, they give the impression of an abundance of water because they provide water all year round. However, old domestic wells cannot be round used during the summer (Caillot, 2005). In some villages, the aquifer is no longer productive enough to fill the overhead tank twice a day (Coste and Ploumpidis, 2007). In some other villages near Sorapet, even if irrigation tubewells belch out water, not enough water is available for villagers at the tap—due to an inadequate storing capacity of the overhead tank. Two further conclusions can be drawn from these facts: on the one hand, reliance on deeper underground resources and on modern techniques of water provision (tubewell supplying water to overhead tanks; private tubewell; hand pump) is on the rise while, on the other hand, tubewell depth may also have to be increased. Under such circumstances, the idea of certain NGOs reviving traditional techniques for supplying domestic water is inapplicable in this alluvial area. Both domestic water and water for irrigation in villages in the alluvial plain in the Villupuram-Puducherry region are more and more dependent on underground water found deeper than 20 metres, and thus on new techniques. By contrast, in rocky areas, open wells are still used and ponds still dug for the provision of domestic water in the villages even if bores supplying overhead tanks and hand pumps are indeed set up.

6.2. Selling Water: A Privilege

Free access to underground water goes hand in hand with the free use of water, which allows the owners of pumps to sell the water they do not need, thus creating an informal market for water.

(a) Positive Social Impact of Water Markets

On a nationwide scale, some estimate that water markets contribute to more than 60 per cent of India's well irrigated area (Kumar, 2007: 63). For our study area it is difficult to give data but these self-initiated businesses are very common in all villages. They seem to be very highly developed in the alluvial and sedimentary areas of our study (of about 150 tubewell owners interviewed, half sell water), but much less so in the rocky areas where wells do not produce a lot of water, and thus are mainly used by the owners themselves. Thus the data collected in Koratur (rocky area) show that the percentage of the area irrigated using purchased water in our samples of 50 wells is quite low: 6 per cent of the area irrigated by the wells in the sample. In Vadanur and Sorapet, the data related to the area for which each bore owner sells water seem irrelevant and underestimated as such an area is 10 per cent and 20 per cent respectively of the irrigated area according to these data. According to the village land record (*adangal*), 158 bores would irrigate 800 acres, i.e. each bore irrigates 5 acres on average. As bore owners of our sample own 2.5 acres of land on average, 2.5 acres [5 minus 2.5] of land are therefore irrigated by purchased water or 50 per cent of the irrigated area.

The main positive impact of water markets is that it is a way to "diffuse access to this precious resource and bring benefits to all sections of rural people including the resource poor" (Shah, 1993: 56), that means to small farmers who cannot afford to invest in digging a well, and who, by buying water, could save their crops and increase agricultural productivity, since GW irrigation is more effective than canal irrigation and, of course, than rainfed cultivation as already expressed in Chapter 1.

(b) Negative Social Impact of Water Markets

The flip side to the coin regarding water market development is that this encourages people to use more water and thus exacerbate the overexploitation of GW and its declining availability (Shah, 1993; Saleth, 1998). Foremost criticism of this surge in GW use stems from regions where the underground potential has already been largely exploited, essentially Gujarat and Tamil Nadu[6]. Some of the criticism focuses on the social consequences of this depletion of the water table: competition to deepen wells pushes up the cost of well irrigation, leading to a situation of inequity amongst well owners, and between well owners and those without a well (Moench, 1992a; Janakarajan and Moench, 2002). Small and marginal farmers will be left by the roadside due to their inability to reach the water table. Another criticism regards the emergence of "water lords" who hold a monopoly on the precious resource. Water markets are biased against the poor, characterized as they are by an unequal trading relationship in which buyers have little room to negotiate. Thus the inequity of resource allocation and access, and the situation of water buyers' dependency are thereby perpetuated (Janakarajan, 1994). In some situations, such as in North Gujarat, from a socio-economic point of view, the appropriation of GW is the business of the dominant caste (Dubash, 2002), so that if any formal regulations limiting the use of water—by increasing the price of electricity for instance—come into force or informal measures involving environmental constraints, the first to be affected are the water buyers, who are pushed out of the market as bore owners keep their water for themselves (Prakash, 2005: 210). When constraints are applied and limit water availability, there is a sharp decline in water markets with them almost disappearing (Prakash, 2005: 216; Janakarajan and Moench, 2002: 3). When the resource is too degraded and has become scarce—chiefly due to the overextraction of GW and, ironically, to the expansion of water markets in the case of alluvial aquifers—water markets, unable to regulate the use of the resource, collapse. Social differentiation is thus highlighted.

All these criticisms are echoed in the fieldwork carried out around Puducherry-Villupuram and numerous examples could be mentioned. One exception perhaps is the second point (social consequences of overexploitation of water tables) which requires further explanation because of the environmental differences between localities that we have already referred to. The case of Vadanur confirms critics' analysis, since the decline of the water table primarily affects small bore owners, who cannot afford to drill deep tubewells (TW) and cannot use their own TW, while the example of Sorapet, where a physical limit has been reached, prompts us to moderate this criticism. TW depth is reduced thanks to investment risk management: smaller landowners are saved in this way from being excluded from the drilling process. However, given the inconvenience of being dependent as water buyers, small farmers would like their own wells, even when this is not economically viable (Cayon, 2006; Roesch in Chap. 8).

(c) Social Constraints on Water Buyers

Water buyers' dependency on water sellers is a very constraining factor which manifests itself in various ways: buyers complain of not having water when they need it and of having to rely all day on the goodwill of the well owner; the seller holds a quasi-monopoly on the sale of water on neighbouring land; if a buyer changes his supplier, the rejected supplier will not sell him water again in future; these markets are characterised by a lack of or very little room for any possible negotiations about the price of water; caste differences may also be a constraint, with some people preferring to buy water within their own caste when the topography and proximity of land permits; finally, buyers are not always able to choose their cultivation methods since some of these are imposed or forbidden by the water seller, who does not want to provide water for farming methods requiring less water and thereby making less profit for him from the sale of the resource [case also noted by Druesnes (1997) in the same region]: these informal markets are thus of no help in using water more efficiently (see also Kumar, 2007) but quite the contrary, since the purchased resource is mainly used for water greedy crops (paddy and sugarcane). If, as far as the tubewell outlet is concerned, the irrigation of a field involves crossing the field of a third party, this can be problematical for the buyer. When the neighbour does not want to "make way" for water, then the buyer has to find another seller. Another social dimension of unfair trading relations is the link to one's political status, as a farmer from the sedimentary area explained: a bore owner refused to give water to a neighbour who had not voted for him![7]

In the past, when the reasons for acquiring a tubewell were explained, the emphasis in literature was on local power and on the prestige associated with such an infrastructure (Janakarajan, Moench, Prakash), whereas today there may also be the desire to break away from any dependence on water sellers, who use the various facets of their power to control the coveted resource. The reply given by a small farmer from Sorapet to my question "why are you installing a bore when the installation is not economically profitable; why not buy water?", is very clear on this subject: "Firstly, owning a bore carries a certain prestige; secondly, we don't have to rely on the water seller's goodwill when we want water, we can get it whenever we like; thirdly, no more quarrelling with our neighbour if they do not want to make way for our water (from the bore outlet to the buyer's fields *via* the neighbour's fields). And we are not bothered about the price of the bore. We need water, that's all there is to it."

The social inconveniences of these informal markets, coupled with the prestige acquired by owning a bore have thus had a tendency to create competition regarding access to wells. This competition leads to increasing the number of bores (within the limits of the electrical connections available)—but not necessarily the number of irrigators because these new bore owners were already buying water before—and thereby to reduce the size of water markets. By increasing the number of bore owners, the situation has changed from one of buyers being deprived of water, when well owners kept the supply for themselves, to one of general privation progressively affecting more and more people, including bore owners. The positive impact this has had might be visible in water management because some—still rare, though nonetheless encountered at Sorapet—are conscious of the need to use less water and to alter the cropping pattern. The negative social impact of the water market, related to inequality in market relations, has led to an increase in the number of bore owners, and therefore, a decline in water market development[8]—stage 4 of Roy and Shah (2003: 332)—which indirectly has some positive impact as it leads to more awareness about water scarcity and its management.

(d) A Market Beyond Economic Logic

Since I am not an economist, I will not refer to economic concepts in my arguments. I would simply like to give a few examples I find particularly striking as they show that the price of water follows no economic logic, since the cost is not calculated according to the initial investment and the charges for maintaining the pumps.

For Saleth (1998), the factors influencing water prices are: depth of water, cropping patterns, source of energy and electricity charges. In our study area, the last two factors do not affect the price of water since most pumps are electric and electricity is free of charge in Tamil Nadu and nearly the case in Puducherry. The crop to be irrigated is the key element for consideration when expressing the price[9], but there is no obvious relationship between the price of the resource and the depth of water either on a regional or village scale. Indeed, the most striking facts are: firstly, the price of water for paddy—generally one third of paddy production—is the same whether in the rocky area from an open well or in the alluvial area from a tubewell; and secondly, the price of water has not changed over recent years in a village like Vadanur, even if TWs have deepened: for the last 30 years water has been sold at the same rate. Why is this so? One bore owner explained that they (the owners) cannot increase the price of water because if they did the buyers would not be able to afford it.

On the contrary, when water is sold for sugarcane or by the hour—that means for crops requiring little irrigation[10] such as groundnut, tapioca, casuarina, blackgram and cotton—we observe mainly on a regional scale a difference which can be interpreted as being due to the difficult access to the groundwater tables or to water availability: thus in the alluvial plain, the rate per hour ranges from Rs. 10 to 30, with this difference in price also being seen within a same village, such as the case of Sorapet, where it is not at all related to TW depth; in the rocky area, the price of water varies from Rs. 25 to 50 but this is mainly in cases of emergency. For sugarcane, the price of water is one-third or one-fourth of production, or a fixed amount varying from Rs. 6,000 to 9,500 per year, with the price being higher in the rocky area. On a village scale, it appears that social relations are important in determining the rate: in a close family relationship, the price would be lower. The change in the price of water may also result from a fear of being short of water. In Tiruvenainallur, for example, located in the alluvial area, where bores have not been deepened quite so much (they are 30 m, 96 feet, deep maximum), the increase in the price of water (from Rs. 10 to 15 or 25 according to places) is a reaction to the three-year period of drought during which cultivation dwindled to the point where very little land could be cultivated at all. Even though water is now available as before, the price remains stable at Rs. 25 and some water owners reduced their number of water sellers because water has come to be perceived as a resource which is likely to become scarce. This might also explain price variations in Sorapet, leaving aside any relation to depth.

Decisions regarding the price of water also depend on a collective consensus among bore owners at village level. Thus at Sorapet, a farmer clearly explained that before the harvest of monsoon rice, bore owners together decide (without consulting the buyers) on the price of water for the paddy season, according to the rainfall and therefore to how frequently pumps are used. This is hardly a market based on supply and demand, but rather a market that follows the rules of a monopoly with people agreeing together on a fixed price: here we are far from the free market competition that Shah (1993) and Dubash (2002) speak about.

Moreover, we have seen the case of Tiruvenainallur, where the fear of being hit by a water shortage leads to higher water prices, even if today farmers are no longer affected by this.

The price of water is thus determined according to some social norms: it is increased by potential hydrous stress, it is limited by the estimation the water sellers make of the economic potential of their water buyers, but the price of water (for paddy and sugarcane) is not fixed according to the cost of investment or of the use of the pump.

6.3. Using Water while Stocks Last

An interesting fact is that farmers use water while stocks last. Very few official limitations on water use exist, the main one however being the restricted availability of the electricity supply that prevents continuous pumping. Nevertheless, some farmers get around this constraint by using a system whereby their pump works on a two-phase instead of a three-phase current which is subjected to a time limit. Let us bear in mind that some indirect regulations on underground water use exist, such as a limitation on the number of new electrical connections, subsidies for drip irrigation, subsidies in Puducherry since 2006 for the cropping of plants other than paddy during the very dry season, and the promotion of new cropping techniques requiring less water, such as the System of Rice Intensification since 2004. Nonetheless, groundwater irrigation is still supported by government incentives, since electricity for agricultural use is free of cost.

This official limitation on the use of groundwater makes local environmental characteristics the principal elements limiting the extraction of this resource which is managed by millions of independent individuals on a nationwide scale. Some constraints are man-made, such as the lowering of water table levels, and others are related to hydrogeological characteristics—which may be accentuated, moreover, by human action—such as the low water storage capacity in the rocky area and the spatial inequality of aquifer potential. As long as farmers are able to dig more deeply they will do so. The wealthiest reach depths the potential of which they do not exploit, installing the motor much higher than the depth of the pipes. If no limitation comes into force and the volume of water exceeds the owner's needs, then an informal market emerges, which certainly allows more people access to the resource but which, at the same time, both encourages the massive use of underground water and thus the decline of the water table, and remains extremely unequal in its trading relations. Any repercussions of limitations on the exploitation of the water table (lowering of level, lowering of aquifer yield) are first and foremost felt by those buying water, who end up being excluded from the market, whereas the pump owner does not particularly limit his own consumption of this precious resource. Where drought conditions affect groundwater levels, bore and well owners react in just the same way, digging ever deeper, keeping the water for themselves and ultimately reducing the number of buyers, while no doubt pushing the price of water up at the same time. Barely modifying their own cultivation practices, they leave the process of over exploitation of water tables to continue unchecked.

Despite the massive drop in the quantity of available underground water, farmers, whether buyers or sellers of water, well or bore owners, make no attempt to save this resource for future use. Their prime objective in using groundwater is not to use as little water as possible, i.e. to economise water in a sustainable way. On the contrary, the tendency is: "if it's there, pump it". T. Shah (1993: 163-164) already observed similar behaviour in Gujarat regarding the use of the open well at the end of the 1980s. Furious pumping took place as soon as the river began to dry up, since farmers knew that this meant a major decline in the yield of the well. Thereupon farmers were "trying to claim as much of the remaining water as possible while no doubt wasting a good deal of it. The competitive pumping must, to a great extent, have hastened the pace of depletion of the aquifer" (*ibid.*: 160). Another tendency is to dig deeper if possible and to sell less water to neighbours without altering one's own cultivation practices (some changes are starting to be seen however). As regards bore owners in particular, their practices seem to be influenced by the return on investment and the fact that they exert some local power. A short-sighted argument has been voiced: "Why think about tomorrow if we need water now?", with some farmers adding that their children will study rather than cultivate the lands. This answer shows a certain degree of selfishness; neither themselves nor their descendants will manage the problem, they will leave it up to someone else.

This practice of using water as long as it is available, without trying to save any, might stem from practices applied by farmers over generations, i.e. the use of open wells and temporary lakes (tanks), as O. Lefebvre (2006: 95) appropriately noted it in his M.A. report: "farmers have always used water on an annual and cyclic basis while stocks last. Thus water from open wells and tanks was used up to dry out the latter. Farmers are not used to managing water reserves on an inter-annual basis, but on the contrary to using all the water available." Farmers have not fundamentally changed their habit of using as much water as possible, when changing over from well irrigation to tubewell irrigation, especially as there has been no official limitation. The continuous supply of water through the year over the last thirty years may also give the impression of an unlimited resource, while the deepening of the bore gives the illusion of an endless resource. Even if the various water tables are declining and the aquifers less productive, most irrigators have not yet changed their irrigation practices in the study area. Is it due to their conception of groundwater, their perception of groundwater storage? To fully grasp these practices, we suggest attempting to understand local knowledge related to this underground resource (Chap. 7).

NOTES

1. We did not depict the lithology of the area on the diagram, because large local differences exist, as Fig. 7.3 (in Chap. 7) shows.
2. Nevertheless, tubewells—using artesian water—have been in existence since 1877 in the Territory of Pondicherry for first industrial then agricultural purposes, up to 100 m deep (see chap. 2.1). In Gujarat, the arrival of submersible pumps dates from the 1970s (Kumar, 2007:104).
3. Auroville Water Harvest website mentioning "a tremendous decline in the level of the water tables, up to 45 m over 30 years in some locations"; Bentz (2001: 69) mentions that bores have deepened by 100 m over 20 years, attaining a depth of 200 m.
4. Personal communication by an hydro-geologist from the Agriculture Department, Puducherry.
5. Ballukraya and Shaktivadivel 2002: "An area may be classified as over exploited if one or more of the following hydrogeological features are encountered:
   - When groundwater levels do not recover their original levels on a long-term basis.
   - When discharge exceeds recharge by an appreciable volume.
   - When shallow aquifers progressively become unproductive.
   - When ground water is pumped from depths greater than that to which adequate recharge currently takes place.
   - When the natural gradient is reversed and remains so throughout the year."
6. Inversely, for regions where the potential is still under-exploited, such as in Nepal's Terai, corresponding to Stage 1 of Roy and Shah's figure (2003), the development of such markets is being encouraged (Shivakoti *et al.*, 2005: 148).
7. Information reported by my colleague I. Prabhakar.
8. Noted also on a nation wide scale by Mukherji (2005), cited by Kumar (2007: 63).
9. For paddy, the price of water very often represents a share of the crop with the production risk thus being shared between the water seller and water buyer. It can also take the form of a fixed amount of grain. In some villages (such as Sorapet) the season itself is taken into account as less water is required during the rainy season, even though crop growth is longer (6 months for *samba* paddy, which grows during the monsoon, versus 3 months for the second paddy—called *kuruvai* here—grown from February to April). The price is also higher during the dry season since farmers are aware of the lesser quantity of tubewell water. Thus, the price is set at 4.5 bags for *samba* paddy and 6 bags for *kuruvai* paddy. In some very rare cases, the quantity is negotiated and reduced to three bags if tank water is used. For

sugarcane, farmers are paid in cash by the sugar factory so the price of water is always expressed in cash, either as part of the tonnage (1/4) or as a fixed amount per cultivated area.

10. For these crops, a total amount can also be set, such as Rs. 2,000 to 3,000 for groundnut, or Rs. 1,500 to 2,000 for cotton.

REFERENCES

Adicéam, E. 1966. *La géographie de l'irrigation au Tamil Nadu*. Paris: Ecole Française d'Extrême Orient.

Balasubramanyam, S. 1971. *The Early Chola Temples: Parantaka I to Rajaraja I: A. D. 907-985*. Bombay: Orient Longman.

Ballukraya, P. N., and R. Saktivadivel. 2002. "Over-exploitation and Artificial Recharging of Hard-Rock Aquifers of South India, Issues and Option." *Annual Partners' meet 2002*, Institute of Rural Management, Anand, 2002, pp. 14.

Bentz, B. 2001. Histoire sociale—Histoire technique: les tanks du bassin versant du Kalluvelli (Inde), Mémoire du Master of sciences DAT (Développement Agricole Tropical) du CNEARC, Montpellier.

Bhatia, B. 1992. Lush fields and parched throats. Political economy of groundwater in Gujarat. *Economic and Political Weekly* 27: A142-169.

Caillot, L. 2005. Accès à l'eau et vulnérabilité en zone rurale. Rapport de Master de l'Ecole Supérieure Européenne, IFP, ESE-IER.

Cayon-Glayère, N. 2006. Rentabilité de l'agriculture irriguée dans un village de Pondichéry, Mémoire en vue de l'obtention du Diplôme d'Ingénieur en Agronomie Tropicale, CNEARC, Montpellier.

Coste, N., and N. Ploumpidis. 2007. "Bittersweet water. Irrigation practices and modern challenges in South India. Trilingual DVD (Eaux douces, Eaux amères-Thaneer: inippum kacappum). Coauthors: K. Anupama, O. Aubriot, Y. Gunnell," Puducherry: French Institute of Pondicherry (Publications Hors Série 5).

Druesnes, N. 1997. *Révolution doublement verte. Les biovillages à Pondichéry*. CIRAD (Programmes Economie, Politiques et Marchés-doc. de travail No. 29).

Dubash, N. K. 2002. *Tubewell Capitalism: Groundwater Development and Agrarian Change in Gujarat*. Delhi: Oxford University Press.

Government of Pondicherry. 1999. *Water resources and their management, Pondicherry Region* (water for life, save water, save Pondicherry). Department of Agriculture.

Government of Tamil Nadu. 2002. *Groundwater Perspectives. A Profile of Villupuram District, Tamil Nadu*. State Ground and Surface Water Resources Data Centre, Water Resources Organisation.

Janakarajan, S. 1994. *Trading in groundwater: a source of power and accumulation*. Madras Institute of Development Studies (Working Paper No. 123).

Janakarajan, S., and M. Moench. 2002. *Are wells a potential threat to farmers' wellbeing? The case of deteriorating groundwater irrigation in Tamil Nadu.* Madras Institute of Development Studies (MIDS working paper No. 174).

Kumar, M. D. 2007. *Groundwater Management in India. Physical, Institutional and Policy Alternatives.* New Delhi: Sage Publications.

Lefebvre, O. 2006. Dynamiques de gestion de l'eau dans un contexte de surexploitation des eaux souterraines. La région de Pondichéry et de Villupuram au Tamil Nadu, Inde du Sud, Mémoire présenté en vue de l'obtention du Mastère Spécialisé Développement Agricole Tropical, CNEARC, Montpellier.

Marius-Gnanou, K. 1991. "L'irrigation et les mutations socio-économiques récentes dans la région de Pondichery (Inde)," in *Eau et aménagement dans les régions inter-tropicales, T.2 (Espaces tropicaux, No. 3).* Vol. 2. Edited by P. Vennetier, pp. 133-160. Talence: CEGET-CNRS.

Moench, M. 1992a. Chasing the Watertable. Equity and sustainability in groundwater management. *Economic and Political Weekly* 27 (51-52): A171-A177.

Mukherji, A. 2005. "The spread and extent of irrigation rental market in India, 1976-77 to 1997-98: what does the National Sample Survey Data Reveal?" in *Water Policy Research Highlight*: IWMI-Tata Water Policy Programme.

——. 2007. Implications of alternative institutional arrangements in groundwater sharing. Evidence from West Bengal. *Economic and Political Weekly* 42: 2543-2551.

Narayanamoorthy, A., and R. S. Deshpande. 2005. *Where water seeps! Towards a new phase in India's Irrigation Reforms.* New Delhi: Academic Foundation.

Narendranath, G., S. Uma, and K. R. Reddy. 2005. To free or not to free power. Understanding the context of free power to agriculture. *Economic and Political Weekly* 40: 5561-5570.

NCRWC. 2002. "Issues of social justice: scheduled castes, scheduled tribes and other backward classes. An unfinished national agenda." National Commission to review the working of the Constitution. http://www.humanrightsinitiative.org/publications/const/issues of social justice scst obc. pdf.

Oliveau, S. 2003. "Atlas of South India 1991," cybergéo, http://www.atlasindia.parisgeo.cnrs.fr.

Prakash, A. 2005. *The Dark Zone. Groundwater Irrigation, Politics and Social Power in North Gujarat.* New Delhi: Orient Longman.

Ramasamy, C., K. Palanisami, K. Lokanadhan, and M. Anjugam. 1999. *Coping Behaviour to Growing Groundwater Scarcity in Tamil Nadu.* Coimbatore: Water Technology Centre, Tamil Nadu Agricultural University.

Roy, A. D., and T. Shah. 2003. "Socio-ecology of groundwater irrigation in India," in *Intensive Use of Groundwater. Challenges and Opportunities.* Edited by Llamas, M.R. and C.E. Lisse, the Netherlands: Balkema Publishers.

Saleth, R. M. 1998. "Water markets in India: economic and institutional aspects." in *Markets for Water: Potential and Performance.* Edited by D. A. A. M. W. R. Easter K.W., pp. 187-205. Boston: Kluwer Academic Publishers.

Shah, T. 1993. *Groundwater Markets and Irrigation Development: Political Economy and Practical Policy.* Bombay: Oxford University Press.

— —. 2007a. "Groundwater: a global assessment of scale and significance," in *Water for Food, Water for Life.* Edited by D. Molden, pp. 395-423. Colombo: IWMI.

Shaheen, F. A., and R. L. Shiyani. 2007. "Energy costs and groundwater withdrawals: results from an optimal control model for North Gujarat". *Indian Jounral of Agricultural Economics* 62: 64-79.

Shivakoti, G. P., D. L. Vermillion, W.-F. Lam, E. Ostrom, U. Pradhan, and R. Yoder. Editors. 2005. *Asian Irrigation in Transition: Responding to Challenges.* New Delhi: Sage Publications.

Vaidyanathan, A. 1999. *Water Resource Management. Institutions and Irrigation Development in India.* New Delhi: Oxford University Press.

CHAPTER

7

Knowledge about Groundwater and its Implications in Water Management

Olivia Aubriot

In an attempt to understand why the prime objective of farmers using groundwater in the region studied is not to economize water for future use, but rather to continue to pump it as long as it is available—to which obviously there are exceptions—let us look more closely at the local knowledge about the underground resources.

7.1. Do New Techniques Imply New Knowledge and Practices?

The use of groundwater for irrigation is not a recent phenomenon and is associated with knowledge handed down from generation to generation. It seems reasonable, given the use of new techniques for extracting water, to wonder what kind of new knowledge is associated with these techniques and this "new resource" (deep waters), whether or not they can guarantee sustainable utilization and whether they interfere with the process of an integrated resource management.

(a) Development of Technical Ingenuity

Let us briefly review the enormous changes brought about by these new techniques for extracting groundwater. They have been the cause of major transformations in the use of land and indirectly in the link farmers maintain with groundwater. They have effectively enabled: (1) the easy extraction of large quantities of water; (2) land which did not previously benefit from any water to be irrigated, (3) access to the resource when neither rain nor surface water is available, (4) aquifers not directly dependent upon rainfall for their production to be reached. The result has been an intensification of the cropping calendar for land which has been able to benefit from this irrigation, and a net and gross increase in the area irrigated. Farmers' fascination with this resource with its many advantages explains their frantic efforts to find and utilise this water which occupies an important place in their agriculture.

Farmers, therefore, use all their ingenuity to access it, especially in areas where the resource is less available. To increase the quantity of pumped water, where the outflow from the well is not satisfactory, well owners may mix the water from two bores, or from a well and a bore in underground pipes (a rare case observed in the alluvial area); or in rocky areas, they may transfer water from one well into another thus combining waters from two places. In both cases underground pipes are used to minimize water loss during transport. Janakarajan and Moench (2002: 26-29) give some other examples of mechanisms by which farmers have overcome the problem of a low yield from wells or bores: in the Noyal basin in Western Tamil Nadu, where the hard rock makes continuous pumping impossible, most farmers use compressor technology which allows them to operate pumps even when there is little water. As the flow is not strong enough to be used directly, they store the water in cisterns and pump again for irrigation purposes. In more than 1/3 of their sample wells, water has to be pumped twice: once from the ground to the storage well and then again for use. In the Vaigai basin in southern Tamil Nadu, farmers have developed an organizational solution since some owners of adjacent wells use their pumps at the same time to channel water from both wells and to irrigate a single plot of land whether it belongs to them or to a water purchaser (Janakarajan, 1994: 7).

The level of ingenuity developed by farmers is again evident in another type of solution which uses the pumping force and pipes to transport water over long distances: in this way farmers get around local hydrogeological constraints by looking for water where it is most available. Janakarajan (1994: 7) gives one astonishing example of a person (a local politician) who irrigates his land with a well set up at a distance of 11.5 km from the River Vagai. In the same village, about fifteen farmers pump water from wells on wetlands and store the water in wells on dry land, then pump the water again to irrigate plots near the latter wells. This means that they have to pump the same water twice, but since electricity is free of charge, this costs practically nothing (*ibid.*: 10). In the villages chosen for our in-depth study, this double pumping was observed in the rocky area, but seems to be quite rare[1]. We also noted that water is transported from a village (TVN) situated in the alluvial area to land belonging to a village in the rocky area (SRVP) by a small group of farmers from TVN (See Fig. 7.1). In 1986, one of these farmers even built and installed 600 m of cement pipes to irrigate the land he owned at SRVP. In 1989, another farmer set up a 500 m network of underground pipes supplying 12 outlets, which today irrigate 30 acres of land belonging to 60 different people (those to whom the water is sold). One farmer from SRVP even bought some land in TVN to install a tubewell and supply water to his fields. Since the 1990s Poly Vinyl Chloride (PVC) pipes have been installed by other owners. A total of twenty hectares (50 acres) of the rocky area are irrigated in this way in the dry season thanks to the groundwater from the neighbouring village situated in the alluvial area, using only 4 bores.

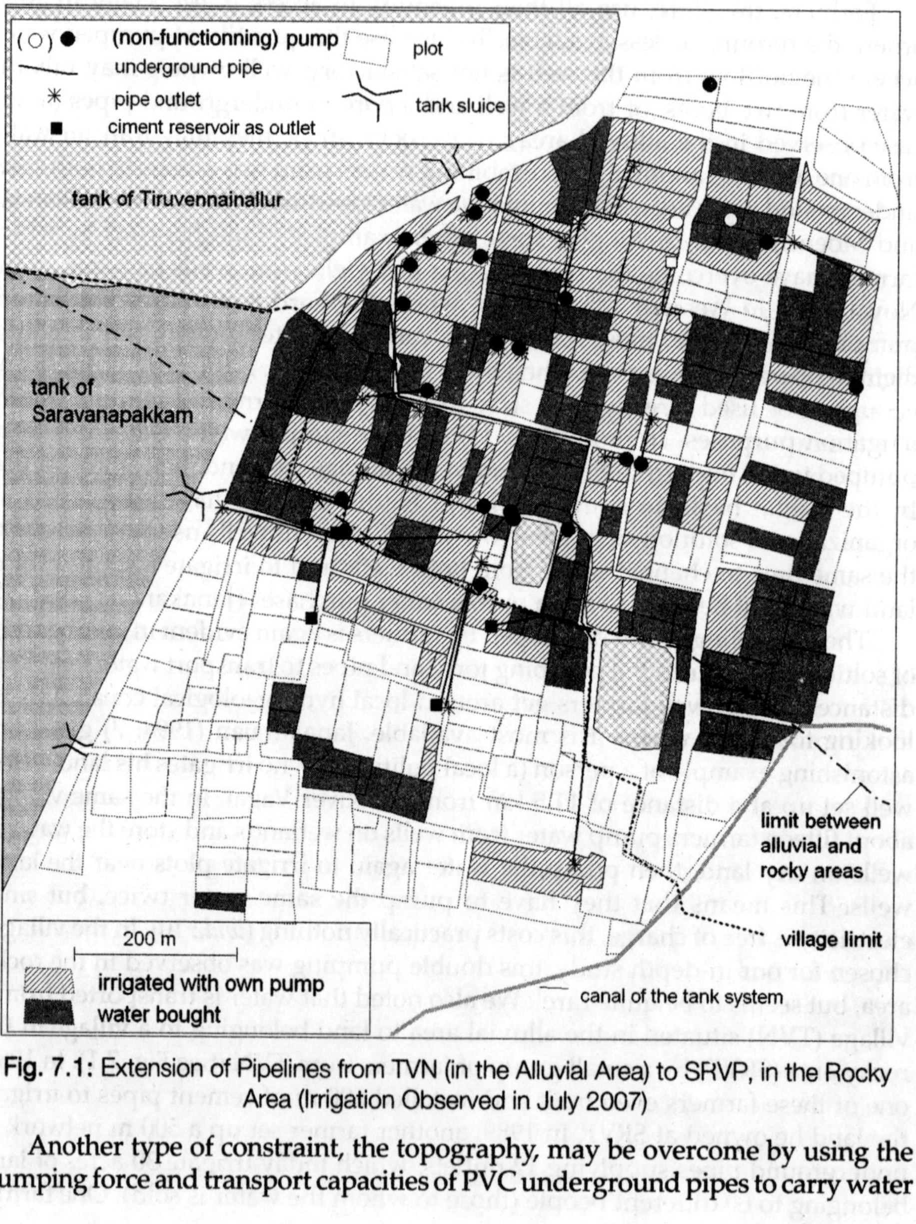

Fig. 7.1: Extension of Pipelines from TVN (in the Alluvial Area) to SRVP, in the Rocky Area (Irrigation Observed in July 2007)

Another type of constraint, the topography, may be overcome by using the pumping force and transport capacities of PVC underground pipes to carry water to higher land. As in the preceding examples, underground pipes are used to reach not only land further away, but also land situated at a higher level. In the village of TVN, underground pipes are only used for transporting water to the parcels situated at a higher level (1 metre), not for minimizing losses due to infiltration[2], since the pumped water circulates through canals when possible. At KOR, owners transport water from *nanjai* wells to *punjai* land, about 12 metres higher and 700 metres away. The example of KOR is certainly remarkable for the fact that these owners use tank water to replenish their *nanjai* well, diverting water from the irrigation canals *fed by* the tank. Thus some *punjai* land (approx. 10 ha) is irrigated using water from the tank through wells situated in *nanjai*. We shall return to this example later (Chap. 7.3).

Lastly, some people install underground pipes to not depend on collective canals for transporting their water, or in order to avoid crossing the fields of a neighbour who refuses to let the water through. Due to the various advantages offered by pipes (to carry water some distance and to upper fields, to limit infiltration and the waste of water during its transport and to avoid crossing neighbouring fields or using the common canal), these structures are used more and more, the only visible sign of which being the outlets[3] henceforth scattered over the landscape (see Fig. 7.2 representing the various outlets and pipe installations). They make up a multitude of invisible, underground mini networks, such as in Fig. 7.1, representing the case of Southern Tiruvenainallur, in the alluvial area. Here two areas can be seen: one with ready access to GW and where underground pipes are not used on an intensive basis, and the other, bordering the village of Saravapakkam (SRVP), where pipelines carry water to the rocky area. We can compare this pipeline architecture with the one analysed by Dubash (2002: 176-181) in North Gujarat in two villages (Ratanpura, V.1, and Paldi, V.2) where the tubewell irrigation architecture differs: in V.1 the pipeline network of a tubewell may be as much as 4.5 km long, carrying water to water buyers' fields, and since there are very few tubewells, water markets have been extensively developed in this village; on the other hand, in V.2, pipelines only stretch to the boundary of TW owners' land (they can nevertheless sell water to their neighbours) and are of a maximum length of 1 km. These differences in water market "architecture and thickness" are due to various historical and social reasons, which also explain the degree of the institutionalization of exchange. In TVN, the thickness of water markets is due more to environmental constraints: in the area where there is good access to GW, the number of bores remains high, the number of water buyers low, the latter being bore owner's neighbours, whereas in the southern part where the number of bores is low, the number of water buyers is high and the latter are located far away from the bore owners. In both cases, even if the thickness of water markets is different, the choice the water buyer has regarding the water seller is very limited: in the first case, one can only ask a neighbouring bore owner, in the second case only very few bore owners exist.

If earlier techniques made groundwater irrigation very restrictive, both because of its low profitability and its cost in man hours, and because of its dependence on local environmental conditions (precipitation for replenishing groundwater and topography as regards the use of water), new ones provide some freedom from these constraints and from the very localised aspect of groundwater irrigation. However, from both a geographical (since pipes are only several hundred metres long and, very exceptionally, more than a kilometre) and numerical point of view, it is very unlikely that groundwater irrigation could really be done on another scale than a very local one, since very few owners are willing to invest in so many lengths of pipe and high-power pumps[4].

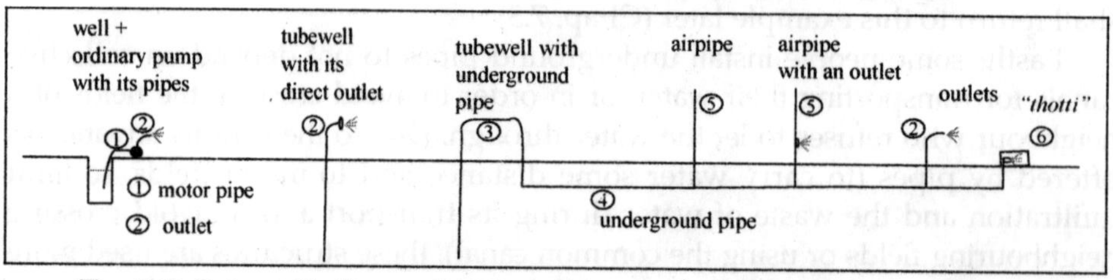

Fig. 7.2: Schematic Representation of Various Outlets and Pipe Installations

(b) No Measure of the Quantity of Pumped Water

One interesting element of GW use through tubewells is the shift in reference relative to the quantum of groundwater extracted: in the past, when human or animal power was used to draw mechanisms, the yield with this equipment was calculated by peasants in terms of *pillaiyars*, or hundreds of buckets, representing the capacity of the recipient that had pulled up the water. The amount of water extracted was estimated in this way, even though the *pillaiyar* cannot be considered a fixed unit of measurement, as is made clear by Adicéam (1966: 260), since the capacity of a bucket is variable. Farmers today count the number of pumping hours, which are generally known, especially to buyers and sellers of waters when water is priced on an hourly basis. They also know whether or not the pump is operating at its full capacity. There is, in fact, a local method for estimating the flow by indicating the length of the jet at the opening of the tube (small spring, long spring) or by showing the size of the jet by pressing one's fingers more or less together at their extremity. This is an empirical method developed by each farmer according to his motor in order to adapt irrigation times to pump flow rate: if at full force, half an hour per field is required, at half force, it would be an hour. This method, however, is not related to any measure of the quantum of water pumped, nor does it give us any information about such quantum since the flow rate can differ from one pump to another, according to various factors: pump power, depth, yield of the aquifer at that particular place.

The use of mechanical pumps and bores means that the volume of water lifted is no longer measured, the only reference being pumping time, which does not determine the volume of water drawn, since the yield is variable according to bore characteristics and farmers do not know how much water they pumped with their bore. The case of open wells is slightly different. The notion of the volume of usable water is indirectly present, even though it is not quantified. Farmers see the resource and the level of well water falling as they pump and can check the time it will take for the well to fill up again. From observing the water level and the time of recharge, farmers know empirically how many hours they can pump for and to which area they can bring water. Yet, even if these estimations are made, no method exists to quantify the quantum of water used.

Even if the number of pumping hours needed to irrigate a field may be evaluated, this is not the indication a pump owner gives to the man he employs to ensure that his parcels are irrigated: he usually stipulates that the water must reach all four corners of the plot and, for paddy, he may dictate the exact height of water needed, expressed in inches or by part of the height of a bund (bunds measure one foot). Except for paddy where heights of water are given, there is therefore no measurement for quantifying the amount of irrigation water. Moreover, farmers have no idea of their pump's flow rate, as no measures are taken at installation time or even after. The absence of measurements (of pump flow or quantity of water used) may in fact curb the development of better GW management since the data relative to improving the effectiveness of irrigation are expressed by scientists in terms of height and volume (a given volume per ha). The engineers from the Department of Agriculture in Puducherry, for example, who are trained to calculate the total amount of water required and to be applied, mention to farmers that the proper amount of water is 1,200 mm for 5 months' paddy cultivation and 1,800 mm for one year's sugarcane, a total of 4.8 and 7.2 million litres per acre respectively. According to the season, however, the irrigation requirement for paddy cultivation varies: during the *samba* season (August to January) this amounts to 542 mm, for *navarai* (also called *kuruvai* in some villages) (January to April) 776 mm and for *sornavari* (April to July) 1,020 mm (GoP, 1999: 8). In order to reduce water consumption, engineers have asked farmers to reduce the frequency of irrigation and the level of water in the paddy fields.

First and foremost, farmers do not understand these figures, especially where crops other than paddy are concerned, since there is no local idea of the quantum of water. Secondly, the absence of any method of measuring how much water is pumped prevents any sound advice being given, as there are differences in flow between pumps and sites. Thirdly, due to variations in water consumption depending on the soil, these quantities are only theoretical: in the village of Manamedu, in the alluvial area, O. Lefebvre (2006: 86) measured a variation of one to three in crop requirements according to soil type (from 800 mm in clay to 2,200 mm in sandy soil, and for three paddy crops from 2,000 to 6,300 mm respectively). Moreover, it is not at all rare to hear farmers trained for such programmes say: "We can't keep to these instructions. We understand that the engineers are asking us to use less water. Some people try to use less, they try a bit, but they do not follow the instructions fully."

To find common ground between farmers and engineers in order to represent the quantum of water used is rather a difficult task, especially for crops other than paddy. In such conditions, the drip irrigation technique in fact reflects a need to reduce water consumption in fields of sugarcane or banana, and the government policy (in TN and Puducherry) in promoting such techniques through subsidies is to be encouraged. Regarding paddy cultivation, which requires flooding the field, the height of the water level in the rice fields and the numbers of days without watering can be given, as is done in a programme such as Tank Rehabilitation Project of Pondicherry (TRPP). However, the technique System of Rice Intensification (SRI) may also help in this respect, as it does not require systematic flooding and can theoretically help to reduce water consumption from 30 to 50 per cent. This technique, originally from Madagascar, was introduced in India in the year 2000. It was first tested by specialists from various networks (farmers' associations, organic farming networks, popular agricultural journals, NGOs) who encouraged its application, whereas Tamil scientists remained unconvinced. Yet since 2004, implementation of SRI has been encouraged by various state governments, including Puducherry and Tamil Nadu, especially for the Cauvery delta (Prasad *et al.* 2005). The number of farmers applying this method is on the increase.

Box 7: *Nanjai* and *Punjai*: Land Terminology Losses its Meaning Because of Changes in Groundwater Use

The terms *nanjai* and *punjai* are very interesting when studying the changes in man's relationship with his environment due to the massive use of underground water. These terms designate paddy-fields and rainfed or "dry" cultivable lands respectively. According to the Dictionary of Contemporary Tamil, two factors characterise *nanjai* land: paddy cultivation and its irrigation using surface water. Indeed *nanjai* (written as *nancey*) is a "paddy cultivable land getting irrigation facility from river or tank"[5] (Subramaniam, 1997) showing implicitly that, traditionally speaking, a well was not considered a source of water for paddy. However, the English translation, "wetland", loses most of this exactitude. Another terminology is used by farmers instead of *nanjai* and *punjai*: *kalani veli* (plane space planted with paddy) and *kolli veli* (other fields), where the presence or absence of wet paddy in fields is a factor of discrimination. This distinction between wet paddy-fields and other land is found in a large number of rice

(Contd.)

Box 7(*Contd.*)

growing societies, as Y. Abé (1995) has shown, when comparing numerous languages across Asia: in such societies, the terminology reflects a classification of the environment based on the presence or absence of wet paddy, showing the cultural importance of this crop. Besides, in Tamil Nadu, it is linked to tank or canal irrigation.

The British maintained a dichotomy between irrigated and non-irrigated land in their cultivated land tax system, with the names "wetland" and "dry land". Wet land corresponds implicitly to the command area of a canal or a tank, the *ayacut*—an administrative category[6]. Today still, whatever the changes, land categorization for land tax is made according to this same classification, both in terms of denomination and of geographical demarcation. Thus, according to the revenue department, in Sorapet village there are 102 ha of "wetland" and 244 ha of "dry land": yet today all of it is irrigated exclusively by groundwater, and some of this so-called "dry land" is actually irrigated paddy-fields!

In what way could this distinction between *nanjai* and *punjai* be relevant today? Why do some villagers still use it, what does it represent? Though it no longer helps to distinguish irrigated paddy-fields from other land—which was the original purpose of this terminology— or to distinguish irrigated land from rainfed land as the English terminology would indicate regarding the land tax classification (wet and dry land), it nevertheless represents a difference in soil. *Punjai* are more sandy or loamy, whereas *nanjai* are clayey, cultivated mainly with paddy and sometimes sugarcane since, according to farmers, it is difficult to grow other plants in such soil. This clayey condition may be due to wet paddy cultivation and to the accumulation of sediments carried by canals over the centuries. When farmers want to transform *punjai* land into a paddy-field, they add clay to the soil, with this clay coming from the tank, then the plot is irrigated with GW. What should this type of land be called then? The answer provided by farmers is: "*nanjai ki punjai*, *punjai ki* nanjai", "it is *nanjai* and also *punjai*, you can put *nanjai* crops as well as *punjai* crops, there are two usages for the soil (*irupadu manu*)."

Here we can see a fourth use for these terms: to distinguish crops according to their irrigation needs (not only wet paddy from others). Those that can be cultivated in rainfed conditions are referred to as *punjai* crops whereas paddy, sugarcane and banana need additional water and are referred to as *nanjai* crops. Because of the ambiguity of these *punjai* and *nanjai* terms, of the diverging definitions according to those you question, today most villagers prefer to use other terms for crops cultivated in an area previously called *punjai*: *manavari* designates rainfed crops and *iravai* irrigated crops, this term especially being used for groundnut.

It is worthwhile examining how a vernacular terminology, formerly related to the classification of the landscape, has become ambiguous and may soon become obsolete, due to major changes in land use and to the surge in the use of mechanized GW irrigation.

(c) Aquifers: Knowledge not Shared by All

To pinpoint the exact position of a well or bore, farmers often call in a water diviner who captures information using a pendulum or through the vibrations of a twig (cf. the film by Coste and Ploumpidis, 2007) or even from the speed of rotation of a coconut[7]. They determine the place where water is most abundant, as well the direction of the groundwater flow. It is not, therefore, rare to hear peasants say that the flow of underground water circulates at such and such place and in such and such direction (in the alluvial region studied, generally from west to east in the direction of the slope, i.e. also the slope of the aquifers). On a village scale, great variations in groundwater availability are often mentioned, with land situated above a GW spring occupying the best position.

Yet those really mastering the quest for water are the drillers (either manual or mechanical) who give advice on the depth to be drilled according to the presence of sand. As bore owners in the alluvial or sedimentary area are present during the boring process, they know the lithology of the position of their wells perfectly well (cf. Fig. 7.3). Those who have sunk a bore manually have a more precise knowledge of this, since the extraction of earth from the subsoil is slower during this type of process. Drillers making a manual bore in fact observe the structure of the soil very attentively while extracting it. On the contrary, where mechanical drills are used to dig deeper, no particular attention is paid to the first metres of soil.

Bore owners are well aware that water is available in layers (at the level of sandy subsoil), and that if the level of water in an aquifer sinks to the point where the source dries up, they will have to go and find a deeper aquifer if they want water. They explain that in their village there are two or three main layers. It is rather difficult to determine whether this knowledge has been gained through 25 years' experience with tubewells or over an even longer period, given that tubewells were set up in artesian aquifers in Pondicherry[8] at the end of the 19th century—which would not be surprising—or through campaigns to encourage the awareness of problems of managing underground water, such as those organised by the TRPP, or perhaps a mixture of both, since the information provided by such campaigns confirms farmers' local knowledge and gives them a more regional dimension.

Lithology is certainly indispensable in determining the depth for placing filters through which water from the aquifer filters into the bore. But how does this knowledge of lithology help farmers to situate the depth of the water-table or to decide at what depth to install their motor? For in fact, the piezometric level (indicating the pressure of the water table) of confined water tables may be higher than that of the sand bed, making the water level in the bore higher than the filter zone. Farmers realize this, since the motor is necessarily immersed, and in such cases they install it above the lithological level. I was unable to grasp the explanatory model farmers use to explain this phenomenon, but they pointed out that the water pressure makes the level higher there.

How then are water levels estimated and how do farmers systematically determine them even though they are subject to fluctuations? Farmers make a precise measurement of the level on two occasions: the first is when they decide that their bore is not producing enough water and want to measure the water level to decide whether to lower the engine; in this case they use a string to measure the depth. The other occasion is when repairs to the motor require the excavation of tubes; the level is then calculated when the tubes are being counted and examined for traces of humidity. Any other method is entirely empirical, deductive, and hypothetical as far as the result is concerned: one method refers to the pressure of the water as it comes out of the pipe and the other is an extrapolation from the depth measured in a neighbouring bore. Now, bore-owners are the first to say that water availability may geographically be very heterogeneous, even between two sites situated a short distance away from each other, which can be easily understood seeing the heterogeneity of the lithology (see Fig. 7.3). Reports by farmers of fluctuations in the water level before and after the rainy season or over the course of time, though made with great certitude and precluding any doubt as to the result, must be treated with caution, since farmers do not systematically measure the level, but instead evaluate it using very questionable methods (which I was not able to test).

Moreover, there is often great inconsistency between the knowledge provided by bore-owners on the one hand and, that of water purchasers and even agricultural labourers who work on the pumps on the other hand. Many of them do not know the depth of the bore or where the engine is positioned or even the depth of the aquifers. They put forward impressive figures, sometimes greater than the depth of bores found in the village. These exaggerated figures and mental pictures of extraordinary depths given by people with limited resources reflect first and theremost big bore owners' frantic quest to place their bores at a great depth—a depth which serves mainly to enhance a bore-owner's social status (Chap. 6)—and secondly, a collective preoccupation with seeing the level of tables sink so that water has to be sought even deeper, even if the depth at which water is currently drawn is not common knowledge.

7.2. Recharging Groundwater: Empirical Beliefs or Imported Knowledge?

(a) Upper Aquifer

Though great disparity can be observed regarding knowledge of deep water tables, this is not the case for the upper aquifer and its recharge. This ancient knowledge shared by all socio-economic classes of the population is undeniable as it concerns fluctuations in the level of the shallow aquifer: (i) this water table is recharged with rain and surface water: infiltration into the ground by

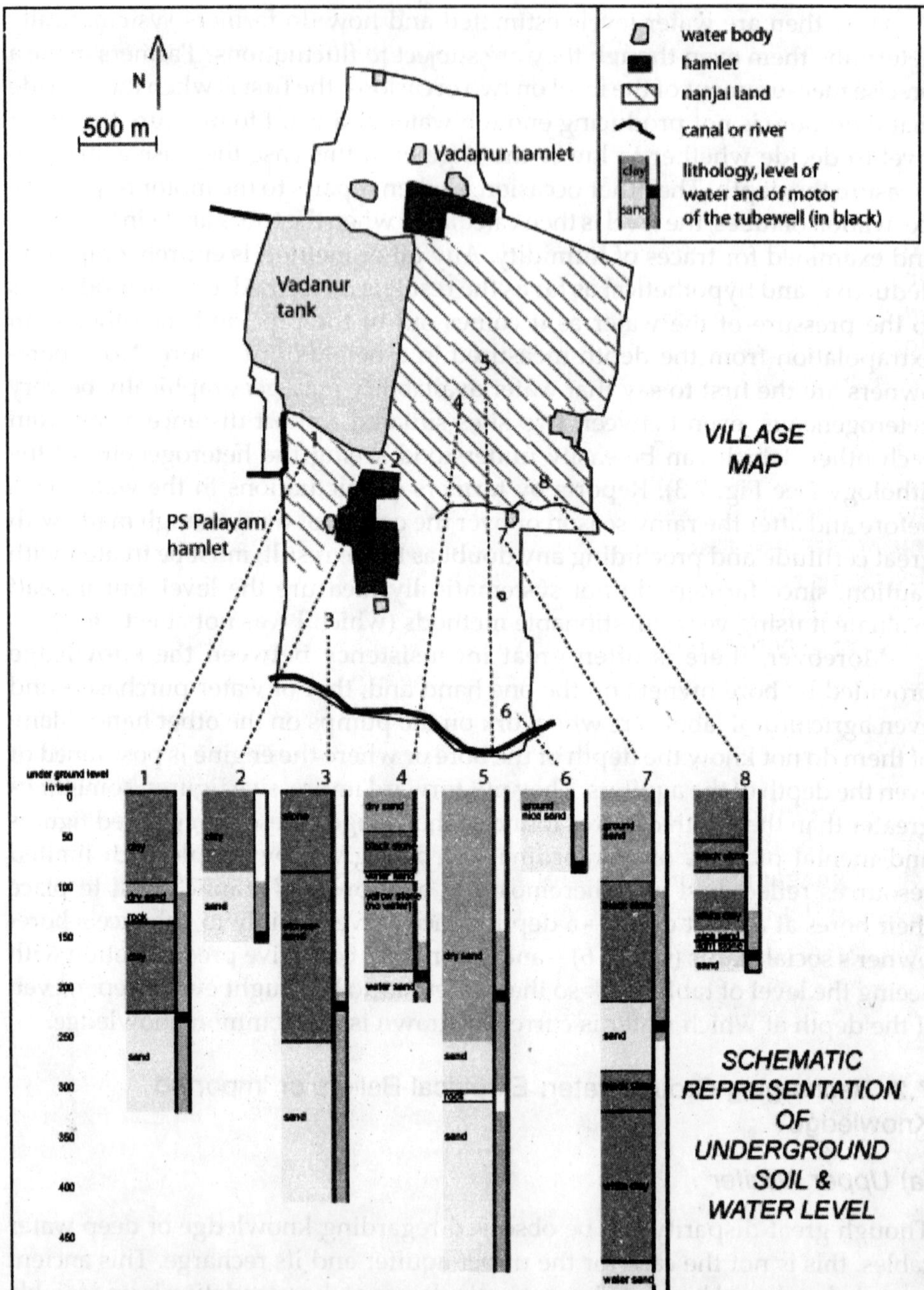

Fig. 7.3 : Spatial Variation in Lithology and Groundwater Level in Vadanur
(From Interviews with Tubewell Owners)

water bodies, water streams (rivers, irrigation canals, drains), irrigated fields and rain waterlogged land; (ii) the shallow aquifer provides a place for storing the resource whose reserves gradually diminish after the rainy season. It is a well-known fact that the availability of this groundwater decreases during the dry months to rise again during the rainy season; and (iii) it is this infiltrated water which replenishes wells.

We hear people talking of "*uthu*" (spring) to explain the provenance of water. However, this widely used and very common term covers definitions and representations which are in no way consistent. First of all, there is the "*mel uthu*" (upper spring) and other deeper springs (*alathu uthu* = very deep spring; *nadu taram* = medium deep; *bhumi ki kile* (underground) = very deep). In the alluvial area, the upper spring is different from one village to another: either there is a good flow of water (*nir ottam*), as in only some places, and it runs in a particular direction, most often that of the slope (it is only then in such *mel uthu* that a tubewell gives a good yield) or, as in other places in the area, there is no so-called *mel uthu*; or else the *mel uthu* is an upper layer of soil into which water infiltrates (throughout the village). In the rocky area the *mel uthu* is the water that fills the open wells; it occupies the empty space and runs out, travelling down from elevations, but not as a flow of water.

Everywhere one hears, "when there is water in the river and in the tank, then there is water in the well". There being water in the wells when there is water in the river or tank is logical to some extent, since the presence of water in the river, which is not perennial, and in the tanks, fed either by the rain or river, means that monsoon rains have been sufficient to generate streamlets of water. Does this mean that there is a belief that the tank recharges the upper aquifer? This is true in the literature. Thus Palanisami and Easter explain that "the management of tank supplies needs to consider the benefits from groundwater recharge" (2000: 165), as "wells depend heavily upon tank water supplies for recharge, under constraints of poor recharge from rains." (*ibid.*: 173). They even suggest the use of some tanks as percolation ponds (p. 174). Janakarajan (2003: 17) also sees environmental functions in the tank "the most important of which is to serve the purpose of recharge structures". Any impact is said to affect *nanjai* wells: "the depth of wells depends on a variety of factors including (…) in the case of *nanjai* wells, the supply in the tanks that recharge them" (Vaidyanathan *et al*., 2003: 20), without any explanation as to why *punjai* wells do not benefit from percolation from the tank[9]. Adicéam (1966: 344) gives the example of the Panaipatti tank in Tiruchirappalli district, whose destruction at the beginning of the 19th century—due to flooding of the Cauvery 20 km to the north—implied the disuse of the wells, as the tank helped to increase the water table. Our findings are more complex in that results differ from village to village according to the topography of the land, the presence or absence of a river in the vicinity and the soil characteristics. Thus the tank in Sorapet which is some distance from *punjai* land is not believed to affect the *punjai* wells, while, in Koratur (rocky area), it is the elevation of the *punjai* fields which supposedly justifies the lack of impact. In SOR again, the tank is believed to recharge only the spring of *mel uthu* present near the drain of the nanjai area, and to affect no other places. In Vadanur (not far from SOR) however, it is believed that *punjai* land adjacent to the tank benefits from water infiltration from the tank.

An important aspect, which is not really documented from a scientific point of view, is the impact of the desiltation of tanks on groundwater recharge. In villages, various opinions are voiced: old men from SOR say that the tank cannot help to recharge the ground water as it has an impermeable bed, and that *kulam* (ponds) are more useful in this respect; a peasant from VAD, on the other hand, criticizes the desilting programme, as it increases infiltration and prevents the storage of water (even if the tank is not used for irrigation there and the storage of water is mainly for recharging the groundwater!).

The common element shared by the various local accounts recorded in our study area is that the level of the upper aquifer fluctuates on a cyclic basis according to rainfall (for its recharge) and dry months (for its drying process). Some well owners note a net drop in the length of availability of this upper aquifer. In Vadanur for example, in the alluvial area, the 50-feet wells are not productive during the summer, whereas it was the case in most summers 30 years ago. This decrease in the length of availability of the upper aquifer has been analysed by some elite farmers as being due to an increase in pumping, though most of the time they accuse the "dry sky".

(b) Accusing the Dry Sky

When we ask why the level of the aquifers has dropped, the first reply given is "the dry sky", the lack of rain. It is true that rainfall analysis (see Box 3) shows that the number of exceptionally dry years has increased over the course of the last 25 years. Even though the annual average has not changed over half a century, this explains the impression of a shortage of rain regularly mentioned by rural population. It is true too that the fluctuations in water levels in the wells (upper aquifer) are directly linked to the alternating rainy and dry seasons, with recharge occurring thanks to the rains.

It is also possible that for some older people, the notion of "dry sky" is linked to other phenomena, especially to the reduction of river spates due to the Sathanur dam. An eighty-year-old man from a village near the river explains, "In those days, there were heavy rains and floods. The water came like a thief into the village. Nowadays there is no good rain, that is why there is a water crisis." Blaming the problems on the "dry sky" may also be due to the fact that old people have difficulty in interpreting the news properly: the same man said that he saw on television that there had been heavy rains in other States, while the soil of his village remained completely dry. It was June at the time, the month in which the south-western monsoon starts in other States with little effect on Tamil Nadu. This "dry sky" interpretation may stem from misinformation about India's various climate zones.

When we pointed out during interviews on the subject that the rains were perhaps not the only factor responsible for the decline in the water table and that there has also been an increased amount of water extracted, linked to an increase in the number of wells and tubewells, three different reactions were observed:

- Total denial of the hypothesis. For these people—who are essentially poor or of limited means and uneducated—the absence of rain is the main reason for the drop in the water table level. They add that if anyone is extracting large quantities of water from the ground in the region, it is the government itself along with the Neyveli Lignite Corporation (NLC). This Central thermal power station supplies electricity to South India and to do that, it pumps vast quantities of water to avoid flood mines (the water is distributed to farmers whose farms are situated downstream). The NLC has thus become the scapegoat so that farmers do not have to blame themselves.
- Acceptance of the argument, but the blame is laid on the NLC[10] in this case too.
- Total acceptance of the argument, with some pessimism as to the lack of awareness of other farmers about the need to alter their practices. Moreover, these farmers who recognise the overexploitation of groundwater find it difficult to change their practices because the social context prompts them to seek water at all costs, not to limit its use.

To blame the "dry sky" is a way of accusing a natural phenomenon rather than questioning anthropogenic action which has increased in these last 30 years. At the same time, the Government is not urging farmers to reduce their use of GW while it puts a lot of effort into recharging groundwater. The recharge of aquifers has become a subject of widespread interest in recent years. The number of development projects aimed at recharging underground layers, including tank rehabilitation, is on the increase. The Government is still in a position of water supplier, whereas it should also be in a position of resource protector through programmes to explain how recharge functions and the need to limit groundwater extraction.

(c) Medium and Deep Aquifers

Opinions are divided[11] regarding the function of aquifers other than shallow ones, with conflicting ideas circulating in the same village—which is hardly surprising given the evident difference in knowledge about the very existence of these aquifers. There is, therefore, in contrast to the upper aquifer, no common, relatively homogeneous knowledge. A single opinion cannot of course be valid for all our study areas, where there is an enormous diversity of situations (even in the case of the alluvial zone; the only one where surveys on this subject have been carried out). Certain contradictions prompt us, however, to attempt to understand how these concepts of GW recharge and fluctuation in GW levels are constructed and disseminated.

The most radical views on how deep aquifers function relate to the fact that the are perceived as 'natural water', the quantity of which cannot decrease (as opposed to "upper sand", which frequently dries out and only receives water from rain and surface water). I heard this opinion several times, but I was literally shocked to hear it voiced by an assistant driller with a far from negligible role in circulating these ideas. When questioned on the coherence of that opinion, given that the level of aquifers is dwindling, he evaded the question and explained that tubewell owners dig so deep (300-400 ft) merely for safety reasons. Such ideas do nothing to encourage effective management of the resource and, on the contrary, may lead to drilling deeper.

Others, often tubewell owners, recognise the decline in water tables and have various explanations for this, their main argument being the necessity, over the last 25 years, of positioning the motor and drilling deeper. Amongst these people, some believe that such aquifers are not rechargeable (with beds of clay between sandy beds preventing the water from infiltrating and from percolating deep down); others believe that since the level fluctuates, as in the phreatic (or upper) aquifer, the water flows progressively downwards.

The idea here is not to validate or challenge these views, since measuring the recharge of aquifers is a difficult and delicate phenomenon, and cannot be the subject of a simple empirical verification. In fact, in order to understand this phenomenon, it is not enough to know the fluctuations in the water table level, as there may lateral inflows of water from recharges taking place in other areas. A strict measurement protocol has to be followed in order to obtain hydrogeological information and data on the permeability, storage capacity, leakage and transmissivity of the various underground layers; GW withdrawals also have to be estimated. Only scientific hydrogeological studies, which are few and far between, would be able to answer this question. One could mention those undertaken independently in the sedimentary area of the Auroville based Harvest project, which show that the overexploitation of the Vanur aquifer impacts the other aquifers. This means that the intermediary beds of clay are not completely impermeable (lecture given in the town hall of Auroville by the geologist Sophie Violette, in 2007). We may, therefore, estimate that that a similar phenomenon exists in the neighbouring alluvial area which has the same geological formations. The information about the alluvial area collected by the Puducherry hydrogeologists does not, however, corroborate this since, according to them (in conversation), there is very little exchange between the beds.

Otherwise, the complexity of the subject lies in the fact that some aquifers may be deep aquifers in one place, then come up to soil level in another place where they are recharged. This is the case with some aquifers in our study area, especially the cretacious and tertiary aquifers of Vanur and Cuddalore respectively. The Cuddalore aquifer—used in Manamedu—is recharged by infiltrations which take place nearly 200 km farther to the south-west, in Pudukottai (information given in person by a hydro-geologist of Puducherry Agriculture Department). It is the same for the Vanur aquifer—tapped in Vadanur—which is recharged in places located some ten kilometres north-west. Thus, the fluctuations observed by villagers are not due to a local recharge, linked to the tank, but to a phenomenon taking place elsewhere. A lack of published scientific information leaves the door open to any kind of interpretation. Moreover, local opinions cannot easily be validated or confirmed.

(d) Producing Knowledge According to Vested Interests

It is in the district of Puducherry, in villages not having used tanks for the last 30 years, that we constantly heard "it is better to keep water in the tank—in order to recharge the groundwater—than to use it directly for irrigation". The rationale is as follows: stored water infiltrates the ground progressively over a 4-6 months storage period (or for just 2-3 months depending upon the tanks), recharging the aquifer whose reserve lasts longer so that, if the tank water is used directly for surface irrigation, it is only available for 2-3 (or 1-2) months—much less than the period for using groundwater—while a great deal of water is lost to evaporation and during canal transportation. This idea is systematically propounded by the Tank Associations (TA) in Puducherry and we wonder if it is not linked directly to the TRPP programme. This programme was indeed implemented in order to reduce farmers' dependence on underground irrigation and to reintroduce the use of irrigation tanks; it led to the eviction of encroaching farmers, it rehabilitated tanks and trained farmers about GW recharge from tanks. A farmer from Sorapet explicitly explained that this idea of storing water in the tank has been circulating in the village since 2004, which confirms the hypothesis that this idea of storing water in the tank was introduced recently through the rehabilitation project.

The main elements that make the above rationale—related to the storage of water in the tank for GW recharge—rather disconnected from reality is that many tubewells in the Puducherry district are to be found in deep aquifers. Talk of storing water in tanks for the benefit of the wells means that the tanks have to be active in the process of recharging these deep aquifers. Many members of Puducherry TA reckon that this is the case, justifying this with a fluctuation in the level of their tubewells. They also credit the tank with recharging the intermediate and deep aquifers, just as the upper aquifer does, thus by extension, ascribing to these aquifers a situation observed for generations for the upper one: the aquifers are recharged during the rainy season, then the water table decreases. However, we have seen that, on a regional scale, this point is not widely accepted. Might it be widespread in Puducherry because the tanks have been rehabilitated and can no longer be encroached on? This is probably the case, but if so, why not use it directly for irrigation? I firmly believe that farmers' idea about the recharge of deep aquifers is:

- First of all, imbued with the current "propaganda" linked to the policies to recharge aquifers, which emphasizes the need to use the various, already existing techniques and the necessity to store a maximum amount of water in the ground, but which barely stresses the conjunct urgency to limit the use of underground water. These policies use the terms "save water" and "tank as bank", which reinforce the idea of storage, not of minimizing utilization.
- Secondly, appropriated and circulated by owners of tubewells, who sell underground water; thus they have no interest in other farmers using tank water, free of charge. Those who buy water explain in effect that they prefer using tank water but, as there is no guaranteed supply, they do not want to risk being without water and being unable to buy it from tubewell owners, who might easily refuse to sell it to them. Buyers do not, therefore, insist upon using tank water. Members of the tank association in VAD, for example, specifically told us that no one wanted to use tank water when the offer was made, thus proving that water is stored in the tank, and of course, they did not spell out the social reasons behind this choice... !

Preserving water to recharge the aquifers is an argument put forward by a particular section of the population in their own interest—in this case, tubewell owners—and against whom other villagers have difficulty in taking a stand in view of the global context of formally advocating the recharge of aquifers, of the absence of scientific data and of knowledge about aquifers only being produced by tubewell owners (as seen above). Two elements lead me to this conclusion: firstly, outside Puducherry, the idea of keeping water in the tank in order to recharge GW is not widespread as we have already said; secondly, in solemn discussions about the process, it is by no means rare to become embroiled in a mix-up. In the case of SOR, for example, only bores located in areas with a good spring (near the drain, the *"mel uthu"*) are able to benefit from tank water as it is the only place where there is a sandy layer. However, the same persons are adamant: it is only when the tank is full or there is heavy rain that bores on *nanjai* land benefit from it; "the water flows into the sandy layer under the tank and will flow wherever there is sand and in the direction of sand availability". Interviewees, moreover, evaded the issue when they said that the aquifer recharge is not only a question of sand beds, but also of the drop in rainfall, since the last proper rainfall dates back ten years (interview, July 2006).

In addition, farmers in favour of the idea of storing water in tanks legitimise this practice by declaring that it is being done on the request of the government that wants to recharge the aquifers. Indeed, they are all members of the Tank Association, which gives them a certain legitimacy at village level to hold this view and maintain this idea. However, this is not the message conveyed at public meetings by the Chief Engineer, who encourages farmers to use more tank water and less groundwater. Development agents involved in the TRPP project always attest to the fact that re-establishing collective surface irrigation is complicated. There are several reasons for this: irrigators' agricultural calendars are now totally different due to GW utilisation (collective irrigation calls for a baseline agreement on the agricultural calendar), the autonomy achieved by owning a tubewell does not make a return to additional constraints particularly appealing to these farmers, and the absence of water availability holds buyers back from taking any risks. Finally, since projects for recharging tanks recommend fish farming in the tanks (see Chaps. 7.3 and 9 below), a conflict of interest can be observed regarding many tanks between those who raise fish and therefore want to keep water in the tank for as long as possible, and those who want the water flowing out of the tank to supply their rice fields.

Here groundwater utilisation is associated with so many social, economic, political and environmental issues that relevant knowledge about the benefits of storing water in a tank can easily be manipulated by some villagers in their own self-interests—in the case of selling of water—and the lack of scientific information and of awareness of how aquifers function, along with the complexity of hydrogeological phenomena, prevents any protest movement from taking place. Moreover, the confidence expressed by bore owners regarding aquifer recharge is not conducive to discussions and thereby other points of view are not voiced.

7.3. Tank and Wells: Competitive or Complementary Infrastructures

(a) Competitiveness

The increase in the number of individual pumps since the 1970s has been described as a source of competition with regards to the tank use and management in many situations, and as one of the main reasons for the decline in tank use (*see* Chap. 1). Many cases corroborate these assumptions in Puducherry district, where the water supply of many tanks is not guaranteed and the disuse of tanks over the last 30 years is in fact linked to the advent of submersible pumps (*see* figures in Chap. 2).

In some villages, such as Sorapet, where only two crops of paddy are possible on *nanjai* land *versus* diversified cropping throughout the year on *punjai* with the help of tubewells, preference goes to *punjai* land: when people have the choice, they prefer to sell their *nanjai* land and buy *punjai* land. Thus, in such villages, irrigation using tubewells has reversed the situation: in the past, *nanjai* were considered the best as they were the only ones to be irrigated and to enable rice cultivation, whereas now *punjai* are preferred because crop diversification is possible, whereas only paddy gows on *nanjai*. Nevertheless, in most places, the preference is still for *nanjai* land, which is more expensive.

A new type of competition has evolved in the use of tank water either for irrigation or for fish farming (and also for recharging GW): with tank rehabilitation programmes, many NGOs suggest that tank associations raise fish on an intensive basis in the tank. Consequently, conflicts of interest emerge regarding tank water use among those who want to use the water for irrigation, and those who either want it stored in the tank for recharging groundwater (cases described above) or wish to keep it in the tank for raising more fish (case of Manamedu, where the group rearing the fish is mostly composed of landless S.C.). In such a situation, there is no direct competition between well and tank irrigation. Yet, since masters of well irrigation (and experts in the notion of recharge by tank) and people rearing fish have a common interest, even if the conflict of interest is clearly shown to be between two functions of the tank (surface irrigation and fish rearing), fish breeders also indirectly receive the support of tubewell irrigators (because of the function of groundwater recharge by the tank). Fish farming is not new, though its intensive practice is a recent phenomenon, implying a certain investment on the part of fish farmers and therefore, an expected return investment. Thus, the rules for informal management of tank water are no longer valid—where priority went to surface irrigation, and the tank was drained at a fixed date in order to catch the remaining fish. Given the fact that fewer farmers use tank water, that more and more people believe in its function of recharging GW, and that landless people can take part in raising fish (case in Manamedu), the conflict over storing water in the tank or using it does not only involve farmers. The multiple functions of the tank, and especially the new practice of fish farming, may indirectly accentuate competition between tank and well irrigation, and also concern landless people.

(b) Conjunctive Use

The concept of conjunctive use is of particular interest to us here, as it is linked to the interrelation between surface water and groundwater at local level. The conjunctive use of tank and well irrigation can be of two types: either the tank and the wells are used during the same irrigation season, which could be called simultaneous conjunctive use, or well irrigation is used on plots that have used tank irrigation when it was available, which could be referred to as deferred conjunctive use.

As Vaidyanathan notes for the Peryar-Vaigai as well as Palar basins, there are wells in the command area of nearly all tanks, either system or non-system tanks with well densities varying from 2 to over 200 for 100 ha (2001: 39, 84). Moreover, "it is conceivable that the shortages in tank water may have been made up by the spread of conjunctive use of groundwater" (*ibid.*: 84). Yet it is difficult to measure the conjunctive use of tank and well irrigation:, firstly, because both the frequency of well use and the area irrigated by well should be integrated; and secondly, because well density does not reflect the intensity of well use (see below and Kumar, 2007). Moreover, it is not clear if conjunctive use is a notion to be adopted at plot level or at tank command area level: at this latter, any use of well water is then called conjunctive use, whether it is once or whether a plot is irrigated only by a well (which is exclusive well irrigation at plot level). If the idea is to make this a quantified concept of conjunctive use, the area exclusively irrigated with GW should therefore be specified. The limit between the use of well irrigation as a competitive (when it becomes exclusive irrigation and leads to abandoning the tank) or a complementary structure to tank irrigation (when it supplements tank water) is indeed a very tenuous one.

Palanisamy and Easter (2000: 152) point out that in many places, groundwater has to supplement tank water supplies, as tanks are often—in more than half the period (according to the Government of Tamil Nadu, 1984, cited by these authors)—insufficient to produce a good paddy harvest. They also note that "the farmers in the tank command area have an incentive to use tank water first and use well water only after the tank water has been exhausted" (2000: 166), which means practising a deferred conjunctive use. These authors then propose strategies for using surface and groundwater to "increase the availability of tank and well water" (p. 173), relying on simultaneous conjunctive use: farmers should use well water when water is still available in the tank in order to increase the period of time tank water is available; farmers could take advantage of the wells whose recharge period would be longer if water remained in the tank longer. In addition, they advised farmers from the study area (Ramanathapuram district) to coordinate tank water release according to the rules for operating the sluices, taking into account the rainfall: leaving them closed for two days following a daily rainfall in excess of 60 mm and closing them on a rotational basis, on alternate weeks (p. 153). Finally, new water management rules need to be defined. NGOs and research programmes have a role to play in such a challenging situation.

(c) Conjunctive Use and Recharge of Groundwater

One reaction to the recharge process is an increase in GW use since more water is supplied. This had already been reported by T. Shah in 1993 (pp. 154-155) in connection with the Gujarat government pilot programme, where various structures were built to increase the fresh water recharge and to fight the pressure of encroaching sea water: instead of limiting GW use, farmers reacted by irrigating larger areas with wells.

The concept of recharging GW using tanks, associated with that of conjunctive use, is rather interesting, as it leads to different interpretations according to the various sites, and thus results in varied practices. Two of our village studies are of particular interest in this respect, since the same argument (tank water recharges GW) is given to justify opposing practices: using tank water for irrigation and keeping water in the tank. Thus, in TVN, because using tank water for irrigation is said to recharge GW, farmers practice conjunctive (simultaneous and deferred) use as will be described below; by contrast, in VAD—a case analysed above as being motivated by vested interests—in order to recharge GW, water is kept in the tank and the latter remains unused. A third example could be mentioned: because the tank is said to not impact GW recharge and as its supply is not guaranteed, conjunctive use and GW recharge by the tank are concepts which are not developed in SOR.

In Tiruvenainallur, where farmers practise simultaneous conjunctive use, the tank is used for recharging GW: farmers explain their practice as recharging the wells thanks to surface irrigation. This might be the case as the upper aquifer is the only one to be tapped. However, conjunctive use is not homogeneous. Let us develop four dynamics observed in this village (the data were collected in 2000 by F. Cesselin, 2001).

- In the southern part of the TVN irrigation network, where the aquifer yield is low (Fig. 7.4), farmers tend to use tank water as much as possible, whereas those relying on the same tank sluice in *Terkuvali* area (Fig. 7.5), but with fields located more to the north and having ready access to groundwater, tend to irrigate more often with tubewells than with tank water. The lower yield of the aquifer combined with secure access to tank water, which recharges GW, may explain such simultaneous conjunctive use in the southern part. While the aquifer's reliable yield in other parts may explain the lesser development of simultaneous conjunctive use.

- Here again (as in Chap. 3), tubewell density does not show the actual extent of GW use and of the conjunctive use of tank and GW. Indeed, let us compare tubewell density[12] in various parts of the village (Fig. 7.6) to the cropping pattern (Fig. 7.7). On the one hand, in the southern part irrigated by the third sluice, this density is half (0.14/acre) that of the northern part, and only two seasons of paddy are cropped against three in the northern one. On the other hand, the tank command area as a whole includes the same density of TWs (0.22 TW/acre) as other rice fields which were previously irrigated using sources (such as the area locally called *Chiteri*) or even as *punjai* land, yet the cropping pattern differs in each of these parts. TW density is not related to the conjunctive use of tank and GW or to the intensity of the cropping pattern.

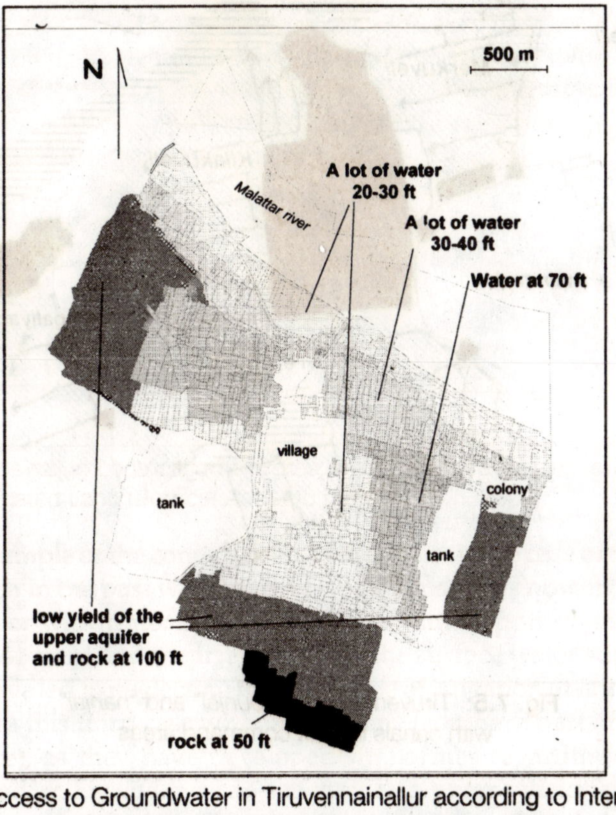

Fig. 7.4: Access to Groundwater in Tiruvennainallur according to Interviews with Farmers (*Source*: Adapted from Cesselin, 2000)

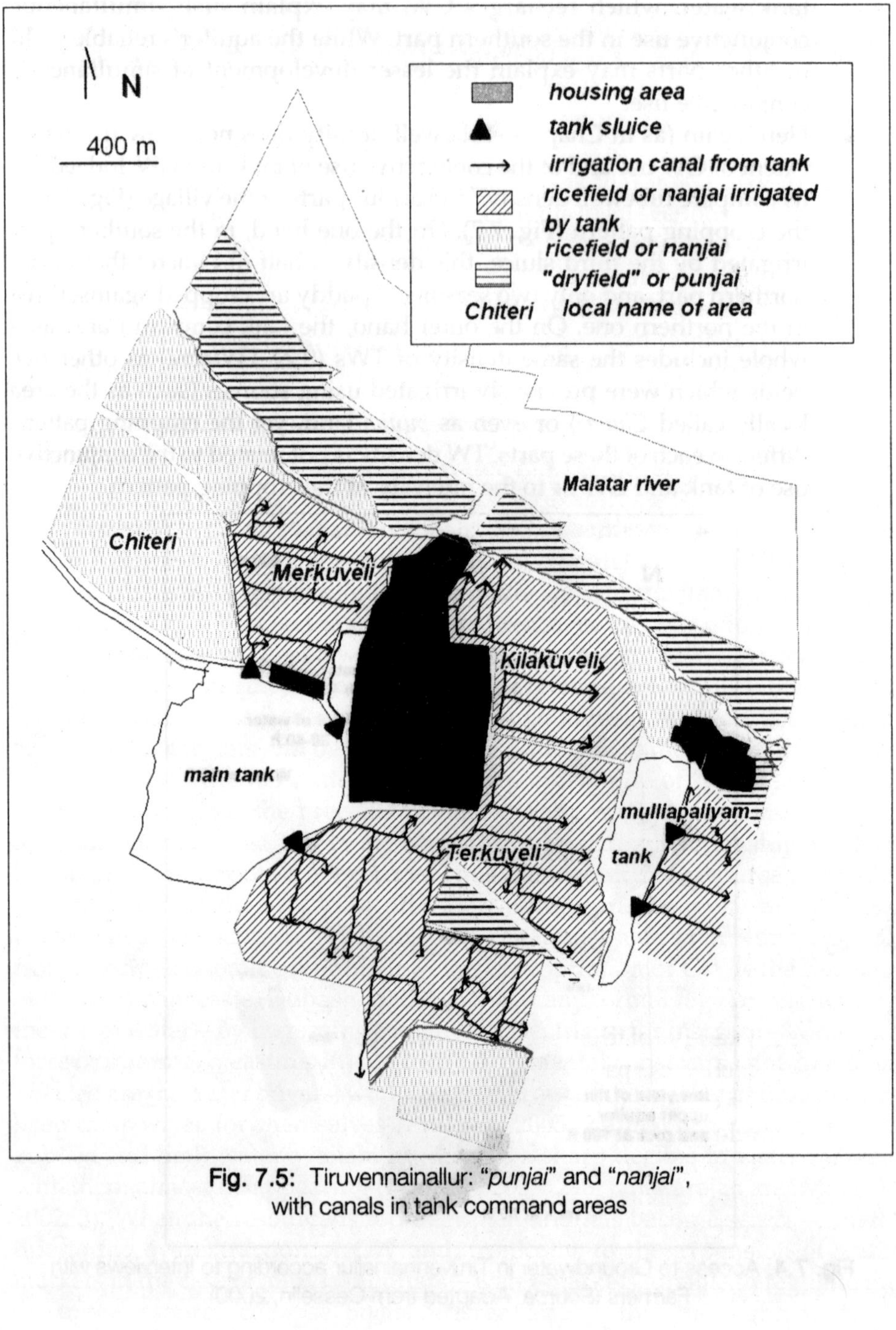

Fig. 7.5: Tiruvennainallur: "*punjai*" and "*nanjai*", with canals in tank command areas

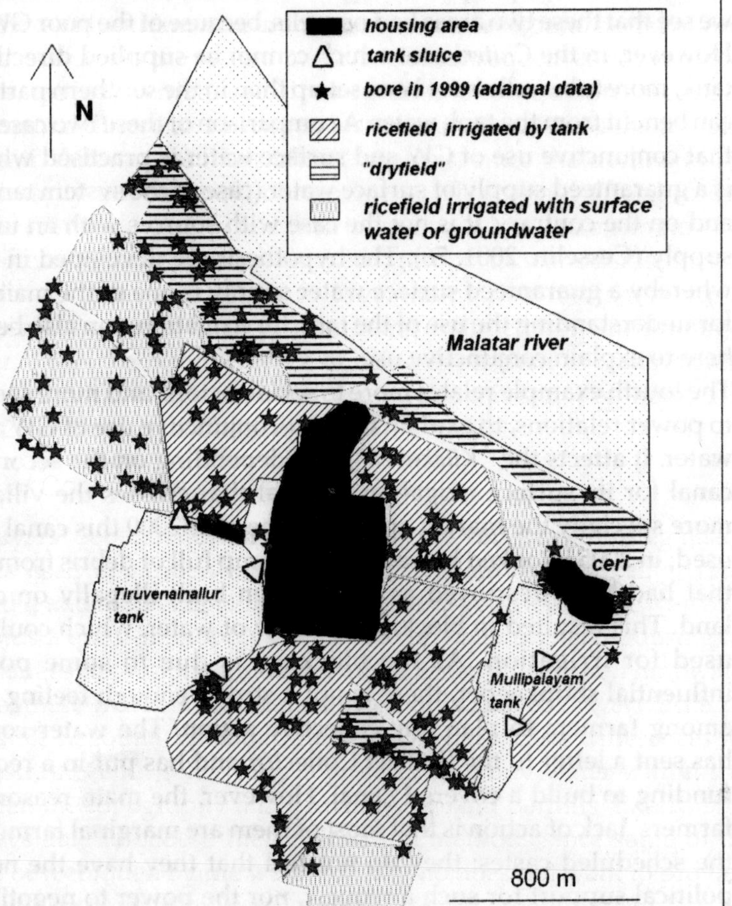

Fig. 7.6: Tiruvennainallur: Localization of Bores, with Reference to Various Types of Irrigated Land (*Source*: Adapted from Cesselin, 2000)

- A third example of the conjunctive use dynamics is the case of the *Chiteri* area, which in the past was irrigated by sources but is now irrigated by tubewells or with water pumped from the tank supply channel. Here TW development and the irregularity of the surface water supplied by sources have led to neglecting the latter type of irrigation. It is interesting to compare this third case with the first one (southern part of the third tank sluice), as they have developed similarities regarding access to GW. Indeed, in Fig. 7.4, we can see that the aquifer's yield is poor and that access is qualified as difficult. Moreover, in Fig. 7.8 taken from the

1950 cadastral map representing the old unmechanised wells (*yattam*) we see that these two areas had no wells, because of the poor GW yield[13]. However, in the *Çhiteri* area which cannot be supplied directly by the tank, more tubewells have been set up than in the southern part—which can benefit from the tank water. A comparison of these two cases reveals that conjunctive use of GW and surface water is practised when there is a guaranteed supply of surface water (case of the system tank water) and on the contrary, it is not the case with sources with an unreliable supply (Cesselin, 2001: 76). The hypothesis we developed in Chap. 4, whereby a guaranteed surface water supply is one of the main criteria for understanding the use of the tank for irrigation, can also be applied here to explain conjunctive use.

- The fourth example relates more to social factors, and more specifically to power relations, to explain the non-conjunctive use of GW and tank water. It affects the "*Kilakuveli*" area depending on the second sluice canal for its surface irrigation, a canal that crosses the village, and more specially the market area. If, in the year 2000 this canal was still used, in 2005 it looked like a rubbish dump full of debris from a house that had been destroyed as it had been built illegally on common land. This resulted in blocking the flow of water, which could not be used for irrigation. As the blockage is due to some politically influential shopkeeper, there is not a strong enough feeling of unity among farmers to plan any collective action. The water committee has sent a letter to the District Collector and has put in a request for funding to build a covered canal. However, the main reason for the farmers' lack of action is that most of them are marginal farmers, from the scheduled castes: they do not feel that they have the necessary political support for such a request, nor the power to negotiate with the bore owners who sell them groundwater throughout the year. Those, who might back such a request, are in fact not so much interested in the release of surface water, since they are bore owners and can use their tubewell water, and even sell it. This last example illustrates how tank water irrigation, even if it is recognised as being of some use in recharging groundwater, is fragile as soon as an obstacle upsets its working routine.

We can conclude from these examples that, in the case of the use of the upper aquifer (recharged by surface irrigation), the productivity of the aquifer (which is not homogeneous through the village) and a guaranteed surface water supply are the two main elements that go to explain the conjunctive use of groundwater and tanks or on the contrary, the competition between them. A guaranteed water supply is nevertheless a combination of environmental and social factors, as it depends in many respects on human action (supply rules and management, cleaning of canals, etc.).

In the case of an aquifer with a low yield, a guaranteed supply of surface water will be determinant for conjunctive use. When the aquifer produces a good yield, the tendency is to rely more on GW. But when there is a disruption in tank water use, the social situation around the groundwater market becomes so unbalanced that it is difficult to organise any collective action towards reusing the tank for irrigation.

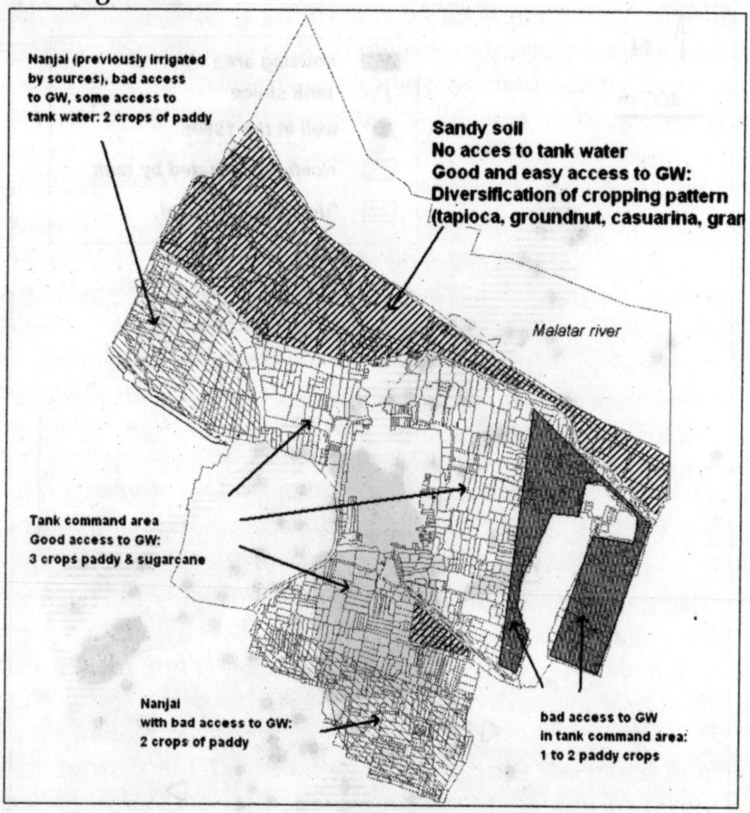

Fig. 7.7: Cropping Intensity, Type of Land and GW Access

(d) Filling the Wells with Tank Water in Order to Irrigate Punjai Land

Another rather peculiar case is that of Koratur, in the rocky area, where some farmers bring tank water to *punjai* land, *via* long pipes and wells located in *nanjai* and filled with tank water: since 1978, big farmers who have land in *nanjai* and *punjai* have been installing long underground pipes (about 600 m long) running from the well in *nanjai* to *punjai* land. Some have specifically bought land in *nanjai* in order to set up a well and irrigate their *punjai*. The well in *nanjai* is naturally filled by underground water but also by tank water diverted into it. A powerful submersible pump then drives the water up to the *punjai* land or to a second well from which the water will again be pumped. This type of investment is of interest to big farmers since *punjai* land requires less ploughing and tolerates more diversified crops than *nanjai* land. On a village scale, there is a gain of about 10 ha on the irrigated area using tank water, which nonetheless amounts to one tenth of the *nanjai* area.

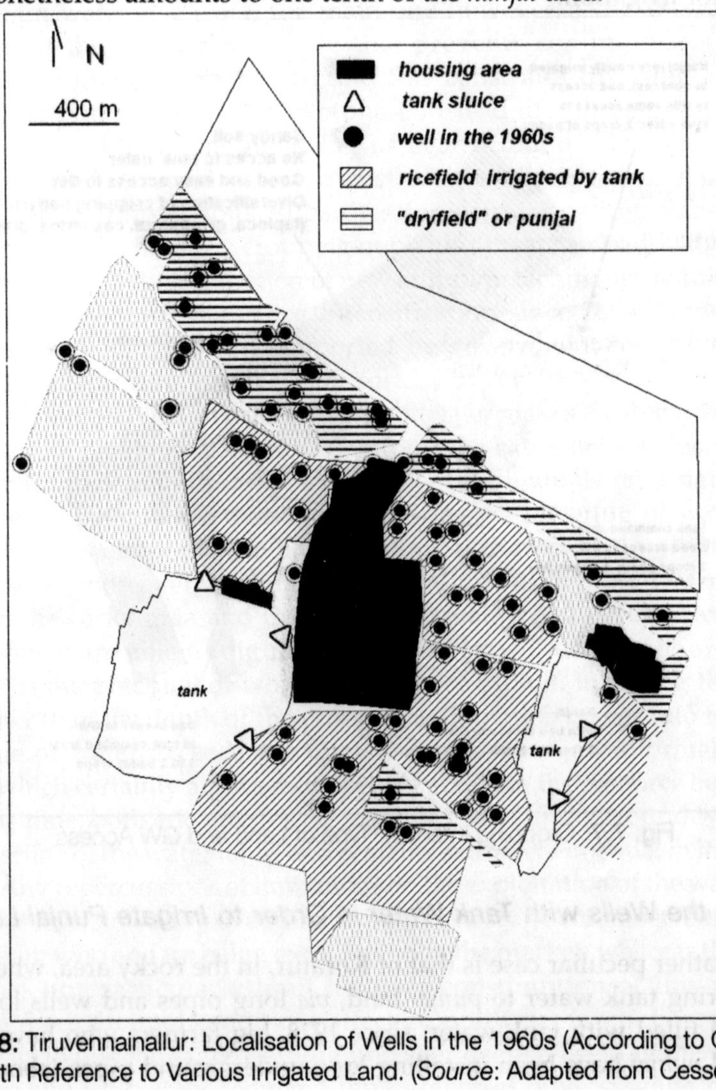

Fig. 7.8: Tiruvennainallur: Localisation of Wells in the 1960s (According to Cadastral Map), with Reference to Various Irrigated Land. (*Source*: Adapted from Cesselin, 2000).

Given the extent of tank water use on *punjai* land, people diverting tank water to *punjai* land are very interested in tank water use: these farmers have taken the opportunity to create a water user association to curry support from the District Collector in order to have any tank encroachers removed and to coordinate sluice opening times more effectively, in short to be actively involved in tank management. Unfortunately we do not know to what extent this extra cropping area—achieved, above all, through wells located near the tank—impacts the irrigation of rice fields downstream. The informal rule for water distribution is indeed to first serve fields upstream, then those downstream. Under such conditions, simultaneous conjunctive use may comprise some competitive aspects, especially if the rules for water distribution have not been redefined. It nevertheless has an impact, as identified by farmers, on the length of tank water storage and therefore on the area of the second cropping.

7.4. Conclusion

In Chapter 6 we described how the absence of any limitation on pumping and the flexibility of this irrigation technique have led to a real quest for groundwater, intensified by the social prestige acquired by owning a well or a bore and by water buyers' willingness to get rid of the bore owner's dependency by acquiring their own bore. This latter condition is certainly borne out, especially in the alluvial area, where water markets are more developed than in the rocky area. However, the example of Sorapet, where an environmental limit has been reached, has shown how, because of the high number of bore owners, GW supply problems have prompted some bore owners to change their cropping pattern on their *punjai* land (so-called "dry land" yet fully irrigated nowadays by GW). In such conditions (i.e. environmental limit, many bore owners affected by GW supply issue), some positive though limited changes, have taken place regarding water use and management. Nevertheless, the lack of a guaranteed water supply from this non-system tank and the belief that GW is not really recharged by the tank, does not lead to the reuse of tank water for paddy irrigation.

The concept of GW recharge through tanks discussed in Chapter 7, is an interesting one, since it has become very common in recent years due to government programmes regarding GW recharge infrastructures. And of course, such programmes have been welcomed by farmers for whom GW holds a primary place and who have observed a drop in its level. Nevertheless, this concept met with various reactions, especially as far as tank water use is concerned. The main reason for this diversity of reactions is the lack of information about the local and regional hydrogeological process that leads to various interpretations. The interpretation given by bore owners often takes precedence over others since, as we have seen, they are the main ones who have access to some empirical knowledge about the fluctuations in GW levels. Besides, their interpretation can affect tank management according to their vested interests, rather than those of most common people. Competition between storing water (for GW recharge) and using it for irrigation can, moreover, be heightened by new financial perspectives introduced by intensive fish farming, an activity which requires storing water for longer periods.

The interest in water (either surface—SW—or groundwater) is so diverse and sometime so contradictory, and the decision-making process so linked to social and political influence, that even if a certain hydrological relationship between SW and GW is recognised by users, it does not lead to any integrated management of these two types of resources on a village scale at least. On the contrary, individual interests dominate all others: either the use of more tank water at the expense of other farmers (KOR), or the tank being left unused or the canal blocked at the expense of water buyers (VAD and TVN) or water being stored in the tank for fishing at the expense of irrigators (MAN). Moreover, we have seen through the example of TVN that the dynamics of water management may differ from one place to another within a same village according to environmental factors (guaranteed SW supply, aquifers yield): this heterogeneity of environmental factors on various scales (village, region, etc.) is one of the difficulties in implementing integrated water management. Another fact to be taken into account is that water use also has to be understood at a hydrological scale. Since we did not adopt a quantified approach, we therefore did not estimate the water balance. Yet knowing that water use in one place affects water availability in another place—for example more SW diverted upstream of a watershed leaves less SW for downstream tanks, but may help to recharge some aquifers, or the use of water from one aquifer has some impact on the water level of this aquifer—we can conclude that any SW and GW integrated management solution must take into account: on the one hand, the influence of local water use on the regional water balance and, on the other hand, the heterogeneity of access to SW and GW on a local scale. This heterogeneity justifies why a homogeneous solution cannot be applied on a State or even regional scale. If this is the case, it must be recognised that it could not be fully equitable.

In the hypothesis whereby a solution is taken which integrates GW management for the mitigation of GW depletion (such as limiting GW use, or SW and GW integrated management), one of the requirements to render such management acceptable by farmers is to provide information about the local and regional hydrological situation. We have seen in Chapter 7 how this knowledge is far from being shared by all farmers, leading to various interpretations of the phenomena. To share information about the hydrological and hydrogeological workings of the water balance, on a local and regional scale is one of the requirements for any participatory management to "function properly". Attention has already been drawn to this requirement (Moench and Moench, 2003; Kumar, 2007). Yet with current policies emphasizing the GW recharge function and therefore adopting a supply side stance, such information is really needed in order to shift over to the demand management side, as recommended by experts (Moench, 1992; Molden, 2007).

NOTES

1. Our knowledge may be distorted as we worked more in the alluvial area than the rocky one.
2. The utilization of PVC pipes is going to be developed, because the governments of Puducherry and Tamil Nadu have been giving subsidies (since the 1990s) for the purchase of these pipes, with the intention of minimizing losses due to infiltration during the transportation of pumped water. In Puducherry, the Resource Conservation Scheme grants 75 per cent subsidies (100 per cent in some cases) with a maximum of Rs. 30,000 per farmer; in Tamil Nadu, under the Drip Irrigation Scheme it is possible to obtain subsidies (which have dropped from 90 per cent in the mid-1990s to 20 per cent since 2005) and under a centrally sponsored Integrated Scheme for Oilseeds, Pulses, Oil Palm and Maize (ISOPOM) since 2004, a pipeline can be subsidized by 50 per cent.
3. These outlets are either tubes coming out of the ground from which the water is diverted directly into the field or cement constructions called *"thotti"* in Tamil, with a certain capacity, through which water from the underground pipes transits before pouring out into the fields through a hole.
4. One may mention striking examples observed in the Cauvery Valley where farmers have organized themselves into an association to set up pumping stations along the river and to transport, over a distance of several kilometres, water pumped from beneath the riverbed and raised onto land previously rainfed, all of which is forbidden on paper.
5. Translated here from Tamil by Vanuatu Subramanian, French Institute of Pondicherry.
6. The term *ayacut* comes from *ayah* (tax) and *kite* (holding) (Subbarayalu, 2001) showing that it was originally linked to tax, not to irrigation.
7. A water diviner we met discovered his talent through a book in which he learned how to use the coconut technique.
8. Let us bear in mind that Puducherry district was the first to test such equipment (the first tubewell was installed in a textile factory), and more than 500 tubewells had already been installed according to Chatterton who visited the area in 1902 (Adicéam, 1966: 256).

9. Interestingly enough, the tank's function of recharging wells is not expressly mentioned in his earlier book on tanks (Vaidyanathan, 2001).
10. This type of explanation may also be due to the fact that farmers actually see hydrogeological engineers making field studies to verify the hydrostatic level of the region's aquifers without any results ever being made public—just as we saw them in the southern part of Puducherry district, 40 km from NLC as the crow flies.
11. T. Shah (1993: 160) was also confronted with some inconsistencies in the knowledge acquired about aquifers in Gujarat : "We inquired if the farmers had a theory of their own to explain how the aquifer of the village behaved. Several were offered but most of them turned out to be inadequate though credible."
12. TW localization refers to the data registered in the cadastral list updated by the Village Administrative Officer. Tubewell density (number of TWs per unit of area) was calculated using a GIS that measured the surface of the given areas. However, as fieldwork reveals that some tubewells have not been counted, these numbers may be underestimated (Cesselin, 2001).
13. Whereas the *punjai* especially near the river had a high density of wells (0.18 wells/acre). In this alluvial village, the existence of wells was not due to *punjai* or *nanjai*, but to the easy access the upper aquifer and to its productivity.

REFERENCES

Abé, Y. 1995. *Terres à riz en Asie. Essai de typologie*. Paris: Masson.

Adicéam, E. 1966. *La géographie de l'irrigation au Tamil Nadu*. Paris: Ecole Française d'Extrême Orient.

Cesselin, F. 2001. *Irrigation par tank et motopompes* à *Tiruvenainallur au Tamil Nadu: compétition ou complémentarité de techniques?* Mémoire du Master of Sciences DAT (Développement Agricole Tropical) du CNEARC, Montpellier.

Coste, N., and N. Ploumpidis. 2007. *Bittersweet water. Irrigation practices and modern challenges in South India.* Trilingual DVD (Eaux douces, Eaux amères-Thaneer: inippum kacappum). Coauthors: K. Anupama, O. Aubriot, Y. Gunnell, Puducherry: French Institute of Pondicherry (Publications Hors Série 5).

Government of Tamil Nadu.1984. *Particulars of revenue collection in different years.* Ramanathapuram district.

Government of Pondicherry. 1999. *Water resources and their management, Pondicherry Region (water for life, save water, save Pondicherry)*. Department of Agriculture.

Janakarajan, S. 1994. *Trading in groundwater: a source of power and accumulation.* Madras Institute of Development Studies (Working Paper No. 123).

— —. 2003. "Need to modernize the tradition. Changing role of tanks in response to scarcity and variability (draft paper)." Paper presented at the Market Development on Water and Waste Technologies Through Environmental Economics, Delhi, 2003.

Janakarajan, S., and M. Moench. 2002. *Are wells a potential threat to farmers' wellbeing? The case of deteriorating groundwater irrigation in Tamil Nadu.* Madras Institute of Development Studies (MIDS working paper no. 174).

Kumar, M. D. 2007. *Groundwater Management in India. Physical, Institutional and Policy Alternatives.* New Delhi: Sage Publications.

Lefebvre, O. 2006. Dynamiques de gestion de l'eau dans un contexte de surexploitation des eaux souterraines. La région de Pondichéry et de Villupuram au Tamil Nadu, Inde du Sud, Mémoire présenté en vue de l'obtention du Mastère Spécialisé Développement Agricole Tropical, CNEARC, Montpellier.

Moench, M. 1992. Drawing down the Buffer. Science and Politics of Ground Water Management in India. *Economic and Political Weekly* 27: A7-A13.

Moench, M., J. Burke, and Y. Moench. 2003. "Rethinking the Approach to Groundwater and Food Security," in *Water reports* 24. Edited by FAO. Rome, 51 p.: www.fao.org/DOCREP/005/Y4495E/y4495e00.htm.

Molden, D. Editor. 2007. *Water for Food, Water for Life: A Comprehensive Assessment of Water Management in Agriculture.* London and Colombo: Earthscan and International Water Management Institute.

Palanisami, K., and K. W. Easter. 2000. *Tank Irrigation in the 21st Century. What Next?* New Delhi: Discovery Publishing House.

Prasad, C. S., P. K. Basu, and A. Hall. 2005. Assessing SRI as a Process: Evidence from India. *Water Policy Research* (http://www.iwmi.org/iwmi-tata) 22:7p.

Shah, T. 1993. *Groundwater Markets and Irrigation Development. Political Economy and Practical Policy.* Bombay: Oxford University Press.

Subbarayalu, Y. 2001. *Studies in Cola History.* Chennai: Surabhi Pathippakam.

Subramanian, P. R. Editor. 1997. *Kriyavin tarkalat Tamil akarati: Tamil-Tamil—Ankilam. Dictionary of Contemporary Tamil, Tamil-Tamil-English.* Chennai: East West Books (Madras) Pvt. Ltd.

Vaidyanathan, A. Editor. 2001. *Tanks of South India.* Delhi: Centre for Science and Environment.

Vaidyanathan, A., K. Sivasubramaniyam, and S. Mariasusai. 2003. *Conditions and Characteristics of Well Irrigation under Palar Basin, Tamil Nadu.* Madras Institute of Development Studies, Working Paper No. 179.

CHAPTER

8

THE PROFITABILITY OF IRRIGATION: BEYOND ECONOMIC ANALYSIS

Marc Roesch

As soon as we engage in conversation with the farmers of South India we hear them say that agriculture is too risky, often unprofitable. They claim that it is rice which poses a problem. With this crop they feel they are losing money almost every year. The expenses involved in cultivating this crop (manure, pesticides, labour) are not covered by their income. Amortising the investment for irrigation seems impossible to them. This amortizing must come from elsewhere. The low profitability of rice is often stressed in many articles in scientific magazines, ministerial reports, as well as in the press (Balasaheb, 2008; Bhatia M.S., 2006; GOI, 2007).

However, new wells continue to be dug and rice always represents a little more than half of the land irrigated in Tamil Nadu (56 per cent in 2005-06 India-stat 2007). The next two most important crops are sugarcane (10 per cent) and coconut (9.1 per cent). The other crops represent much smaller percentages (fruits and vegetables 8.7 per cent, groundnut 6 per cent, etc.).

The question which arises is: if rice does not allow one to bring in the sums necessary for the amortizing of an investment in an irrigation system, what about the other crops? How are the drillings of the latest wells (which number continue to increase) financed? How are the loans contracted for this irrigation repaid?

Most of the farms in Tamil Nadu are very small-scale (less than 2 acres). However, it has been known for a long time that in small-scale family farming the decisions are not based solely on economics, but on a whole complex combination of social and cultural elements. To tackle the question, we have thus favoured a qualitative analysis of the farms and their economic evolution (livelihood, income, asset, indebtedness). We have traced the economic evolution of the family. We classified the farms according to their capacity to make a living (or not) from agriculture. We established a typology which enables us to identify the explanatory factors.

In the first part of this chapter we will present this typology. It confirms that small-scale farms which primarily cultivate rice cannot live from agriculture. One or more members must work outside the farm to allow the family to live. On the other hand, the farmers who cultivate sugarcane seem to have a better livelihood.

Also, in the second part we will make a comparison of the profitability of various crops to identify those which would be likely to make an investment in irrigation profitable. We will see that very few crops can really amortize this investment. The question is thus: why and how do small farmers continue to want to invest in irrigation?

In the last part we will present how the investments are financed and how they perceive the question of amortization. We will show that it is while being involved in debt that the farmer finds the required funds, and they repay by "financial juggling". The investment and its "amortization" are not considered according to the crops which the owner thinks of cultivating, but according to a whole range of outside assistance which they can mobilise, and the agricultural or non-agricultural income of the family.

We will conclude with what such a perception of profitability of agriculture and more particularly of irrigation implies.

This work is based, on the one hand, on two series of surveys carried out in 2006-07. One by Nicolas Cayon in Sorapet (Puducherry Territory), one of the villages studied in the project "Management of Water" of the IFP (*see* Box 1 and Cayon, 2006) and the other by Marion Maignan in two villages close to Mailam (Villupuram district, Tamil Nadu) (Maignan, 2007). The surveys in Sorapet were carried out among 95 farmers cultivating between 0.2 and 30 acres of land[1]. Their objective was to identify the various agricultural production systems in west Puducherry. The surveys carried out near Mailam in the villages of Aathikuppam and Needi, 40 km from Puducherry, wanted to analyse the degree of indebtedness of the farmers owning less than 5 acres. They represent more than 90 per cent of the farmers in Tamil Nadu (see Fig. 8.1). These villages were selected in collaboration with the most important microfinance NGO of west Tamil Nadu among the four other villages that it recommended to us as representative of the agricultural villages of the area where it is active.

These villages have been benefiting from the support of this NGO for nearly 10 years (SHG implementation, access to banks, training). The survey near Mailam concerned all the 156 farms from the two villages with less than 5 acres. Following this survey, 28 households were selected among the 156 in order to represent all the categories of land, crops and extra-agricultural activities present in the two villages. Twenty-seven of these families are from the Vanniar castes[2], as is 75 per cent of the population of these two villages. The objective was to identify, through qualitative talks, the diversity of family resources (land, work, and asset). The survey was about the evolution of their farms, the budget of the households and their capacity to manage their resources[3].

In addition, the more global reflection relating to the investments presented below has been based on several surveys conducted for 5 years in Tamil Nadu in about fifty villages west of Puducherry and in the northwest of Chennai within the framework of a research programme carried out by the researchers of the "Labour Finance and Social Dynamics" programme of the IFP on the impact of microfinance in a rural environment.

8.1. Fragile, Small-Scale Farming: A Typology

In Tamil Nadu, more than half of the farms had less than 1 acre (4.047 m²), and nearly 75 per cent less than two acres (0.8 ha) in 2001[4]. Because of the size and resources of these farms, mechanization is not developed very much. Ploughing and harvest are mostly done respectively with a pair of oxen or manually but can also be carried out by hired material. The farmers having more than 10 acres (4 ha) are generally mechanized.

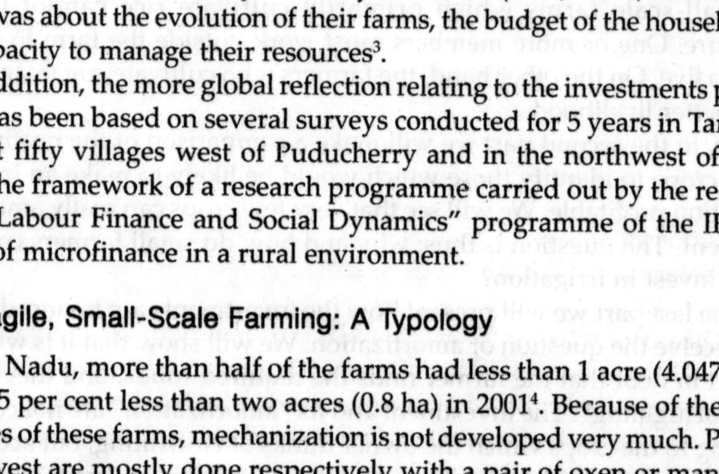

Fig. 8.1: Farm Structure in Villupuram District and all over Tamil Nadu (from NSSO, 2007)[5]

The distribution of farms in Villupuram district is similar to the farms in Tamil Nadu.

The 28 farms of the selected Aathikuppam and Needi villages are located in a rocky area where bore wells and open wells are found[6]. For each case we have collected information on the crops, families, extra-agricultural activities, on their concerns, their enterprise, their successes and their failures. We estimated their level of income, their level of debt. For each one we have noted down their "life story". For these small-scale farm holdings, we wanted to know, what were the factors which made it possible for some to live off their crop, while others were obliged to seek work outside the farm. Through these "life stories" we could identify several levels of livelihood. We included other factors in relation to agricultural and non-agricultural activities. A synthesis of certain criteria were selected to describe the characteristics of these farms, their lifestyle and management mode. These criteria are:

- Crops and areas.
- Water supply for the crops (open well, open well-*cum*-bore well, bore well, buying water).
- The off-farm activities (agriculture coolie, small shop, employment in town).
- Livelihood: for this criterion we divided the farmers into:
    - Those who manage to live "normally"[7] from their agricultural activities alone (type A).
    - Those who combine farm activities and non-farming activities. They say that they reach a satisfactory living standard (which means above the poverty level and by saving capital) only because they work outside the farm. Agricultural income is insufficient to go beyond this poverty line (Type B).
    - Those who live below the poverty line, in a precarious balance or in a phase of recapitalization. We have subdivided this group between those who continue to raise crops (type C) and those who stopped agriculture (type D).

These criteria allowed us to classify the 28 farms into 4 groups:

Type A (7 farms): Small scale (between 2-5 acres), they cultivate paddy and sugarcane, the paddy area being 1-2 acres and that of sugarcane always higher than that of rice[8]. These farms are characterised by their capacity to live only from agricultural income. The story of their life reveals that these farmers were able to accumulate capital thanks to the growing of sugarcane. From the profits of this crop they bought land or a team; one of them bought a tractor. They all have a bore well which allows them to irrigate the sugarcane all year round. Without being "rich farmers", these owners are able to meet their needs and to improve their livelihood year after year.

Type B (7 farms): Between 1 and 4 acres, these farms each have two or three crops at the same time, among them paddy, groundnut, sugarcane (maximum one acre), cotton, black gram or casuarina. The losses for a bad harvest one year are compensated by the profits of another crop the same year. These farmers have at least one family member who has a regular, external, paid activity (employment in a factory, a service, a trade...). Agriculture makes it possible to ensure basic needs (food and small domestic expenditures), and the extra-agricultural activities ensure the needs related to the social life or the children's education. It is thus the income from these paid activities which will make it possible for them to have an acceptable "living standard". They all are equipped with open wells or bore wells. Very often these farms cultivate little or no sugarcane because they do not have the appropriate land, the land is difficult to access, or the engineers of the sugar refinery refused their request for plantation for various reasons.

Type C (9 farms): Very small farms (between 0.3 and 2 acres) all cultivating paddy on a major portion of the land. The remainder is covered with a little cotton, or groundnut together with cotton, black gram, etc. These farmers do not cultivate sugarcane either because the land is not appropriate, or because they do not have the authorization from the mill (not enough land), or because they do not have access to water all year round (no bore well, no neighbour able to sell water to them). Rice is partly self-consumed. Observations on their budgetary management show that these families are in a precarious situation. Their off-farm activities are irregular (they are primarily agricultural or construction coolies) and are hardly sufficient for a living. During the good years, the income from their crops enables them to repay their crop loan and to have some income from agriculture. During the bad years, they are obliged to go into debt to repay their agricultural investments. They are always deeply involved in debt and do not manage to break the cycle of debt/repayment. Four of them buy water and say they lose money with agriculture. In spite of these losses they continue to cultivate. The other five have an open well. They are at the edge of survival. As soon as they have some income (good harvest, remunerative work through migration, on a building site) these farmers try to buy land (but land is scarce) or to rent some. They also try to improve their water supply (digging a well for those who do not have any, deepening of the existing wells, drilling in the open well).

Type D (five farms): All small-scale farms (between 0.13 and 1.8 acres) which do not irrigate any more or gave up agriculture temporarily. Two years ago, four farmers among these five cultivated paddy, the fifth cultivated groundnut together with cotton. They all have an open well. But each one was confronted with different problems. Three out of five farmers gave up agriculture because of a motor problem, a disagreement on the use of the well, or because of a well filled up by an "enemy". The two others cultivate only during the rainy season rainfed crops such as groundnut, sesame or black gram, the well being unusable (motor problem for one and well which changed owner for the other). All hope that one day they will be able to cultivate again and irrigate part of their land. Meanwhile they work as coolies.

The typology demonstrates that agricultural income by itself seldom suffices to ensure a decent standard of living. Only the farms which have at least an acre or more of sugarcane manage to bring in enough income to rise above the poverty line. Families which are at the limit of survival and enter a spiral of decapitalisation (standard C and D below) have the following characteristics:

- They have less than 2 acres of land (and less than one acre of sugarcane).
- They have irregular, insufficient, non-agricultural activities to meet the family's needs; they are generally farm labourers or construction coolie.

The farms which buy their water are particularly vulnerable (type C). But this does not mean that those which have a well or bore well make a profit from it.

The question which arises is why the growing of sugarcane is a crucial factor, what is its profitability and its position compared to the other crops, and what is the importance of the water cost in this profitability?

We will analyse below the profitability of the crops to understand this large number of farmers on the brink of survival and why they continue to invest in the drilling of wells.

8.2. Compared Profitability of the Various Crops: Concept of Profitability

"To the farmer ... the difference between total output and cash input are the important criteria in his choice of crop cultivated" (Epstein, 1962: 222).

In the Tamil language, the equivalent of "profit" is used only for trade. To employ this word during discussions with farmers would often make them smile. During our interviews, the farmers say that a crop is financially interesting if after harvest (and the sale of the produce) "there is something left for them after all has been paid". Neither the "costs" of family labour (that is considered free), the financial costs of the loans, nor "opportunity costs" of this unpaid labour are taken into account. Also not taken into account is self-consumption, (subsistence farming), nor the by-products which a farmer will not have to buy (straw, groundnut leaves, etc...).

The decision to plant such a crop, to buy water or to invest in irrigation will not be made after a traditional "cost/benefit" economic reflection. It is, above all, the condition of the land and access to water. The farmer often does not have an extensive choice of possible crops. Secondly, the crop will be evaluated according to the risks which it will bring. The investment (cost of renting land, seeds, manure, labour) is one of the greatest risk factors. The farmers have no savings and most of the time they have to borrow to invest. This loan makes them dependent on the lender, not only economically, but also socially. It is only afterwards that a more economic reasoning sets in. Then they will look at decisions made according to the criterion: "Is there something left over for me or will it cost me money?" This is a characteristic common to the so-called family farms. To make economic calculations based on a business model to decide the choice of crops and the investment in irrigation does not make much sense, under these conditions, because it does not allow one to understand the decisions and choices of the farmers [C. Gladwin (p. 45) and P. Barlett (p. 137) in Barlett, 1980].

(a) Gross Profit and Net Profit

Generally, the gross profit is distinguished from the net profit. The *Gross profit* is calculated as sales minus all costs directly related to those sales. These costs can include manufacturing expenses, raw materials, non-family labour, selling, marketing and other expenses. For the *Net profit or Net income,* is what remains after subtracting all the costs (namely, business, depreciation, interest, and taxes) from a farm's revenues. Net income is sometimes called the bottom line, also called earnings.

For family agriculture, to make the calculation of the net profit is of no interest, because it remains too theoretical, too farfetched. Indeed, family budgets and those related to the crops tightly overlap. For example, a farmer takes a loan from a co-operative bank or moneylender. To some extent, this loan is used for agricultural investments and consumption needs. How to distribute the financial expenses (considerable) between the crops and the family consumption? The farmers do not make this distinction and will not account for these financial expenses when they take stock of their land.

The calculation of the gross profit is not easy to calculate, for several reasons:

- The estimate of the produce value:
  - The farmers sell their produce throughout the year. The prices vary not only according to the time of year, but also according to the nature of the relations between producers and purchasers (social bond, financial bond, etc.). Because of these bonds, a farmer can actually decide to sell his produce at a lower rate than the minimum price applied by the Indian State[9]. It is thus impossible to have an

exact idea of it. For our calculation, we will choose the "minimum price". One part of the produce is self-consumed, should one evaluate it? At what price? In order to calculate and compare produce, self-consumption is not taken into account and the minimum price is applied to all the produce.

- The by-products (rice straw, groundnut leaves, etc.) are partly or completely used on the farm. They have a commercial value. But as they are not sold in the majority of cases, we will not take into account these products in the economic calculation.[10]
- The cost estimate: the family labour should be evaluated with its "opportunity cost"[11]. But the job market is irregular and very uncertain. It varies widely according to the villages, the season, the nature of work. In the calculation we will take as "labour cost" the price of the external labour reported by the farmer during the interviews.

The profitability will be calculated by counting only the cost of the paid labour, as the farmers told us.

(b) The Cost of Water

Before entering into details of the cost-benefit analyses, we should give a progress report on the cost of water. The cost price of water is extremely variable from one farmer to another, from one village to another. Below we present some examples of this diversity (least expensive to the most expensive):

- *Water from the tank or channel: quasi-free*. The only cost is that of the labour for the maintenance of the channels and the water pipes up to the plot of land. System maintenance generally intervenes outside of the crop season. This off-time work is not calculated by the farmers and it is separate from the peak periods. One can thus consider that this cost is quasi-null. This type of irrigation accounts for approximately 27 per cent of the land irrigated in Villupuram, district where our study is located (NSSO 2005-2006 in India-stat 2007).
- *Open well equipped with pump*: This type of well, traditional in the area, continues to be dug. It is recharged by surface water through the aquifer and the farmer has the guarantee of finding water there at the end of the rainy season. This water is normally sufficient for a second crop. In Villupuram the open wells feed 50 per cent of the irrigated surfaces (NSSO 2005-2006 in India-stat 2007).
- *Bore well*: To drill is expensive and hazardous, therefore risky. It varies from Rs. 50,000 to 200,000, even more, according to the depth and nature of the land to cross to reach the water table. A submersible pump costs

around Rs. 45,000 and its maintenance approximately Rs. 500-1,000 per year. This system accounts for 23 per cent of the irrigated land in Tamil Nadu (NSSO 2005-2006 in India-stat 2007). It is in constant progression.

- *The purchase of water*: Certain farmers buy water from owners of wells or bore wells. The rates are negotiated on a case-by-case basis, while being based on local practices. In the majority of cases in the area under investigation, water is paid in kind, the cost being 1/3 of the harvest. Sometimes the amount is fixed in advance, approximately 5-7 bags of paddy/acre. When a farmer needs additional water, at the end of the season for example, the price can be fixed per hour of use. The prices vary from Rs. 20-60 per hour according to the village.

In calculations of profitability and depreciation below we will limit ourselves to two scenarios, that of "free" water (which comes from a tank, a bore well or personal open well, we will count Rs. 500 maintenance/season) and paying water at a rate of 1/3 of the harvest.

(c) Poverty and Income Levels

The official statistics fix at Rs. 13,250 the poverty line for a family of four people in a rural environment.[12] According to our observations, a woman agricultural coolie earns approximately Rs. 35 per day for 6 work hours, and a man Rs. 50-80 for 6-8 work hours. The various cross-checking made during interviews with the coolies (agriculture and construction) reveal average remunerations of Rs. 7,500 per year for a woman and Rs. 15,000 per annum for a man, which is a total of Rs. 22,500 per annum for a family when the couple works. This figure can be regarded as a minimum so that a family can live decently. At this level of remuneration families cannot accumulate capital, they do not save. In fact, our surveys show that 90 per cent of the households which are on this level of income are involved in debt, and half are involved in debt for more than one year of income (Guérin *et al.*, 2007, Roesch *et al.*, 2006). To determine if a crop allows a family to live off its work we must compare the "profits" of the various crops with these incomes.

(d) Results of the Compared Profitability

Table 8.1 presents the different gross profits from the principal crops. One will find the details of the calculations in the Appendix.

To be closest to the way of calculating by the farmers, we considered that family labour is free. The farmer does not remunerate these people; he considers that it is not an expenditure.

The figures given below are averages calculated per acre for one year. For sugarcane, it is an average for 12 months of crops. We present the results for the first year of installation, then those of the following years. For paddy, it is the total for two seasons of crops (the majority of the farmers cultivate two seasons, if they have enough water). For groundnut, it is the result of only one season. Very few farmers cultivate three seasons of crops; the level of water is often insufficient, even null, for a third rice season (very much water intensive), in open wells as well as in bore wells. On the other hand, sugarcane and casuarina can support a rationing of water at the end of the dry season. For the casuarina (3 to 5 years of cultivation), they are the results for the whole cycle divided by 5. For this crop we did not present the results in the case of purchased water, as this case almost never presents itself.

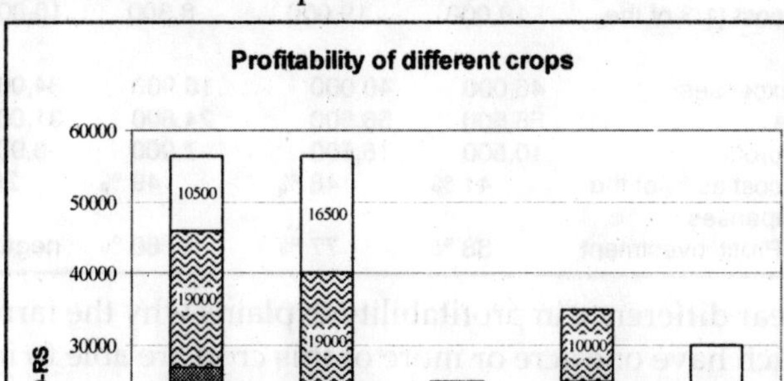

Fig. 8.2: Comparative Profitability of Different Crops (for 1 Acre of Land)

Note: We have put the "poverty line" at about Rs. 22,500 per year. To reach this income (with the condition of not having to pay for water) one needs 1 acre of sugarcane, 2 acres of groundnut and 4 acres of rice. If one pays for the water, one needs 2-2.5 acres of cane and 4 acres of groundnut.

These results of Table 8.1 show that sugarcane brings the greatest profit. It is 50 per cent higher than casuarina (which has the disadvantage of generating income only after 3 to 5 years). It is double that of groundnut while presenting less risk. However, it requires a high initial investment (Rs. 20,000-27,000) which represents the equivalent of an annual income. This need for investing is compensated by the fact that the banks, the input suppliers and the "lenders" are more open to credit requests from farmers growing sugarcane. The expenses related to the credits come as a deduction from the gross profit; we will look at this point later.

Table 8.1: Comparative Profitability Table (for 1 Acre of Land)

		Sugarcane 1st year	Sugarcane ratoon	Groundnut	Paddy (total 2 seasons)	Casuarina/ year on 3 years
When own water is used	Expenses without water cost	27,500	21,500	8,600	24,900	9,500
	Income	56,500	56.500	24,800	31,000	30,000
	Gross profit	29,000	35,000	14,200	6,100	20,500
When water is bought	Expenses without water cost	27,000	21,000	8,600	24,900	
	Water cost (1/3 of the yield)	19,000	19,000	8,300	10,000	
	Total expenses	46,000	40,000	16,900	34,000	
	Income	56,500	56,500	24,800	31,000	
	Gross profit	10,500	16,500	7,900	-3,900	
	Water cost as % of the total expenses	41 %	48 %	49 %	29 %	
	Gross Profit/investment	38 %	77 %	86 %	negative	

This very clear difference in profitability explains why the farmers of the A and B types which have one acre or more of this crop are able to increase their capital gradually and improve their livelihood.

The question of water: The profitability of the crops remains fragile. The availability of water is the principal element. If they must buy water, profitability falls from 50-64 per cent[13]. For rice the profitability becomes negative.

If water is purchased, the profitability of groundnut is greatly reduced, but the investment at the beginning is lower and the rate of profitability remains interesting (86%). In addition, it is a crop which is appropriate for the sandy areas where rice cultivation is not possible and that of sugarcane more marginal.

It can seem surprising that the farmer continues to grow rice in spite of very low profitability. Without paying for water, in a normal year the rate of profitability is 24 per cent and one would need 3 acres of rice grown for two seasons to reach an income equal ... to the poverty line. The figures presented are averages. Paddy has a very strong inter-annual variability and is particularly sensitive to insects and climate. The average output used for the calculation above are 5,500 kg/ha in the 1st paddy season, and 4,600 kg in the 2nd season. The average rice output given by the statistics is clearly lower. In the district of Villupuram they vary from 3,000-3,500 kg between 1990 and 2000 and around 2,500 kg between 2001 and 2006 (http://www.tn.gov.in/deptst/agriculture.pdf). In fact, paddy is very often a crop which loses money.

There are several reasons why the farmers continue to grow paddy:

- First and most important is the social status. A person is not really considered by his entourage as a farmer if he does not grow paddy. There is real social prestige attached to rice production, to nourish his family with one's own rice rather than someone else's or market one (Deshpande, 2007).

- The rice fields are not appropriate for some crops and all farmers are not authorised to grow sugarcane. Some farmers have no other choice than to grow rice.
- The output figures are an average. The production variations from one year to another being inportant, the farmers always hope to have a good harvest.
- Paddy can be sold after 3 months of culture, at a relatively stable price fixed by the government. It therefore represents an earlier money income than the other crops. For families which have daily difficulties finding the money necessary to live, not being obliged to wait for a late harvest (and thus to borrow while waiting) can be a good reason to prefer a crop which is harvested earlier.

On the other hand, from the results on Fig. 8.2, it is surprising that the farmers do not all plant sugarcane or casuarina. We can found some reasons if we have a look on the sugarcane production conditions.

(e) Sugarcane is a Crop under Regulation

We saw previously that sugarcane had a determining effect on the livelihood of the farms. We should examine for a moment the organisation of this production, its financing, the advantages and disadvantages for the farmers.

For this production the purchase price is theoretically fixed a year in advance; the farmer knows to whom he will sell and when. He is thus able to anticipate his income. The only uncertainty is the output, this uncertainty is weak. Parasitism is not very important; the varieties have stable outputs if the farmers follow the technical recommendations. This crop is closely monitored by the engineers of the sugar mill; they advise and control the farmers.

It is generally possible to envisage a minimum production. In the event of a serious problem (drought, flood, etc.) it is possible to obtain a postponement of the loans from one year to another if the harvest has been bad.

If the farmer is thus able to anticipate his income, his entourage (not only family, friends, but also money lenders) can also have an idea of the amount. They know his plot, his technical skills, as well as the approximate date of payment for the harvest. Sugarcane thus offers to the lenders a safe margin of repayment much larger than other crops. For the other crops, the volume of the harvest and the amount of produce are much more uncertain. By promising to repay at the time of the cane harvest, the farmer who grows cane can obtain credits more easily.

On presentation of the authorization to plant delivered by sugarcane mills, the agricultural banks offer a credit of Rs. 7,500 (€ 125) per acre of cane with a subsidised interest rate. This credit accounts for approximately 1/3 of the production costs (see below). Repayment to the bank is directly assured by the mill at the delivery of the harvest.

This crop also presents two considerable disadvantages:

- The order to cut is issued by the engineers of the mill. Now, a harvest delayed for a few weeks after maturity can lead to production losses, in weight as well as sugar content and thus to less income.
- It is necessary to wait a little more than one year after the plantation or the previous cut to be able to receive the income from his crop. A family must then be able to invest in the crop and live for one year while waiting for this income. It is, therefore, necessary to borrow (sometimes at 5 per cent per month) and to pay the interest monthly. When one does not have any other source of income, this waiting is long and expensive.

But there are other factors which are taken into account:

- The mills receive more requests for authorization to plant cane than they are able to grant. The engineers of the mill can choose their farmers. Their criteria are a combination of criteria dictated by the mill (land, availability of water, technical skills of the farmer, proximity to the mill, accessibility to the plots, etc.), and of the political and social pressures (in particular caste membership);
- The availability of water. Sugarcane must be irrigated all year, but many wells do not have water all year (particularly the open wells). The deepening of an open well or borewell is a necessity for the crop. In some cases, a cane producer will buy water at the end of the season by negotiating his contract in hours, at a fixed price or by a fraction of the harvest.

The growing of sugarcane is a powerful stimulant to dig wells or drill borewells. Especially since the farmers also see this investment a means of irrigating rice. The question for them is to find the required funds.

8.3. The Investment in a Well and the Question of Amortising

According to the Census[14], 1.2 per cent of the wells in Tamil Nadu are financed by the bank and 94 per cent on "savings". However, in all the examples of financing of a borewell or open well which we found (not only through the surveys evoked previously, but also through those done at the time of the surveys on the impact of microfinance), irrigation is financed by a combination of funds obtained from the family circle (in the form of gifts or loans with or without interest), of credits from money lenders or pawn brokers. The surveys confirm the very small proportion of financing by the banks. On the other hand, the share of "savings" is also very low. The main funding source is credit.

The cost of an open well or a bore well varies enormously according to the depth and the nature of the land. Most of the time, the prices given by the people interviewed fluctuate between Rs. 50,000-120,000. For a farmer, it is very difficult to find this amount from only one source and to find the whole amount at once. He is, therefore, obliged to "juggle with different sources" and he must "juggle" for several years in order to pay the investment.

(a) Debt Juggling

In Tamil Nadu, 90 per cent of the people are permanently indebted (India-stat, 2007; GOI, 2007; Roesch *et al.* 2006; Roesch *et al.*, 2007) and this credit can cover investment and big expenditures, (health care, ceremonies, housing), as well as consumer expenditures.

The families are permanently juggling with a multiplicity of debt sources; depending on the importance of the need they can resort to several options available at the time; none of the loans have the same interest rate; it is not uncommon for a high interest loan to be repaid by another loan at a more advantageous rate or quite simply by another loan when the creditor claims his/her due.

In order to finance their wells the farmers will combine different sources of financing when they have to pay off the investment. The banks lend if the borrower has serious guarantees (land or tractor to be pawned). This is, of course, not the case for the small farmer who has only one or two acres to present as guarantee. In the order of priorities they turn to:

- Initially to relatives to ask for a free loan,
- Then pawn the jewels, theirs or those of their relatives (between 2 and 3 per cent interest per month),
- Take a credit at the SHG *via* their wife (between 2 and 3 per cent interest per month. The possible maximum amount in a SHG is Rs. 20,000),
- Financial Company (3% per month),
- Pledge the land, and
- They turn to a money lender only as a last resort (3-5 per cent interest

per month). According to our surveys, most of the time only the private lenders are able to lend amounts higher than Rs. 20,000 (Roesch, *et al.* 2006).

Repayment is made in three ways:

- The money lender and the pawn broker claim a monthly payment of the interest and a one-time payment of the capital. This monthly interest payment is made from the wages of the husband, the wife or the contribution of the migrant children or those living in town. If the wage income is insufficient the farmer will take another loan to pay this interest.
- The payment of the monthly installment including capital and interest claimed by the financial companies and the SHG will also be taken from these same sources or in the event of insufficient resources on the day of repayment, by a new loan.
- The repayment of the capital (in the case of the money lender and the loans from relatives) or claiming pawned jewels will happen when the farmer has a major source of income: from migration, sale of sugarcane, good groundnut harvest, gift of a family member.

We present in Table 8.2 an example of financing a borewell at Rs. 70,000.

- Loan from relatives at 0 per cent: Rs. 20,000.
- Loan from a SHG or a finance company at 3 per cent per month over 10 months: Rs. 20,000.
- Loan from money lender at 5 per cent per month: Rs. 30,000.

Table 8.2: Loan Structure

	Amount of monthly installment	*Remainder to be paid at the end of 10 months*
Relatives	0	Rs. 20,000
SHG or finance companies	Rs. 2,600 to 2,060 (declining balance)	0
Money lender/pawn broker	Rs. 1,500/Rs. 900/month	Rs. 30,000
Total	Rs. 4,100 to 3,500/month	Rs. 50,000

A family able to pay a monthly installment of more than Rs. 3,500-4,000 (that is a regular income of more than Rs. 5,000 monthly) may find it beneficial to borrow as much as possible from the finance companies or the SHGs.

Those having few regular resources will have to call upon the money lender or pawn broker for a part or all of the funds to pay the interest at the level of their monthly income. In very many cases, the farmers take a loan from the SHG or from a finance company (if they can present guarantee or collateral), and they call upon the money lender to find the funds for the payment of the instalments. If one takes the example above, for a loan of Rs. 50,000 from a money lender the interest will be Rs. 2,500/month, Rs. 1,500/month for a pawn broker. These sums are closer to the income of a household whose members are agricultural coolies or temporary migrants. This allows them to wait for the sale of the harvest, outside assistance or to find another lender proposing a lower interest rate, in order to repay the capital partially or completely.

We end up with a paradox: the weaker the extra-agricultural income is, the more the farmers are obliged to call upon the most expensive lenders, those who claim only the monthly interest.

(b) Amortising in the Framework of the Family Budget

For the farmers, the loan for the well belongs to the category of "family" loans like repairing the house or financing a ceremony. It is an expenditure necessary for the livelihood of the family in order to have an income (in the case of sugarcane) as well as food safety or social status (in the case of rice). The investment will not be reasoned solely according to the "profitability" of the irrigated crop.

Above all, the income from a crop must cover direct investments, and crop expenses. From this fact, the farmer will consider that a crop is "advantageous" if, after having paid all the expenses directly related to the crop, there "will remain something", even if this "something" is lower than the effective cost of the amortising of the well. The decision to invest and make the loan payments related to the well is in the same category as the other family credits. They are done according to available resources and expected income.

This reasoning is used for the installation of a new well, but it is even more so in the case of deepening. If water is insufficient for a second crop season, the farmers seek to finance the deepening without finding out if the profit made on the second crop makes it possible to repay the investment. When the farmer can pass from a season of irrigation of rice to a year of sugarcane the investment has a good chance of being advantageous. But if the investment makes it possible to make a second rice crop (considering the gross profit of paddy close to zero on average), it is probable that it cannot be repaid with the income of the rice crop.

Therefore, the choice to dig a well does not rest on a purely economic calculation of profitability of the irrigated crops, but on a complex set of economic elements (access to credit, access to donations, agricultural and extra-agricultural income, other ongoing credits, etc.) and social (need for food safety, social status, relationship with "lender", etc.).

Conclusion

The analysis of the profitability of the crops and its link (or the absence of link) to the way of financing irrigation demonstrates a very narrow overlap of economic and social elements.

The ownership of a well clearly improves the social status of its owner. He is not dependent anymore on his neighbours for his water supply; he can cultivate his rice and other crops without being in a socially vulnerable situation. Moreover, he reduces his vulnerability by being less dependant on pluviometric risks and by increasing the range of possible crops. He can schedule the planting of his crops and diversify production. If he can cultivate sugarcane he increases his income dramatically.

Improvement of social status, reduction of vulnerability, security of production and increase in income are many elements which can lead a family to priorities the digging of a well.

But this acquisition requires a real effort and the ability to take risks on the part of the farmer. The first effort consists of mobilising all the financial resources and social capital to find the funds. Second is to ensure monthly repayments which result from this and therefore, for many farmers that means seeking additional financial resources outside the farm.

On the one hand, there is reduced vulnerability with respect to climatic risks, but on the other there is a technical (drilling is not always productive) as well as economic risk, by increasing the current debt. The great diversity in profitability of the irrigated crops has as a consequence a great diversity in the return on investments possible for a farm. An uncertainty exists regarding the possibility of using the income of the irrigated crops for the repayment of the loans taken for the well. Therefore, a farmer who invests in a well projects himself into the unknown. To the uncertainties in his capacity to repay his credits is added that of the availability of the water resource, of his capacity to manage the well (technically, economically and socially, see the examples of the farmers of the D type).

Considering the many farmers with less than 5 acres who launch themselves on this adventure, it is very likely that the advantages of such an investment are more psychological and social than economic. If the farmer only reasoned in economic terms, many would have given up cultivating rice or even agriculture. But do they really have a choice?

APPENDIX

Cost and Income Per Acre

The figures below are a rough estimate of the price and quantity of the different element of the calculation of the expenses and income. The informations were given by farmers and technicians during the two surveys conducted in Puducherry Territory (Cayon, 2006) and Villupuram district (Maigna, 2007) South India. We select the most frequently mentioned figures.

Casuarina: During the first year groundnut is cultivated between casuarina rows. Groundnut income is low, most of the interest of this production is to cover the soil, improve the fertility and provide some by-product (leaves, seeds). Due to the uncertainties of the groundnut production, we include the cost of this production in the casuarina crop expenses.

Paddy: Paddy is cultivated during two seasons. The differences in temperature, humidity and parasitic pressure between the two seasons oblige the farmer to utilise different varieties. In the example below we use the data on Chinapoony variety for the first season (*Samba*) and ADT 37 for the second season (*Sornawari*).

Sugarcane: the data concern the first year of cultivation (instalation of the crop) and the second year cultivation based on the ratoon. Sometimes the farmers are allowed to produce a third year with ratoon before planting rice. The third year expenses and income are close to second year's data.

Expenses		*Price (Rs.)*	*Quantity*	*Total*
Casuarina				
Ploughing		50	12 days (d)	600
Casuarina seeds		2,000	1 kg	2,000
Groundnuts seeds		40	25 kg	1,000
Tree nursery	Sowing	30	6 d	180
	Weeding	30	33 d	990
	Fertilizer	420	0,3 d	126
Transplanting	Men	70	14 d	980
	Women	40	14 d	560
Groundnut sowing	Women	40	10 d	400
Year 1				
	Weeding	30	97 d	2,910
Groundnut harvest		15	10 d	150
Splitting seedling		60	10 d	600
Groundnut top harvest		40	6 d	240
Canal maintenance		70	8 d	560
Fertilizer	Super	250	2 bag	500
	Potassium	230	1 bag	230
Field preparation	Hoeing	1,100	8 d	8800
	Earthing-up	40	7 d	280

(Contd.)

(Contd.)

Expenses		*Price (Rs)*	*Quantity*	*Total*
Pruning	Men	70	18 d	1,260
	Women	40	19 d	760
Year 2				
	Weeding	30	17 d	510
Fertilizer	Men	70	6 d	420
	Women	30	6 d	180
	Complex	420	3 bag	1260
	Potassium	230	1 bag	230
	DAP	490	1 bag	490
	Urea	240	2 bag	480
Year 3	Earthing-up	40	7 d	280
	(no expenses, only family work)			
Water				4,000
Total Expenses				30,976
Income	After 3 year			100,000
Groundnut				
Land preparation through external contractor		1,600	1	1,600
Labour		30	100 d	3,000
Seeds		40	24 bag	960
Fertilizer: Potassium		230	1.3 bag	299
Super		250	3 bag	750
Complex		420	1 bag	420
Insecticides: Florafix		28	1 bag	28
Rogga		23	1 bag	23
Hisulfur		20	1 bag	20
Harvest		25	60 d	1,500
Total				8,600
Water 1/3 of an average yield				8,300
Well maintenance (if no water is bought)				500
Income 35 bags x Rs. 710				24,800
Paddy				
	Samba season (July to January)			
Nursery	Ploughing	50	1	50
	Seeds	16	5 kg	80
	DAP	490	0,2 bag	98
	Complex	420	0,12 bag	50,4
Land Preparation				
	Ploughing	400	3 d	1,200

(Contd.)

(*Contd.*)

Expenses		*Price*	*Quantity*	*Total*
	Organic manure	400	3 cart	1,200
	Complex	420	2 bag	840
	Canal maintenance	90	3 d	270
	Zinc sulphate	100	0,5 bag	50
	Fertilizer	90	2 bag	180
	Levelling	90	6 d	540
Transplanting	Women	50	20 d	1,000
Weeding	Women	50	30 d	1,500
Fertilizer	Urea	240	3 bag	720
	Potassium	230	2 bag	460
	Neemcake	330	1 bag	330
	Amonochlorate	340	1 bag	340
	Men	90	2 d	180
Insecticide	Monochromotapas	480	0,25 bag	120
	Poltin powder	160	0,5 bag	80
	Men	10	8 d	80
Harvest		155	20 d	3,100
	Total			24,924.3
Water cost	1/3 yield for both season			10,000
Income		1st season	35 bags x 560	19,600
		2nd season	30 bags x 380	11,400
	Total			31,000

Sugarcane	*Ist crop cycle*	*Ratoon crop*
Land preparation through external contractor	2,600	1000
Labour	5,400	5,200
Seeds, fertilizer, insecticides	9,500	5,300
Harvest	10,000	10,000
Total	27,500	21,500
Water	19,000	19,000
Income 50 t x Rs. 1.130	56,00	56,00

NOTES

1. 1 acre = 4.047 m^2
2. Vanniar are considered middle class (MBC) and mostly involved in agriculture (Kanthimathi *et al.*, 2008).
3. Since the surveys were made in 2006, the prices relate to this period. However, the price of fertilizers and pesticides has increased. The selling price of rice has not changed in spite of the increase in the price of rice in the towns and global market. As for the cost of labour, it is also constantly increasing: the development of the labour offer on construction sites in towns or on roads has strongly affected the volume of agricultural labour

available; the programme set up by the Indian Central Government to guarantee rural employment, the National Rural Guarantee Scheme, offers daily wages above those practised in the villages and thus raises the salaries. (Under the National Rural Guarantee Scheme, one member from each of India's 60 million rural households is guaranteed 100 days of work each year). It seems that the current trend is a deterioration of the profitability levels presented.

4. Department of Economics and Statistics, 2007.
5. NSSO: http://www.indiastat.com and Statistical Handbook.
6. From a survey conducted among 156 farms drawn randomly in two villages of Tindivanam district (Tamil Nadu, India), 28 farms of 0.2-5 acres were chosen randomly. Each farm was subjected to an in-depth study (social and economic), of its life cycle, management mode, etc.
7. By "living normally" we mean feeding oneself, housing, proper management of loans, timely loan repayment, sending one's children to school, meeting social expenses without falling into over-indebtedness or living in visible poverty.
8. One of the farms cultivates cash crops (vegetables and flowers).
9. Every year, the Federal Government publishes a minimum price for government agency to buy rice from farmers. This rice is meant for the ration card distribution programme for poor populations. At the same time, it helps to avoid a fall in prices (Dorin and Landy, 2002).
10. When a farmer makes the "decision to cultivate", the existence of by-products will influence his choice. But this choice will have a social and cultural more than an economic origin; it is more related to the wish to have his own by-products rather than be obliged to beg for them or buy them. It is not really the result of an economical consideration which would compare the production income, by-product included.
11. The cost of passing up the next best choice when making a decision. For example, if a family member works on the family field what will be the "loss of earnings" because they cannot go to work as an agricultural coolie on a neighbouring farm?
12. POVERTY LINE definition, as per Government of India is at Rs. 10 per day, per person, approx. Rs. 276/month = Rs. 13,250/month for a family of four MINIMUM WAGES, as per Government of India, for all the States, about Rs. 40-60 per day per person, average about Rs. 50 per day. For 25 days per month Rs. 1,250/month = Rs. 15,000/year.
13. Profitability loss = (a – b)/a; gross profit without buying water = a; gross profit including purchased water = b.
14. Census of Minor Irrigation of the Minister of Water Resources (2001) http://wrmin.nic.in/mi3census/preface.htm.

REFERENCES

Balasaheb. 2008. "Agricultural Indebtedness: Crisis and Revival". *Economic and Political Weekly* 43 (5): 47-52.

Barlett, P. 1980. *Agricultural Decision Making: Anthropological Contributions to Rural Development.* New York: Academic Press. 378 p.

Bhatia, M.S. 2006. "Sustainability and trends in profitability in Indian Agriculture". *Agricultural Economics Research Review* 19 (Conference Print).

Cayon, N. 2006 *Quell avenir pour la revolution Verte en Inde du Sud? Etude de cas à Sorapet* Mémoire ESAT CNEARC Montpellier France, 179 p.

Department of Economics and Statistics. 2007 *The Statistical Handbook 2007, table 4.9,* http://www.tn.gov.in/deptst/agriculture.pdf

Deshpande, R.S., Shah, K. 2007. "Agrarian distress and agricultural labour". *Indian Journal of Labour Economics* 50 (2) April-June 2007: 256-272.

Dorin, B., Landy, F. 2002. *Agriculture et alimentation de l'Inde: les vertes années (1947-2001).* Paris: INRA Editions. 253 p.

Epstein, T.S. 1962. *Economic Development and Social Change in South India.* Manchester: Manchester University Press.

Government of India. 2007. *Report of the Expert Group on Agricultural Indebtedness.* Banking Division. Department of Economic Affairs. Ministry of Finance Government of India. July 2007. 109 p.

Guérin, I.; Venkatasubramanian; Roesch, M. 2007. "Ne nous libérez pas ! L'ambiguïté du principe de l'avance sur salaire à partir de l'exemple des briqueteries en Inde du Sud", *Revue Autrepart* 43: 121-133.

Kanthimathi, M. Vijaya and A. Ramesh J. Genet. 2008. "Genetic study of Dravidian castes of Tamil Nadu", *S. Journal of Genetics* 87 (2): 175–179.

Maignan, M. 2007. *Endettement des agriculteurs et et marché financier: le cas de la région de Mailam Tamil Nadu,* stage de 2° année d'ingénieur agronome, ENSAM Montpellier, France 38 p.

India-stat. 2007. *Indiastat.com,* http://www. indiastat.com.

Roesch, M., Helies, O. 2007. "La microfinance: outil de gestion du risque ou de mise en danger par sur-endettement?" *Autrepart* 44: 119-140.

Roesch, M., Guerin, I., Kumar, S., Thanuja, M., Venkatasubramanian, Ponnarasu. 2006. *Impact assessment of the Peblisa project in Tamil Nadu. Report* for the International Labour Organisation. Pondicherry: French Institute of Pondicherry.

PART 3

Tank Irrigation Management Institutions

CHAPTER 9

TANK IRRIGATION MANAGEMENT INSTITUTIONS

P.I. Prabhakar

Tank management institutions cannot be understood outside the village social context, as they are part of the village institutions. Indeed, in India, the village social system is a compendium of institutional arrangements in the village mainly regarding religious affairs (festival organisation, temple maintenance); common resource management (including water irrigation), and intra and inter-castes problems. The interactions and individual relationships between local institutions of a social, religious, economic and political nature underlie the functioning of the village social system. An individual has different types of role to play. The inter-relation between individuals belonging to different institutions is one of the factors that explain their role, their social status and the relationship of power. As caste organisation has been and is still a fundamental feature of the social structure of Indian villages, the caste composition in a village is essential to understand the social relations and the power relations. Let us recall the dichotomy between castes and "outcastes" (called Scheduled Caste or SC under the Constitutional category, and previously called untouchables), either in the social hierarchy or in the spatial organisation of the village. In Tamil Nadu, the residential locality of outcastes is called a Colony, a *"cherry"* in Tamil and more familiarly a *"pet"* in Puducherry, and is situated on the outskirts of any village. The outcastes are still a minority or even absent from the decision-making realm. We will see that local power is indeed based on caste hierarchy and kin group competition, and that the same type of power structure exists in all village institutions.

Rivalry between caste groups, between kin groups and lineages within a caste, prevails in village societies, generating factions. The intrusion of political parties in this rivalry has re-ascertained and is accentuating the differences and disparities that used to exist. Indeed, a political affiliation is generally based on social networks, that means social factions, so that rival political parties reflect rival social groups. And the rivalry is manifested in various forms such as disputes and conflicts between families, kin groups and lineages, augmented by the respective members' political affiliation claim. D.Mosse (1999) in mentioning the Tamil society says, "social dominance does not remain unchallenged by new institutions which also provide the means to advance externally defined development objectives such as equity, democracy and social justice.... the point is that these social changes intersect with local caste conflicts or factional affiliation and with strategies to challenge as well as retain caste power."

We will see how this may also be true of formal irrigation institutions that provide ample space to exercise and sustain the differences that prevail in villages. The last decades of the 20th century witnessed a pervasive policy consensus headed by the World Bank to transfer State management of natural resources by and large to the user community. This resulted in the nationwide Participatory Irrigation Management (PIM) approach in all irrigation-related activities (Hooja *et al.*, 2002). Thus through projects and legislation, the Indian Federal States initiated the formation of water user associations (WUA) at village level. The States drew up stereotyped guidelines for forming these new associations. In the case of Andhra Pradesh (Mollinga, 2001; Reddy, 2005), these newly formed associations have been taken over by the local elites, while farmers find it difficult to adapt to the associations, as this does not suit their already existent system and thus leads to disinterest among irrigators. As a result, the duly established WUAs remain non-functional and no further action has been taken to shape processes within the WUAs, apart from organisation of the elections. The aim and the sustainability of such institutions are then questionable.

The central focus of this chapter is both to prove that caste composition and land ownership by castes explain the management of tank irrigation, and to describe the links between the new irrigation institutions and other institutions, including traditional irrigation institutions. This will be done through a study in three neighbouring villages in Tamil Nadu State (Koratur, Saravanapakam and Tiruvennainallur) and three neighbouring villages in Puducherry Territory (P.S. Palayam, Vadanur and Sorapet).

The first two parts of this chapter deal with informal institutions involved in the tank management of five tanks, namely Koratur (KOR), Saravanapakam (SRVP), Tiruvennainallur (TVN), Vadanur (VAD) and Sorapet (SOR). VAD tank is, however, shared by two villages, VAD and PS Palayam (PSP), which are administratively under PSP village *panchayat*. The third part of this chapter deals with the social dynamics of the new WUAs.

9.1. Informal Local Institutions Involved in Tank Management: Common Features

According to Adam Pain (2004), institutions represent the rules for governing or regulating the behaviour of individuals. For Gary Bouma (1988), they are sets of norms which apply across a variety of specific organizations. Institutions are "complexes of norms and behaviours that persist over time by serving some collectively valued purposes" (Uphoff, 1986, quoted in Norman Messer and Philip Townsley, 2003). Institutions may denote the rules that individuals use to govern specific relationships with one another. In her book on crafting institutions for self-governing irrigation systems, Ostrom (1992:19) used the term "institution" as the set of rules actually used (the working rules or rules-in-use) by a set of individuals to organise repetitive activities whose outcome affects those individuals and potentially affects others. Many scholars use the terms "institution" and "organisation" synonymously. However, here we use these terms at two different levels: an institution is a set of norms that guides the functioning of one or many organisations within it. Institution has thus a broader meaning than organisation. Organisations may acquire special status and may become institutions if they satisfy people's needs and meet their expectations over time. To mention only some of the activities that fall under informal caste and village organisations: kinship roles in life-cycle ceremonies, system for monitoring punishments and resolving disputes, labour exchange, etc. The few informal village organisations are the village temple management committee, the irrigation committee, the labour management committee, etc. These organisations run according to the religious, cultural, economic and political aspects of village institutions. All those that do not have hard-core written rules but are governed by beliefs and unwritten rules transmitted by oral tradition might be termed informal organisations and institutions. Some of the formal institutions are the elected village panchayat, the water user association (WUA), the parent teachers' association, the local political party association, and the registered youth and welfare association, all of which are either registered or have some legal status. These formal institutions have written guidelines, or even rules enacted by-laws for the running of the organisations within them.

(a) Traditional Institutions

Tanks are ancient irrigation structures. The informal institutions that manage them and the common resources they contain (water, grass, fish, trees, etc.) can be referred to as traditional as they refer to cumulated knowledge based on principles accepted by the community and deeply rooted in the past (or at least believed to be). Tank management institutions have long been in existence, and have adapted to changes over time to various social, cultural, economic and political situations. These informal institutions have become extinct in some villages, while they exist in other villages in one form or another. It is, therefore, difficult to offer a general description of the situation, and to say if a specific function has become completely extinct.

Another point we would like to emphasize is the fact that the terminology may interfere with the understanding of the organisation since it refers to traditional terms, even if the function or the structure of the institution have undergone significant changes. We will give examples where an institution referred to by a traditional term is not necessarily traditional.

The informal institutions that were/are managing the irrigation tanks in the study villages are: kambakaran, vaikal maniyam, natamai, periya natamai, seetukarar, kambatham, miras. Some formal administrative institutions may also be involved, such as *thoti, maniyakarar*.

Kambakaran: This term literally means "stick man". Men holding this position are always from the servicing castes, the SC, belonging—in our area—to the *Paraiyar* caste (one of the servicing castes in the Hindu caste system) which is at the lowest level of the caste hierarchy. The persons in-charge of *kambakaran* functions change every year on a rota basis, according to the lineage division of the caste. In some villages, they are the heads of the lineages and the title is passed down through generations, while in others they are chosen among the lineage members. The number of *kambakarans* in a village varies from village to village. For example, in Koratur and Tiruvennainallur there are four, but only two in Saravanapakam.

With regard to irrigation tanks, they guard the trees in the tank and, open and close the sluices to irrigate the command area. As for system tanks, when water is released into the system, the *kambakaran* regulates the head sluice through which the water reaches the tank. If there are any blockages in the upper reaches of the supply canal, these will be removed by *kambakarans*. These are the duties to be undertaken by the *kambakaran* under the orders of the *natamai* (leaders) and *periya natamai* (big leader) in the village.

The *kambakarans* also undertake the task of cutting firewood for wedding ceremonies for the other castes in the village, of digging burial pits or of bringing wood for cremating the dead, of sometimes cremating and burying the dead (under a different title—*vettiyan*) and of undertaking other menial jobs during the temple festival. When he has no such duties to perform in the village, the *kambakaran* is also a wage labourer or a farmer (in recent times some of these caste elites have been able to own land).

For the services rendered by the *kambakarans* regarding the irrigation tank, farmers in the tank's command area give them a share of production (bunches of harvested paddy crop or money for the sugarcane crops). For other services in the village, *kambakarans* are paid in cash or kind.

Thoti: The *thoti* carries out menial jobs in the village. He is a village servant, associated with the revenue administration, as he is assistant to the Village Administrative Officer (VAO) or *maniyakarar*. Interestingly enough, in the villages studied in Puducherry, *thoti* was synonymous with *kambakaran* and they performed the functions described above. Thus, in Vadanur, when the *maniyakarar* managed the tank affairs (until the mid-1980s), the two *thoties* would follow his orders and follow the supply canal to check the flow and ensure that water arrive in the tank. In Sorapet the planks of the surplus weir (*kalingal*) would be operated by the thoti under the instructions given by the *natamai* (village leader), whenever the tank was full. They would also maintain other common property belonging to the village, such as temple land and trees along the road. This shift in responsibilities from the *kambakaran* to *thoti* may find its roots in the formal associations the French Government set up, with farmers using the money collected for these associations to pay the *thoti* (see Box 8).

Natamai and kothukarar. Natamai are the heads of those castes making up the majority in the village. As Vanniars and SC are numerically predominant in the study area, in each village there are invariably *natamai* for these two castes. Yet the *natamai* of high castes are the informal leaders in the village. The numerously significant caste is divided into lineages (*kothu*), and the *natamai* are generally the heads of these lineages. In PSP, such heads are called *kothukarar*. There are four from the Vanniar caste, though the term natamai is not used. In Saravanapakam there are seven *kothus* from the Vanniar caste, but here the heads of the *kothus* are called *natamai*.

The function of the *natamai* and *kothukarar* is to manage and take decisions in various domains: on the one hand, on religious, kinship and social issues for their respective caste, and on the other hand, for overall village common resources. As regards the religious domain, they organise temple feasts and festivals by collecting *vari* (tax) from the families of their respective lineages. They have an honorary leading role to play in the life-cycle ceremonies of the families in their respective lineages. They also resolve disputes about familial and civil issues within the caste. The management of village common property, such as tanks, trees, fish, grass, threshing floors falls to the *natamai*.

The *natamai* of the landowning castes are actively involved in village common affairs, whereas the SC *natamai* play a role within their residential locality. They do not have a larger stake in the decisions regarding village affairs, except in Koratur, where they have participated in village meetings. This can be attributed to their relatively long-standing status of landowner in this village. In general, the SC *natamai* take orders from the other *natamai* about the services to be carried out by the SCs in the village during feasts, at times of death, at weddings (*see kambararan*), which means that a hierarchy exists among *natamai*.

Box 8: Formal irrigation associations in the past: *caisses communes* and *syndicats agricoles* during the French era

Overview from Girod (1938: 85-102) by O. Aubriot

Formal associations were set up in Pondicherry by the French government according to a decree signed in 1859. They brought together rice field owners from one village or from several villages who had a common interest, mainly collective irrigation canals. The aim of these institutions was initially to maintain and to improve these canals. Another decree passed at a later date introduced new functions such as the protection of rivers and the maintenance of waterways used for agricultural activities. The associations were given French names, *caisse commune* and *syndicat agricole*, the latter being an institution created in 1911 which offered the group of landowners the possibility of managing common projects, of applying for a loan and of collecting unlimited contributions from members, unlike the *caisse commune* which imposed a maximum contribution per unit of land. *Syndicats* took over from a few *caisses communes*: there were 15 *syndicats agricoles* and 65 *caisses communes* in Pondicherry in 1938, with the *syndicats* being set up in wealthier areas. A *caisse commune* committee was made up of four landowners with an administrative officer —an "irrigation supervisor"— as president, showing the importance of the administration in such institutions. Girod (p. 91) proposed grouping together several *caisses communes*, as they were very often organised on a village basis, some having 3 to 10 members for only 3 to 6 ha. We can conclude that the existence of such a small organisation proves the importance of the village as a social unit.

Revenue came from contributions made by members and from the sale of products from tanks such as fish and grass. From 1911 onwards, additional revenue came from the tenancy of trees along tank embankment and canals. According to the legal text, this money was earmarked for the maintenance of irrigation canals (tanks and supply channels being the colonial government's responsibility). However, the associations used it to pay those distributing water to rice fields, the "water distributors", who were one per distribution canal. Their salaries represented nearly 80 per cent of members' contributions and 23 per cent of the associations' budget. The *caisse commune* budget earmarked for canal maintenance was thus reduced and not even used (the institutions "didn't even have the idea of carrying out any work" so Girod (p.92) complained), whereas the *syndicats* had greater revenue and maintained their canals. Nevertheless, any neglect was also due to poor administration (p. 93): the control of the institutions was the responsibility of the Revenue Department that had neither the finances nor qualified staff to carry out a verification of the works; and the

(Contd.)

Box 8 (Contd.)

Public Works Department, in charge of other hydraulic projects, was not aware of the people's needs.

Three points of comparison can be made with traditional irrigation institutions in Tamil Nadu. First of all, water distributors came under the orders of the irrigation supervisor, an administrative officer—even if water distribution was not an association's official function! This might be due to the second point; the fact that the water distributor was paid, in cash, by the association: as president of the association, the administrative officer had to supervise water distributors, who amounted to about fifteen per officer. Though not paid in kind—contrary to the landowners' practice today in some Tamil Nadu *tiis*—water distributors were paid indirectly by landowners, through the institution. These two first characteristics (water distributors reporting to the administration and their salary being paid by the formal institution) may explain why these water distributors were called "*thotti*" in Puducherry case studies, a term referring to an administrative function. The third point of comparison is related to the sale or the tenancy of natural resources from the tank: it was already a means of setting up a common fund, but this was not used for religious purposes as is currently the case in informal institutions in Tamil Nadu.

Compared to formal associations today, the main differences are first and foremost that today there is a strong incentive to involve institutions in tank management and maintenance, whereas in the past the French colonial government did not demand their involvement; and secondly water distributors, when still in existence in Tamil Nadu, are managed by *natamai* and paid in kind by all landowners, not by the WUA.

Girod, P. 1938. *L'agriculture et l'hydraulique agricole dans l'établissement de Pondichéry*. Pondichéry: Imprimerie moderne.

Periya natamai ("big natamai"): Where this title exists, the man belongs to the dominant caste, which is the Naidu caste in the villages studied (case of Saravanapakam and Koratur). There are very few Naidus, but they are the biggest landowners in the villages. The *periya natamai* are the decision-making authority on temple issues and village affairs. When the other *natamais* of the village are not able to solve certain problems, these difficult cases are taken to the *periya natamai*, who by consulting other *natamais* and elders of the village resolve them. Inter-caste disputes and issues are settled by the *periya natamai*. In the absence of the *periya natamai*, the *natamais* of the village collectively administer affairs. After discussions with the other *natamais*, the big natamai pass guidelines and take decisions on the use of the village's public money (*ur podu kasu*—see below for details). In the case of Saravanapakam, until 1996 the *periya natamai* was also president of the village *panchayat*: informal and formal leadership was in the hands of the same person.

Maniyakarar or Village Administrative Officer (VAO): This post is an administrative one, under the State Revenue Department. In Vadanur, a Brahmin who was also a big landlord in the village, held this post until the 1980s. In addition to the village administration, he also looked into managing the Vadanur tank. With the assistance of the *thoti*, he kept an eye on the supply of water in the canal leading to the tank, and also supervised the sharing of tank resources (fish and water) between the two villages and among farmers in the command area. The current VAO adheres to the official duties assigned to him. Moreover, he is not a native of the village, thus he does not take part in issues related to the irrigation tank.

Vaikal maniyam: It literally means "canal accountant". This institution was only found in Tiruvennainallur amongst the villages studied. The main tank in Tiruvennainallur has three sluices, and each sluice (See Fig. 7.5) and its corresponding canal has a representative called *vaikal maniyam*. The *vaikal maniyam* has specific responsibilities, such as irrigation canal maintenance, regulation of water distribution and sluice maintenance. They also have to organise the cleaning of the distribution canal once a year. In order to manage and maintain the canals, the *vaikal maniyam* has to use his abilities to the full and organise farmers in the command area to clean and repair the canals. This contribution can be through physical work or in cash. The *vaikal maniyams* are usually elites, respectable persons who command the respect of the other farmers in the command area of the specific distribution canal. The *vaikal maniyam* is considered a prestigious position in the village of Tiruvennainallur, where the post is not necessarily passed down from one generation to the next.

Villagers of Tiruvennainallur feel that the *vaikal maniyam* system in the past (until the creation of the WUA in 2004) was good and proper. All villagers respected the charisma and the attitude of a *vaikal maniayam* to serve for the common good of the village. They had no vested interest. Villagers accepted their decisions and suggestions. The *vaikal maniyam* summons the farmers in the command area of certain sluices, by ordering the *kambakaran* to make an announcement around the village by playing the drum and shouting out the message. The farmers dutifully assemble and clean the main distribution canal. Individual farmers clean the other branch canal that passes through their land. One person per plot of land in the command area was the rule for cleaning the distribution canal. As regards the activities in the tank such as desilting and bund strengthening, one person from each household in the village had to come forward to carry out the task.

The traditional village institutions described above illustrate their roles and responsibility in irrigation tank management. However, if different vernacular terms are used to define some institutions, their functions may be almost similar (such as *thotti*, *kambakaran*, or even *nirkhati*). On the contrary, the same institution's name may cover various functions (the *kambakaran* may be encouraged or not to water the fields, by closing the sluice everyday or only once in the season, etc...).

(b) Informal Common Fund

The *ur podu kasu* (village common fund) is an informal account generated through temple *vari* (literally means tax) from the village households and through auctions of produce from the village trees, tank fishes, thorn trees and grass. The village periya natamai manages this informal account, or in villages with no *periya natamai*, one of the *natamais* of the village takes responsibility for it and manages the money. This practice is ongoing in all the villages studied.

Generally speaking, the *ur podu kasu* is used for temple festivals, village feasts, to host dignitaries who visit the village for common purposes. As there are usually separate temples for the *ur* (caste residential area) and for the Colony, separate amounts are given to each for the temple festivals. Regular disputes break out regarding the money earmarked, because it is not proportional to the money generated on a yearly basis and on the whole, the situation is detrimental to the Colony people. In all the villages studied, the SC *natamai* have to negotiate with other village *natamai* to obtain a share from the *ur podu kasu* in order to organise their temple festivals.

Vadanur tank is common to the villages of Vadanur and PS Palayam, but two *ur podu kasu* are generated, one for each village. A physical demarcation in the tank divides the tank into two: for the trees on the tank bed and on the bund, and the trees on either of the sides belonging to the respective villages. The respective trees are under the control of the *natamais* of Vadanur and the *kothukarar* of PS Palayam. Each village auctions its trees and the revenue generated is saved and managed by the respective *natamais* and *kothukarar*. Regarding the fish, the revenue generated from its auction is shared in a proportion of 2:1 between PS Palayam and Vadanur. This unequal share is justified by the fact there is one share for each village, while the third share covers the costs of PS Palayam people's work of bringing water from the canal to the tank.

Managing and maintaining irrigation tanks requires monetary support. The means of generating funds are practically the same in all the villages studied. The appropriation and management of funds is where social dynamics is most vibrant. The traditional institutions and the roles played by individuals are interlinked with the effective use of the acceptable and sustainable funds generated.

(c) Water Specific Activities

In collective irrigation systems, irrigation management implies specific activities related to water use (acquisition, allocation, distribution, drainage), to the infrastructure (design, construction, running, maintenance), and to the organisation (conflict resolution, communication, mobilisation, decision process) (Uphoff *et al.*, 1985). As the infrastructure is State property, the activity related to its maintenance is the responsibility of PWD in the villages studied (there is no *Panchayat* tank in our sample of 6 villages). However, activities related to water use are organised by villagers.

Until March 2004 (date of the creation of the WUA) in Saravanapakkam, the *natamais* were responsible for matters concerning the irrigation tank. They visited the PWD officials at Tirukoilur and Villupuram to make requests for the release of water from the Sathanur dam and in Rhagavan channel. They attended regular meetings held at the Tirukoilur *anicut* (starting point of the Rhagavan channel). The *natamais* would monitor the water flow into the head sluice of the tank from Rhagavan channel. When the flow dwindled due to blockage of this channel, the *natamais* organized farmers and the *Kambakarans* to clear the blockage. They also made arrangements to clean the distribution canal within the command area. To undertake these activities, and especially to pay workers for cleaning the canals, the *natamai* took money from the big *natamai,* from the *ur podu kasu*. When funds were not available, the *natamais* would organise the farmers in the command area and ask them to undertake the work at no cost. Here we can see the use of the *ur podu pasu* for water specific activities.

In Koratur, the *natamais* gave orders to the *kambakarans* to remain posted at the head sluice on Rhagavan channel when water was released, then to close or open the three sluices of the tank. The *kambakarans* in KOR were also the SC *natamai,* taking orders from the other *natamais*. KOR *kambakarans* were involved in allocating water to the fields, which was not the case of those in Saravanapakam.

In Vadanur during the 1950s, there were three *natamais* from the *Ur* and one from the Colony who had responsibility for the maintenance and management of the irrigation tank. For example, since the water stored in the tank was even used for drinking, there was a strict ban on anybody defecating inside the tank. They also checked that the planks were inserted in the surplus weir of the tank during the Tamil month of *Puratasi* (which begins in mid-August) in order to store water. At beginning of January, they also checked that some planks were taken off to facilitate the flow of water to the tanks below. The elders of the village recollect that there was regular rain and that at the time the tank would fill up regularly with water directly from the Pambai River through the Konamoonji canal, which has since been in disrepair. There was enough water in the tank for two cropping seasons of paddy, the only crop that was cultivated in the command area of the tank—whereas nowadays the tank is rarely filled. Moreover, they also remember that villagers abided by the rules and regulations then.

In Tiruvennainallur during the late 1960s, when the distribution canals in the command area of tank required urgent repairs, the *vaikal maniyam*—who were the big landlords from the caste which was dominant in the village in the past—, would undertake the work immediately using their own *padi azhu*[1] (labourer). Generally speaking, the cleaning of the distribution canal was carried out regularly every year during the month of *Aid* (July-August). Villagers recollect that in many instances the *vaikal maniyam* himself was also engaged in physical labour. The other responsibilities that used to fall to the *vaikal maniyam* were to control the unnecessary use of water from the tank through the sluice; to respect a strict timing in the allocation of water and closing of the sluices; and to give permission to open the sluices. The *vaikal maniyam*, along with two or three people from the command area, would monitor the supply of water and settle disputes. The *vaikal maniyam* would also have regular contact with PWD government officials, by attending meetings and holding discussions with them.

The traditional system of managing and maintaining irrigation tanks in the villages was, in retrospect, considered effective, though a section of the village remained excluded—a criterion that was not considered before. The land owners in villages were few, and the number of farmers in the command area was also limited when compared to today. Hence there were very little confrontations and difference of opinion in the organisation of the management and maintenance by the persons responsible. The roles were well defined, orders were passed by the persons responsible from higher castes, such as the *natamais*, vaikal maniyams, *kothukarar* and *maniyakarar* and the tasks were undertaken by the *thotis* and *kambakarans* according to the rules. There were also farmers in the command area and villagers who had a role to play in times of need with regard to irrigation tanks; for example, in Vadanur tank, when water flows through the system canal, the *kothukarar* (lineage head) organizes villagers to go into the canal and clear what is obstructing the flow of water, which they call 'bringing water'. This practice is said to exist for many years. In Tiruvennainallur, in the past, villagers were organised by the *vaikal maniyam* to clear the supply canal and farmers in the command area to clear the main distribution canal.

Irrigation tanks were not used for irrigation alone. There were multiple uses of tanks, since the source of water in the tanks supported the overall local ecosystem (see Chap. 11). Fishing in the irrigation tank was a tradition. For example, in the Vadanur tank, all villagers and even people from other villages would be allowed to catch fish over a period of about ten days during the months of *Chithirai* (April-May) through *Adi* (July-August) depending on the water level in the tank. During these months, when the water receded below the level of the sluices, implying that the water in the tank could not be used for irrigation, fishing was the only activity that would take place in the tank. The *maniakarar* would send notice through the *thoti*, who would make the announcement using his drum to stop fishing in the village. Then the traditional in-land fishing community from other villages would be called upon to catch the remaining fish that would be shared amongst predetermined shareholders, one of the shares going to the fishing community. The dynamics of the multiple uses of tanks is detailed in another chapter (Chap. 11). The part that follows documents the impact of other uses of irrigation tank on the institutions managing it.

9.2. Development of Informal Institutions

Compared to what has been described above, in our case studies we can find some differences either in the rules or in the make-up of the institution members: is it due to an evolution of the organisation in response to the society evolution or to a specific evolution of each site or to a specificity of the case study that was already existing and that endures? However, let us present a system that was even existing before the system described in the previous part, and that helps in understanding the relation between the water sharing system and the common resource sharing.

(a) The Ancient System of Managing Common Resources: Its Link to Water Sharing

The *pass vari* system—which means water distribution tax—worked in the three study villages of Tamil Nadu (Saravanapakam, Koratur and Tiruvennainallur) until the early 1960s[2]. Under this system there were different taxes for landholders in the tank command area depending on the quantity of land they own, and the allocation of tank water was in accordance to the tax paid. This practice of water allocation was strictly adhered only during water scarce periods in Saravanapakkam, but systematically in Koratur (the tax payers were called the *seetu karar*). In Saravanapakkam, the highest taxpayer would irrigate his land first and then the others will irrigate accordingly to the tax paid. The *natamai* with a major share (because of his bigger landholding) took the responsibility of the activities regarding the irrigation tank. In Koratur, the *kambakaran's* service was utilized to distribute water to fields: depending on the water availability in the tank and on the requirement of farmers, with the advice of the *natamai*, he decided the quantity to put to each field. There were

9 *seetukarar* (tax payer): the (SC) Colony had 2½ *seetu* (share of the tax), the Naidus had 1½, the Vanniars had 4 and the neighbouring village of Kuchipalayam had 1 *seetu* of share. The fact that the SCs held a considerable share shows that they have been important land holders in this village for quite a long time.

It seems, as we didn't find this term in the literature, that share allocation based on *pass vari* became void during early 1960s due to changes in the state policy, as far as the farmers could explain to us. By the time the *pass vari* system ceased to exist, quantified share of water allocation was not made, even at scarce periods, and then every irrigator started to take water from the tank whenever he wanted in a haphazard manner.

At the time of the *pass vari* system, those who have right on water, the *seetu karar*, had also rights over the fishes in accordance to the taxes paid, except in Tiruvennainallur (see below) where the fishermen community had the right. In Saravanapakkam and Koratur, the fishes caught from the tank would be divided amongst the shareholders, that means among landowners. One of the *kambakaran* of Koratur recollects that during the *pass vari* system the SCs' share of fishes and grass used to be promptly given to them by the high caste *natamai*. The *seetu karar* of SC could sell these fishes to whom ever—either to the villagers or to anybody outside as well—and the revenue from this activity was utilized to build temple in the Colony and to organize temple feasts. In Tiruvennainallur, the access rights to fishes has been hold (and is still now) by the fishermen community (Nattar) who had the right to catch the fishes all through the year. Once a year, under the notion of *"yeri kolla vittutanga"* (the tank is left open) villagers could go and catch fishes let remained by fishermen. It is interesting to note that the practice regarding fishing here had been the reverse to that of Vadanur, where first the tank was open to everyone and then fishermen—called from other places, thus outsiders to the village—came to do the catch.

However, the other common resources like trees and grass in the tank, or trees on the bund did not have any significance with the *pass vari*: there were either auctioned and money generated was utilized for common purpose of the village or let to be freely accessed by the villagers. In some villages (Tiruvenainallur for example) the grass, used to thatch house roofs, was open[3] to all villagers.

When the *pass vari* system became in disuse (in the 1960s) all the resources of the tank, i.e. fishes, grass in the tank bed, trees in the tank bed and on the bund were auctioned and managed by the *natamais* as described before. For example in Koratur, the two goddesses temples in the caste village (*ur*) were built many years ago utilising the funds generated through the auctioning of the trees and grasses in the tank.

So in the case of Saravanapakkam and Koratur, the access to fish was that of landowners according to the landholding. Then it evolved to an auction system that means a monetarised system where the economic value of the resources is used to feed a common fund managed by the *natamai*. In Tiruvenainallur the grass and in Vadanur the fishes were common resources in open access for villagers, that have also evolved to an auction system.

(b) A Collective Benefit from the Grazing on Private Lands: Case of the Duck Auction

A very interesting innovation took place in Tiruvenainallur, inspired by the auction system of the common resources. In the year 1976 a "duck auction" (*vathu kuthagai*) was introduced. The bidder who won the auction could let his ducks graze in the command area after the harvest of paddy for three seasons in a year. And the money collected by the *natamai* was used for the cleaning of the distribution canals in the command area. Before the auction was introduced, the duck owner used to pay each individual farmer to let his ducks graze in their harvested land. And each irrigator had to contribute (through physical labour or money) in order to carry out the cleaning the distribution canal.

This innovation is surprising for two main reasons. First of all, the auction is not related to a common resource, but is associated to private lands. The common aspect of this auction refers to the geographical area, especially the tank command area where paddy is cultivated (and then its stubble and felt down grain grazed). The common aspect refers then to the area irrigated by the canal that brings water. This canal is not a common property as such, as it is believed to be the property of the landowners, those who indeed can give the rights to the duck owners to graze their land. Secondly, the villagers were able to come to a consensus for a collective purpose (canal maintenance) even if private holdings were involved. They found a way to pay a service (workers cleaning the canals) without paying out a *paisa*. And ever since this duck auction fund came to exist, the command area farmers' contribution of physical labour or money disappeared. The workers being mostly landless people, this system reminds the difference of work done by landless and landed people: physical and manual one for the first ones, more managerial work for the second ones. It reflects also how the physical work is denied, and culturally considered as inferior, as it is given to outcaste people, low in the social hierarchy.

This duck auction was started by a Reddiyar and then it was taken up by an elderly Chettiar, who was a prominent personality in the village, an affluent person as well as a politically influent one. The duck auction fund was kept by this Chettiar. Initially the auction amount was less than ten thousand rupees. Then the amount increased and in 1996 a joint bank account was created, not on the name of the Chettiar but of his son and of another influential Udayar from the village, who is also one of the Udayar *natamai*. The shift in the money responsibility to these two persons is explained by their affiliation to the same political party and their influential status in the village.

Every year during the month of January a meeting regarding the duck auction would be held, bringing together the people involved in the tank and irrigation affairs (*natamai, vaikal manyiam, kambakaran*), the elders of the village and 40 to 100 farmers and villagers. The review of the duck auction fund would take place, and during the meeting the performance of the *vaikal maniyams* would also be reviewed. These latest had indeed an important role to play in the cleaning of the distribution canals: they were in charge of managing the work and since the creation of the duck auction of paying workers. However, since 2001, due to a factional strife among the Udayar, the duck auction has not been held because some farmer did not want that the *vaikal manyam* or the *natamai*, from the other faction, gives the authorization of grazing on his own land. As a consequence, the duck auction has not taken place, the money in the bank account remains unused and one of the canals—having a long part that crosses the market—is in a very bad shape. For the two other canals, farmers have been cleaning the part of the canal bordering their respective field. However, for the canal crossing the market area, neither the old system of all the farmers participating in the cleaning nor a collective fund coming from the grazing is in place. Moreover, some building destruction managed by the road department (as the building encroached the road) has blocked the canal and no one wants to struggle for this: the responsibility is that of the road department, the rich and influential farmers have tubewells and can even sell their water, so that the situation of no water coming from the tank is not really problematic for them and last, but not the least factor, the canal brings water to the SCs fields, most of them being not equipped with borewells.

The scenario here gives us a good picture how a collective initiative few decades ago has dwindled down to no initiatives regarding the irrigation canals that supply water to fields. Those who took the collective initiatives then are no more; they had a concern to support the marginal farmers who in many cases are SCs who do not have the monetary strength to drill bores and use ground water. However, now with the increase in number of individual borewells among other caste farmers that could be classified as small and even large farmers and that are in many aspects better positioned politically and socially than most of SCs farmers, these tubewell owners do not have the will to revive the collective initiatives. They have a vested interest over the inability of the marginal farmers and are capitalizing on the water that the SCs and other marginal farmers require for irrigating their fields from the tubewell that wealthier farmers own.

(c) Encroachment as a "Shared" Common Land

As many authors have already expressed (Vaidyanathan, Janakarajan, Palanisami), encroachment of supply channels and tank beds is one of the characteristics of the tank use decline as well as the traditional institution decline, since 1960s. As quantified in the first part of the present book, many tanks of the alluvial area and of Puducherry hinterland suffer from encroachment. The process has started a few decades ago, and some tanks like Vadanur tank or Sorapet, have been totally encroached and unused for irrigation for nearly thirty years. The reason given by farmers for the decline of the tank irrigation use and then for the use of the space for cultivation, is the lack of water supply and the failure of rain. Even, the government had allotted, during the late 1960s, land inside the tank of Vadanur to a person who also drilled a bore inside the tank. Then many others started to encroach. The traditional village institutions were not able to control this encroachment.

Due to the TRPP (Tank Rehabilitation Project Pondicherry), the encroachers in Vadanur and Sorapet were removed by the tank association (TA) in 2006 and 2007 respectively, the eviction of the encroachment being one of the first actions of the TAs in this project. In most cases encroachers are from the neighbouring villages, cultivating in the tank bed. The social dynamics that exists in encroachment of tanks is interesting, as it points out the nexus between the landless and the landed, the power of the tubewell owner, the plight of the landless SC, the inability of the authoritative personalities in addressing the issue of encroachment, etc. When the irrigation tank was encroached, not all the landless were able to encroach due to the landowners who also have encroached owing to the proximity of their borewells in their own lands that could be used to irrigate encroached land; there were also encroachers from the neighbouring village situated in the periphery of the foreshore area of the tank. Latter there was also the practice of 'selling' the encroached land to others (Prabhakar, 2008). The encroachment is notably due to a decline in the strength of the village institutions that could not oblige the villagers to respect the rules, the elite themselves not following them. The irony is that these village institutions intervened to resolve the dispute that took place between encroachers for the access to the encroached land: landless people complained to the *natamais* and *kothukarar* about the inequality of area of land encroached inside the irrigation tank by landless and landholders. The village institution divided and allotted land inside the irrigation tank for those who complained. However, during the period of encroachment, the physical structures of the irrigation tank (sluices and surplus weir) were deliberately destroyed for the water not to be stored during rainy seasons, and thus the encroachers to continue their cultivation inside the tank. The encroachment also deprived the grazing land for those families that were only dependent for their livelihood on their livestock.

Again here the losers were those who did not have access to borewells as they had to leave the land or forcefully 'sell' the land to persons who had access to borewell water.

(d) Evolution of the Society: Informal Institutions in Transition

The society is evolving, so that the institutions themselves have also undergone changes. Sorapet is a good example to show how the name given to an institution can remain unchanged even if the institution has completely changed. There, since the *natamai* did not have any heir, the institution did not continue as such, rather it had been transformed into a socio-geographical institution with "two *natamais* for each street" since the early 1980s. Hence nowadays there are ten *natamais* for the five streets of the whole village (colony included). The institution evolved then from one *natamai* to a group of *natamai* that means from an organisation with only one person taking decision to a group of person representative of the castes, or at least of the immediate neighbourhood.

The income generated from the auctions in the tank is divided among the ten *natamais*. The money will be pooled in from the natamais when there are activities undertaken for the village in common, such as the temple feast, drinking water tap repairs, etc. A stipulated amount would also be received by the natamais of the SC Colony who will utilise it for the SC temple feasts. It is not a fixed and predetermined share but an amount that the SC *natamai* has to ask for.

In Vadanur, a big landlord who had an administrative function (*maniyakarar*) managed also the tank. After the shift of his office from the VAD to PSP during late 1970s one big farmer from the Pillai caste living in PSP who owned a majority of the land in the command area of the tank undertook the responsibility of the tank. After his demise, the lineage leading persons from the Vanniar caste as well as Reddiars from PSP who are also big landlords having good relationship with each other took the responsibility over the tank. Among the Vanniar, the leading persons are from one lineage (*kothu*), named *Kumalamuttar*, that had led the vanniar caste in PSP for many generations and draw respect from the other three Vanniar lineages who did not hold land to the same extent than the *Kumalamuttar's* one, that in turn had much less land than the big Reddiar landlords.

Thus, in Vadanur tank, the management of tank shifted from one village (Vadanur) to the other (PSP) and in Sorapet, from one person to a group of elites. Even if the tank water has not been used for irrigation for the last 30 years, the local institution (*natamai* in case of Sorapet and Vadanur, *kothukarar* in case of PS Palayam) has been still involved in the tank resources management like fishes and trees sharing. Moreover in Vadanur, it is in charge of "bringing water to the tank" (*yerikku thani konduvaruvadu*), for the fish cultivation. VAD tank being at the tail end of the Pambai irrigation system, the water supply is not automatically provided. Thus, when the water is released in the system, a big team of villagers from both the villages are organised by the leading *kuthukarars* of PSP to go in the supply canal to the upper reaches and clear the blocks made by the villagers of the upstream tanks. The team of villagers negotiates with villagers of the upstream tanks and some incidences of violence are reported. This activity is said to be an old practice.

In Tiruvennainallur, until 1962 a Reddiar had a command over the village due to his large land holding and his relationship with the officialdom of the State. He controlled all affairs of the village including the irrigation tank, and he would give directions to the *vaikal maniyam*. But after his death, his sons sold most of the lands, which were bought by people from other castes in the village. The locus of power and authority that was held by the Reddiar in the village transferred to big land holding Chettiar and new land owning Udayar and Vanniar. The Udayar *natamais* slowly took over.

In 1992, instead of an elderly Vanniar, the villagers selected two young farmers as *vaikal maniyam* for the command area of the southern sluice: one was a Vanniar from the head reach and the other an Udayar from the tail end of the distribution canal. The choice to change the number was suggested by the newly selected *vaikal maniyam*, as the command area is large. Regarding the caste choice, it has been driven by the activeness of the young farmers who had a record of working for the common cause of the village. Through this case, we see again an evolution of the institution, with more persons involved and from more various castes than before. An interesting point is that the number of *vaikal maniyam* is justified and legitimated through an explanation referring to the four directions (one *vaikal maniyam* for each direction), as if it was a "traditional" justification of the organisation, a traditional way of thinking the organisation, whereas this number is quite recent.

The case of Saravanapakam presents the opportunity to show the process of the decline of the traditional aspects of irrigation institutions. For example, the command area farmers open the tank sluices to irrigate their crops without the help of the *kambakaran*. One of the *natamai* from the village expresses that the *kambakarans'* service is utilized for checking the blockages in the supply canal of the tank and for the village menial activities during the deaths, wedding and festival times, but not for the irrigation activities. The contention of the *kambakaran* is that today they have only the responsibility to open, regulate and manage the head sluice of the tank on the Rhagavan channel at times of water supply but they are no longer involved either in operating the tank sluice nor water distribution unlike the earlier days when, they were responsible also for the distribution of water from the sluices to the fields. The *kambakarans* of Koratur (a neighbouring village) acknowledge that the *kambakarans* of

Saravanapakkam are still active as they are able to collect their share from the command area production, and this share is spontaneously given by farmers to the *kambakaran* without any demand of the latest. The issue in Saravanapakkam is that the *natamai* would like to exclude the *kambakaran* from the affairs of the tank and the struggle of the *kambakaran* is to reinstate their role in the tank affairs.

(e) A Hierarchical Management of Irrigation: From a Centralized Organisation to a More Shared Power of Decision

The responsibilities pertaining to the management of irrigation tanks are operated at two levels in the organizational structure of Traditional Irrigation Institution (*tii*) in village society: one is at a supervisory level and the other one is menial in nature (Janakarajan, 1993).

From our study villages it is understood that the *natamais* had the role and responsibility of maintenance and management of tank related affairs. In the absence of *big natamai* the other *natamais* operated the *ur podu kasu* which was mainly utilized for common purpose of the village such as temple festival, reception of representatives of the administration and also the repairs of irrigation structures. The classification of the organizational structure of *tii* in the study villages reveals four levels (see Table 9.1). The *big natamai,* who is only one, from dominant caste (in big land holding and power) hold the first level of deciding authority and discretion of funds, but he did not exist in all villages. The *natamais* of the more numerous high castes in the village are at the second level and undertake monitoring of the physical works of the tank. At the third level are the *kambakaran,* who do under the command of the higher castes *natamais* regarding the village affairs, including the tank. The tasks carried out by *kambakaran* are menial in nature and they take orders from the high castes *natamai* and the *big natamai* (and not from the SC's *Natamai*). In one village only there was a specific function (*vaikal maniyam*) for the management of distribution canals, the regulation of sluices and the visit to PWD officials, activities that are undertaken by the *natamai* in other villages. This adds an intermediate level. At the fourth level are the *ayacutdars* or the command area farmers. They all irrigate their own fields. In some places they have also to undertake activities like cleaning the distribution canal or bringing water to the tank.

Janakarajan (1993) who studied *tii* in 15 study villages in Palar *anicut* system, found that only seven were effective, that means where the maintenance work of inlet channel, main distribution and field channels has been done periodically in the past 5 years. In this Palar area, the institution name are not exactly the same (for example the specific irrigation functionaries are called *kavaimaniyam*), and there is a well established water allocation called *murai* that does not exist in our area of study.

The historical transect outlined from the 1960s for the *tii* in the study villages shows significant changes. The first level that was prevalent in all the villages disappeared in VAD, PSP, SOR and TVN, but is still existing in KOR and SVP. In the absence of the first level, the *natamai* in the second level collectively carry out the role of the first level. There is also a rhetoric that is often heard nowadays in the villages amongst a section of people, mostly youngsters, few elders and sometimes the *natamai* themselves, that stipulates the position of *natamai* is "*talaiku talai natamai*" (head to head is *natamai*). It means everyone considers himself as *natamai,* and it reflects that nowadays the *natamais* are not esteemed as they were in the old days. This can be explained by two main facts. The presence of popular political parties in the villages has created a sense of diffusion of power centers that were once with the *natamais*. The other fact is that many new land owners have come into existence. As land owning is a sign of power and better socio-economical status, the access to land holding by castes who did not have access before, is also an element of democratization of the land holding power. Nevertheless, the influence of *natamais* does exist regarding most of the decisions on temple affairs, life cycle ceremonies and to some extent irrigation tank affairs.

Tables 9.1 and 9.2: The four levels of organizational structure of functionaries in village social institutions pertaining to the affairs of the irrigation tanks in the study villages, during the 1960s and in 2006

Table 9.1

During the 1960s		Tamil Nadu			Puduchery		
Villages	Traditional terminologies of functionaries	KOR	SVP	TVN	PSP	VAD	SOR
1st level: Deciding	*Big natamai*	1	1				1
authority & financial	*Kambatham#*			1	1		
Discretion	*Maniyakarar*					1	
2nd level: Decision making, working at moderate level, monitoring	*Natamai*	3	4		4	3	4
	Kothukarar			4			
	Vaikal maniyam			4			
3rd level: Executing the decisions made, menial work	*Kambakaran*	2*	2	3 then 4			
	Thotti				2	2	3
4th level: Contributing money, sometimes also carrying out physical works	*Ayacutdar*	√ *	√ *	√ *	√ *	√ *	√ *

 # -That was the title of the Reddiar who owned very big quantum of land in the village.
** -Irrigating fields*

Table 9.2

In 2006		*Tamil Nadu*			*Puducherry*		
Villages	**Traditional terminologies of functionaries**	*KOR*	*SVP*	*TVN*	*PSP*	*VAD*	*SOR*
First level	***Big natamai***	1	1			1x	
	Kambatham#			Nil	Nil		Nil
	Maniyakarar						
Second level	***Natamai***	√	√	√		√	10
	Kothukarar				√		
	Vaikal maniyam			4			
Third Level	***Kambakaran***	√	√	4			
	Thotti				√	√	√
Fourth level	***Ayacutdar***	*	*	*	√!	√!	!

*√–Active role, *–Irrigating fields, ! – tank not used for irrigation, x–Not from the community*

Change with the previous period

(f) Conclusion

The power in the villages was consolidated on the basis of the quantum of land holding and the dominant nature of the caste. In our case studies, the biggest land holder held the authority and power. The dependency relation was prevalent in the village between the big land lords and the other small, marginal farmers and landless. Due to the mobility of labour forces to neighbouring urban centres, to the development of roads and transportation facilities, there was a considerable rise in the economic status of those who were depending on the big land holders in the village. Then, dominant land holders could not draw the same respect that they used to when others were mostly depending on them. There was also considerable change in land holding pattern in the villages, with many marginal and small farmers mushrooming and the wide disparity in the quantum of land holding between big farmers and others reducing. All these factors are some of the reasons explaining the dilution of power in the villages. Consequently to this transition in the social structure, elites and dominant individuals who were reigning in the traditional institutions in the villages had to share their power with the emerging landed elites.

However, the cultural practices of inequality in a village have been maintained in order to retain the big landholders' power and authority embossed by the caste values. Thus, even if SCs have now access to land, they are not involved in the 1st or 2nd level of organization we described above. The cultural remains of discriminative practices that existed in the caste hierarchy under the symbolic notion of pollution are also found in the present day social settings with less rigour and symbolic attributes. Nevertheless, in one of the study villages, Koratur, where the SCs have been landholders for a long time as their important claim over the *miras*—hereditary rights over the shares based on the land holding—proves it (they hold 2½ shares out of the nine shares in the tank command area), they are involved in the decision-making process related to common resources management. As D.Ludden (1989: 87) mentions, the gentry of share holding families, caste and community had transformed not before nineteenth century. In the study villages the landholding pattern has come under change during the middle of twentieth century. The proportionate rights that were held by the shareholding families over few resources of the tank were first transformed into rights for the entire village and then transformed into a right given to one person or a group of persons against an amount of money, the financial benefit being used for collective purpose, especially for temples. It is a transfer of rights using the economical value of the resources and transforming a collective use of the common resources to an economical use of these resources and a collective use of the financial benefit. And with the advent of the statutory local bodies like the village *panchayat* and Water User Association (WUA) these shares of resources are again under dispute, which is the highlight of the current phase of transition.

9.3. Formal Associations and the Participation in Question

Though local informal irrigation institutions exist in villages, development projects have defined formal users associations. Consequently, in Tamil Nadu many irrigation rehabilitation projects and programmes have been undertaken. The first of their kind was a special minor irrigation programme in 1951, followed and paralleled by many others (Palanisami, 2000: 47-52). In the 1990s, Participatory Irrigation Management (PIM) was implemented in India through a "Big Bang" approach, since a uniform structure and uniform working practices are prescribed for all water users associations (WUAs). Though theoretically it is possible for legislation to provide a flexible bottom-up approach (Hooja, 2004), this had become void in the case of the Indian States. In Tamil Nadu, there had also been many attempts to introduce irrigation legislation through acts and amendments (Janakarajan and Rajagopal, 2002). But it is only since the year 2000 that the TN Farmers Management Irrigation System (FMIS) Act has been enacted. This is one of the outcomes of the World Bank's Tamil Nadu Water Resource Consolidation Project (TN-WRCP), implemented by the Public

Works Department (PWD) from 1995 to 2004, during which Farmers' Councils where formed in order to facilitate farmers' involvement in the management of irrigation systems. With the TN-FMIS Act, river basin boards have been created and the Farmers' Councils rearranged into WUAs. About 1,566 WUAs were created during the 2004 elections that were held in 20 districts (TN-WRCP Implementation Completion Report, 2005). A follow up project to the 'Irrigated Agriculture Modernization and Water Resources Management Project', an integrated management of water resources in all 17 River Basins in the State, on 63 sub-basins[4] is currently in progress.

In Puducherry, some formal associations (see Box 8) were actually set up during the French period and the system was in place until 1964, 10 years after Independence. From 1999 to 2004, the European Commission (EC) funded the Tank Rehabilitation Project of Puducherry (TRPP), contributing to 81 per cent of its budget and providing consultancy services to assist with implementing the project, while 6 per cent of the budget came from contributions from local communities. The Puducherry PWD implemented this project with the help of the Agriculture Department and local NGOs involved to mobilize the community and form Tank Associations (TAs). Since the Puducherry government (GoP) did not heed the few recommendations and suggestions made by the technical team which undertook various interim studies, the EC withdrew its support. These recommendations concerned extremely delicate aspects, such as the transfer of the irrigation tank property rights to the TAs. GoP decided to fund the project and extended it for a further three years. In 2007, the Government appealed to the World Bank and other funding agencies to support the project, but has not succeeded in obtaining any funds.

(a) Main Differences between WUA and TA Characteristics

WUAs are formal associations created under the FMIS Act in Tamil Nadu, whereas Tank Associations here refer to the associations created during the TRPP for all the tanks in Puducherry (2 will be presented). Since our three case studies in Tamil Nadu belong to the WRCP project, the WUAs here refer to those implemented through this project. However, the main differences between WUA and TA are:

- TA is a registered body unlike WUA which is a statutory body: that implies that the WUA has a legal status governed directly by a legislative act, whereas the TA has a status of civil society governed by the Societies Act.
- NGOs have been entrusted with the assignment of mobilising local communities and forming tank associations. They have the role of social facilitator among TA members, and between PWD and the TA. A

Community Organizer (CO), recruited by the NGO, is therefore assigned to 2 to 3 TAs. By contrast, in the WUAs studied, PWD officials have organised the elections themselves and have direct contact with WUA members.

- To a large extent, the WUA reaffirms the role of the already existing power-centred village institution with the added possibility of political party. The WUA limits its members to only one category, the *ayacutdars* that is farmers who have land within the tank command area. On the contrary, TAs provide opportunities for an egalitarian setup or they at least theoretically give voice to those who are not used to airing their opinion. Indeed, TA members are the villagers, with representatives on the Executive Committee (EC) according to a fixed proportion of various social categories: *Ayacut* farmers (60%), agricultural labourers/ landless (30%), other groups such as washer (wo)men, shepherds, fisher (wo)men, etc. (10%). In the group of *ayacut* farmers, different categories such as marginal, small and big farmers, and those with and without a well have to be represented. Women should also make up at least one third of the EC.
- Under the guidelines of the TN-FMIS Act, 2000, PWD demarcated the territorial constituencies for the Water Users Associations, each covering around 500 ha. This demarcation was made on the basis of the extent of the command area without taking into account any social and cultural factors. Thus, in the Tirukoilur *anicut* system (where the study villages are situated) WUAs are organised in such a manner that some associations manage just one tank while others manage up to six. Conversely, the TRPP created TAs according to social criteria such as having to share the same tank or tanks belonging to the same village, as seen in the two cases studied in Puducherry: Vadanur tank is shared by PSP and Vadanur villages, whereas the Sorapet Tank association has to manage the three tanks in the village.
- One of the first activities of the TAs under TRPP, encouraged and supported by the government, was the eviction of encroachers, whereas in Tamil Nadu, the creation of the WUA by itself was the main objective of PWD.

Another distinction between the TAs and WUAs studied is that the 2 tanks in Puducherry have not been used for the last 30 years whereas the 3 tanks in Tamil Nadu are still in use, but this is due to the specificity of our case studies and it cannot be generalised.

Under the TRPP guidelines, a list of 16 objectives has been given to the TAs. These objectives are very broad and include various common interests within the village, such as women's empowerment, the village's social development, etc. and are not only restricted to tank or irrigation related affairs, whereas WUAs governed by the bylaws of the TN-FMIS Act only have objectives related to technical and structural aspects of the irrigation tank alone, such as following the rules related to the division of the command area into territorial constituencies, the election process, etc.

(b) Puducherry Tank Associations: Their Social Constraints

With the advent of TAs, village institutions that were managing the irrigation tanks had to compromise with the prerequisite composition of a TA, i.e. inclusion of women, landless people and other socially excluded persons who must take part in decision making. Let us have a look at an issue that came to light when TAs were being set up as well as how two tank associations function, that of Vadanur and Sorapet, and report on how villagers perceive them.

TA Formation and Encroachers' Opposition

At first, the TAs for VAD and SOR tanks had difficulties in becoming properly functional, due in particular to the eviction of encroachers. In SOR, a few village elders wilfully opted not to associate with the TA owing to the pressure put on evicting encroachers, while Office Bearers (President, Vice-President, Secretary, Joint Secretaries) had to be identified by the NGO as persons who had some interest in their tank. On the contrary, in VAD, the leading *kothukarars* of PSP and the *natamais* of VAD on learning the inception of the Tank Rehabilitation Project of Pondicherry (TRPP) in 1999 voluntarily introduced themselves to the project management unit and proposed to implement the project in their villages. In these two examples, the interest of villagers in the project was not the same, and the purpose of the project (with the eviction of encroachers as its first priority) has been an important factor influencing the choice of the Office Bearers and the creation of the association.

Thus, in Sorapet, many big farmers who were cultivating land inside the tank prompted few factions in the village to sabotage the set up of the TA. The lack of interest among village people obliged the Community Organizer to concentrate on the formalities of the TA creation and to focus on financing work in the tanks with money from the project, rather than on the quality of membership. There are 23 members on the Executive Committee to manage the three village tanks: the *Periya eri* (big tank) with a command area of 100 ha, the *Pudu eri* (new tank) with 10 ha of command area, and Aripankulam a storage tank. None of these tanks had been used for irrigation for the last 30 years (farmers in the command area had been using underground water, either from their own tubewells or by buying water from tubewell owners). Among the five Office Bearers, only two were village *natamais*. The fact that members were defined by an outsider and that the other *natamais* did not want to be part of the TA shows that a TA is an *ad hoc* association formed according to exogenous criteria, with no appropriation by the villagers. That might explain that few influential people are involved and that members have very little legitimacy from the villagers' point of view.

One of the SOR TA issues has been the way the *pudu eri* encroachment eviction was managed, since most of the encroachers were Scheduled Caste (SC) families. The only sluice in that tank was repaired and the dead storage area deepened. Yet the SC encroachers opposed de-silting activities in the tank and are still expecting some compensation. The lack of interest on the part of the government in resolving this issue does not help the proper functioning of the TA, which is thus perceived as having a poor status. Owing to this issue, the TA Office Bearers are now suggesting that a separate TA be set up for the pudu yeri and that the present TA could function more effectively by managing and maintaining only the other two tanks. This conflict about encroachment eviction comes as an addition to the ones that regularly oppose SCs and other village castes. This highlights the fact that the village is not necessarily the best social unit for a TA when several tanks are involved, since social opposition between castes may be very strong and jeopardise any eventual consensus necessary for solving problems.

In Vadanur, even though the Office Bearers did organised themselves properly, the process of setting up the VAD TA was also said to present difficulties, both for the NGO and for the influential persons of PSP and VAD who had an interest in the irrigation tank. It took one year to set it up (registered in March 2001). The main opposition came from encroachers who were cultivating land inside the tank, this issue being moreover taken up rigorously by leaders from the *saniathanmodu* lineage that opposes the leaders from the traditionally powerful lineage (the *kumalamuttarmodu*), who were later on elected as the TA Office Bearers (OB). In fact, the NGO representative was manhandled and four of the people involved in the tank management were tied up inside a temple in Vadanur. The Executive Committee members admitted that this task had turned into a very bitter experience as they had encountered the wrath of many villagers cultivating land inside the tank. The social status and the political links of the influential persons of PSP, as well as their approach in dealing with the agitation, have been vital factors in their nomination as members of the TA. The treasurer of the TA is indeed a wealthy Reddiar of PSP, an active member of a political party (Congress), owning the biggest rice mill in the region—he is called the 'auditor' by villagers—and he has good connections with the political leaders of Puducherry territory. He is an influential person and has undertaken many activities for the common interest of PSP, like providing rice to landless people at times of natural disasters, setting up health camps in the village and providing food to school children.

The TA president is also the village *panchayat* president (for the Tamil Nadu part of the village) and has been in this post for nearly fifteen years. The secretary has served as president of the agricultural co-operative society in the village. The vice-president a Chettiar from VAD is a big landowner and one of the ten *natamais* in VAD. All are active members or local leaders of political parties: ADMK for the president, PMC (a regional party) for the secretary and the vice-president, and Congress for the Joint Secretary. Their influential status in their respective parties and the alliance of these parties with the ruling party government have a positive impact in facilitating their activities at the local level and hence in stifling adverse actions.

Out of the 25 executive members of Vadanur TA, fourteen are from PSP, of which the three main leading members of the TA (president, secretary and treasurer). Regarding caste composition from PSP, four members are Reddiars, nine are Vanniars (mostly belonging to the *kumalamuttarmodu* lineage, especially the president and the secretary) and one is SC. Of those from Vadanur, nine are Vanniars and one is SC. One member is from another neighbouring village, with land in the tank's command area. The three women are from the Vanniar caste, two from VAD and one from PSP. The Office bearers from PSP have been the same since the beginning, whereas in VAD changes have been made: a Muslim was the vice-president and a Vanniar was the joint secretary; as their participation in the meeting and other activities were not conclusive, in 2005 a landowning Chettiar and another Vanniar replaced them respectively.

Ever since Vadanur TA's creation there has been some tension between the TA and the encroachers who were evicted from the tank bed. Most of them were landless SCs. The TA leaders assert that the rival lineage (the *saniathanmodu*) had made use of the tense situation to set people from VAD and PSP against the TA. Even in the 2006 *panchayat* election, the *saniathanmodu* members contesting the elections promised that they would allow the landless to cultivate land inside the tank. Indeed, as far as *saniathanmodu* members are concerned, they do not take kindly to the TA since they feel EC members would support unquestioningly the decisions taken by the Office Bearers (OB) who are dominated by persons from the *kumalamuttarmodu* lineage and by the wealthy Reddiar.

Office Bearers are indeed village elites, with political contacts or administrative responsibilities. The EC members from PSP have more important functions within the TA than those of VAD and moreover they control the TA as most of the EC members belong to the same faction (that of the powerful OB). The political affiliation and the historical dominance of the caste and lineage have an effective impact on the functioning of the TA. However, the guidelines for membership of the TA stipulate representation from all categories of people in the village, the existing social system has a mechanism that selects the persons with close affiliation that supports the dominance in some way or other.

Interestingly, even if formal associations set up by the French are described by elders today as a good system, and even if this "conclusive" experience is not so old, it has been difficult creating TAs in most villages. For various reasons —including the suspension of support to these previous formal associations by the Government and the surge in groundwater exploitation on an individual basis (as is the case in Tamil Nadu)—irrigation tanks and their use in irrigating crops in the command area have taken a back seat as far as the interest shown by farmers is concerned: at the end of the 1960s, farmers forfeited interest in the collective management of tanks. After that, most tanks in Puducherry became encroached upon and less used or they were even left unused. Voluntary collective interest in the management and maintenance of the tanks was non-existent in most villages (Sorapet is a good example of this).

Thus, Puducherry case shows that the existence in the past of a formal association does not necessarily help when forming a new association. This can be easily understood as many aspects have changed: encroachment situation, decline in the use of tanks, institutional rules since 1963 the Public Works Department has been authorized to take over the full management of the operation and maintenance of the tanks (Loes Sank, 2003), etc. The existence of these former associations is nevertheless recalled by TRPP engineers, as if it was a success criterion for recreating a village tank association.

Management of Common Resources by TAS

One of the issues for TA sustainability highlighted by the Expert report by the TRPP is the financial sustainability of the TA. A model for generating funds has thus been proposed: TAs are given a three-year lease of usufruct rights over trees and fishes of the tank, have to share the income generated with the *'commune' panchayat* (an administrative unit coming from the French) at the rate of 7:3. As we have seen in the previous section, prior to the creation are TAs, the revenue generated from these resources was usually utilised for basic temple and village needs by the *natamais*. The TA is tacitly obliged to contribute a share to the village common fund.

The VAD TA has planted trees for the most part in the foreshore area of the tank, after it was set up. It is estimated that after ten years the TA will earn about Rs. ten lakhs from selling the wood, and the president declared that this amount would be an asset for the TA for its efficient management of the tank. However, palm trees on the bund continue to be auctioned every year for *toddy* tapping and the revenue is collected by the respective *natamais* and *kothukarar* of VAD and PSP, and the TA has no say in this.

The TA auctioned the fish in the VAD tank for a period of two years after it was set up. During the first year, the TA members decided that fish would be farmed and sold by them. They appointed guards to watch over the fish raised in the tank. They harvested the fish, then transported and sold them in other villages. Only the Office Bearers became involved in this process while many EC members did not. As the Office Bearers were from the landowning category, the labourers who used to work in their field were used to do physical work of selling fish. Owing to the EC members' lack of involvement in these tasks, the following year a public auction was called for, which drew people from far away places.[5] A few members of the TA won the auction by making a collective bid. But after the harvest of the fish, this group submitied its accounts to the TA explaining that they had incurred loss, and the TA decided to compensate for it by returning the money collected during the auction! This provoked the anger of many villagers, even some on the TA's EC and induced a polemic, many people being opposed to this compensation. This also led an EC member to opt for disassociating himself from the EC and the TA, but the TA office bearers managed to quell the growing unrest.

There is also a difference of opinion about the disposal of mud coming from the process of deepening and desilting the tank. Villagers from Vadanur assert that if the soil had been sold publicly, it would have generated a good revenue for the TA. Some of them would have bought the mud and utilized it to cut bricks or would have spread it on their cultivated fields. They condemn the TA for laying farm roads on the banks of the river, which they say would be washed away by floods.

The TA now collects the revenue generated from tank resources that previously went to the village institutions. This has created a frosty relation between VAD and PSP. VAD people claim that the revenue that used to come to their village institution no longer does so. VAD people consider that the revenue goes to PSP thanks to the dominant position of the TA office bearers who are from PSP.

Functioning and Objectives of TA

For the TRPP project, VAD TA is considered a model TA as far as the amount of physical work achieved is concerned (members have managed to regularly desilt the tank; they have laid farm roads in the ayacut and on river banks. They have also taken initiatives and planted trees in the tank bed. The fish were farmed by the TA and auctioned). Moreover, the TA's finances are said to be healthy owing to the work carried out, and it has been able to save some money. The revenue generated by selling the trees is kept by the TA. The members intend to use the money to set up a computer centre in PSP, which is likely to be useful to students from the surrounding villages, including VAD.

The influential status of the Office Bearers of Vadanur TA has undoubtedly helped the association to undertake rehabilitation work in the tank.

As far as the objectives are concerned, the OBs of VAD TA are hoping to store water and increase the water resource[6]. Moreover, by storing water in the tank they feel that salt water intrusion might be prevented—an idea introduced by the TRPP—and that the ground water level would increase. There is indeed frequent mention of the fact that tank water was not used for irrigation for over 30 years. There are many villagers who admit that they have only ever heard of a systematic and organized distribution of tank water but have never seen it at all. In addition, the secretary of the tank association attributes the fact that the water in the tank would not be distributed fairly and that would cause a problematic situation to the non-use of tank water. In order to avoid such a situation, the tank water is only stored to recharge the ground water level. The TA's Office Bearers claim that this benefits many surrounding villages, not only VAD and PSP. Yet farmers who do not own a borewell consider this non-use of tank water for irrigation, even when water is available in the tank, as prejudicial. They assert that, though they could have free access to tank water which they are denied, they have to pay borewell owners for water to irrigate their land. The TA's objective is not to find a solution to sharing tank water as fairly as possible. Generally speaking, a better management of water is not really one of the TAs' priorities.

In comparison, SOR TA, created in 2003, was only able to restore the structure of the main tank to its optimal status. The tank has been desilted and the sluices repaired; this has increased the storage capacity in the tank and hence the tank can now hold water for six months instead of two months before the work had been carried out. But here again, the objective is not the proper management of water. The TA has laid farm roads in the command area of the tank. Due to the lack of cooperation on the part of some farmers, 25 per cent of the roads are waiting to be laid. Banning any encroachment on the tank bed has led to cattle and goats grazing there when the water is drained from the tank. This has settled a problem for people who have cattle and goats, who represent most of the households in SOR.

The Sorapet Office Bearers are of the opinion, regarding how their TA functions, that they could carry out the work mentioned by the CO and assigned by PWD. However, they expressed the difficulty in conducting monthly EC meetings, since members would not turn up on the set date. They also feel that the work could have been done better under their own supervision than by contractors—a discourse regularly heard and which refers to the money that ends up in the contractors' pockets. Another bone of contention that the Office Bearers like picking is that when the project was under EU, the TA was able to do more work than when the project was taken up by PWD which is now limited. Since the withdraw by the EU experts from the project, PWD officials have an upper hand in imposing their erstwhile approach so far removed from the community approach subscribed to by the EU project unit.

The main objective presented by SOR TA is to improve the water resource, store water that could enhance groundwater capacity. The association members say neither can they do any more than this nor could they bring about a major change in the village. The TA Office Bearers lament about the long list of objectives given and monitored by the TRPP—even if the list remains unknown to most members of the association. One of these is restoring villagers' rights concerning the tank. For this, SCs, women and the landless have been admitted into the collective administration of the tank. The TA Office Bearers disagree with such ideas and claim that these categories of members do not attend the monthly meetings held in the village. For their part, women, landless people and SCs explain that they are not used to speaking in front of men and/or upper-caste people and that when they speak, the others do not heed their voices. This explains their indifference to the meetings.

The model the TRPP wanted to experiment with by creating and running the TA reflects a holistic approach for the mitigation of existing social realities. For that, it encouraged the participation of the vulnerable section (such as SCs, landless, women) of the society in the activities of village development and of irrigation tank affairs. However, this is an unfinished task. On paper, TA membership includes all categories of people. But in practice, the influential personalities in the village still take the authority and dominate, not providing the opportunity for vulnerable people to actively participate in the TA functioning. The historically deprived SCs feel that their voices is not accepted by other caste villagers, that Office Bearers remain indifferent to whatever they say. In the case of women, however, even if some are active in the social context through Self Help Groups (SHGs) and village activities, they are often looked down by the elites of the village. Hence the women neither energetically participate nor voice their opinion in contributing to the decisions of the TA which are always taken by OB who are all men. Hence the situation in the village context remains relatively same analogically "old wine in a new bottle": the old village institution with a new name and new superficial structure but functioning in the old way. The social dominance of powerful elites and caste groups always wins but real needs of the needy remain unheard and unaddressed.

Thus, it can be said that the TA was actively up and running during the project period and even afterwards, especially as far as using funds for work on the structure of irrigation tanks was concerned. However, the social objectives that are of an egalitarian nature are not achieved due to the existing social structure and to the caste system in which it operates.

(c) Tamil Nadu WUAs: Competition Through Election and with the Informal Village Institution

WUAs formed in Tamil Nadu according to the guidelines of the TN FMIS Act provided a competitive situation with formal elections and a ballot vote system. It provided a space for politically vibrant personalities to stage their performance in the same way as the local *panchayat* election and system do. However, the persons who were earlier members of the informal village institutions express their reservation about participating in the WUA. Here again this is explained through two examples: WUAs for Saravanappakam (SVP) and Tiruvenainallur (TVN) tanks and Koratur (KOR) and Kothanur (KOT) tanks.

WUA Formation and the Notion of Election

Prior to the creation of WUAs under the Command Area Development Programme run by the Agricultural Engineering Department, it is said that farmers' councils (FC) were set up. The *modus operandi* in setting up these FCs was obviously through political party nomination. The local MLA would nominate local functionaries from the corresponding political party in the village such as the president and the office bearers of the FC. Locally this is referred to as *vivasahigal sangam* (Farmers' association), but very few villagers know of the existence of such FCs.

During March 2004, the elections for the president and Territorial constituency members of the water user association for TVN and SVP tanks (one WUA for both tanks) were held. The command area of both tanks was divided into four governing divisions, and the divisions were numbered one to eight. Nominees were invited to present themselves to the PWD. The main eligibility criterion to compete for the posts of WUA president and members of the Territorial constituency was to be a landholder in the tank command area attested by the relevant legal documents justifying ownership in one's own name. A voter list[7] was then prepared by PWD according to the landholder list given by the revenue department.

There were two candidates from TVN competing for the post of president, both from the same Udayar caste and the same political party affiliation as well. The reasons for them both standing as separate candidates are personal. Both deployed their respective candidates for the posts of members of the territorial constituency (TC). The candidate who won the post of president spent one *lakh* rupees (US $ 2,127) for the elections. He declared that it would have been a question of prestige had he been defeated in the election. Of the four members of the TC who won in TVN, one of them hailed from the traditional post of the *vaikal maniyam* at the western sluice of the tank, and belonged to the Udayar caste. Of the four members of TC, three were Udayar and one was from the Vanniar caste. The four members of TC from TVN and the president were active members of political parties at local level[8]. In SVP, however, out of the four TC members, unanimously nominated, three are *natamais* from SVP and the fourth is a politically influential person from Gokulapuram.

In October of the same year, government officials signed an agreement with the WUA president mentioning the work and maintenance of the tank and canals. In addition, PWD formed sub-committees to address finance, work, water management and monitoring issues.

The Koratur (KOR) WUA is a similar type of WUA since the tanks in KOR and Kothanur have clubbed together to form one association. However, here there are only four Territorial constituencies (TC), three in KOR and one in Kothanur. There were five persons competing for the post of WUA president, all from the village of KOR. The WUA president spent rupees 75,000 (US$ 1830) on the elections. Competition here was between two political parties. The four who opposed the actual WUA president were all from one political party and the president from the other. The WUA president was already the Farmers' council president, showing a certain continuity in the process. There was competition for three out of four posts as managing committee members from the Territorial constituencies. One of the TC members from KOR was unanimously elected.

A general notion prevailing in the study villages, and also to some extent where the WUA has been set up, is that the WUA has paved the way for competitive politics because of the election process of ballot box voting. Owing to its similar nature to that of the election of local bodies, those holding political party membership and a position at local level perceive it as a power structure to compete in the WUA. Those with a dual interest—good state of the irrigation tank as well as maintaining their political position—decide to compete and resort to any possible means to win the elections. The members of the informal irrigation institution who look at the election in a negative light abstain from participating. Thus many people in the study villages express the opinion that due to the election process, WUAs mainly attract those who have a vested interest in enjoying the status of elected member or in making money, not necessarily those who are interested in managing irrigation tanks.

Impact of money on the running of WUAs

The WUA has to make proposals to PWD regarding the work to be done to a tank. The PWD sanctions the work and the amount allotted to it, then the money released by the government is deposited in a joint bank account in both the WUA president and a PWD Assistant Engineer's names; both of them have to sign for the money to be used. Moreover, an initial deposit of Rs. 100 per acre is given to each WUA.

In TVN-SVP WUA, 2 main types of work have been sanctioned: the desilting and cleaning of TVN tank supply canal (for an estimated Rs. 100,000) and the repair of damage in both tank bunds (for an estimated Rs. 40,000). The first type of work was overseen by the president and a TC member, who rented an earthmover for 45 hours (at the rate of Rs. 550 per hour, hence for a total cost of about Rs. 25,000). It is said among WUA members that the remaining amount was pocketed by the persons who did the work and used to bribe the MLA (Member of Legislative Assembly) and PWD officials. One TC member from the rival faction of the WUA president asserts that such practices are those of contractors, and should not be that of an association that could use the money for collective purposes. Interesting enough, even if this TC member is well aware of the misappropriation of the money by the WUA president, he was not able to do anything in terms of reporting it to the higher authorities since both the president and himself belong to the same political party. Thus, it would appear that political affiliation creates a strong social bond that leaves no room for criticizing the acts of members.

The second type of work was undertaken for the TVN tank by this TC member of the WUA who provided his opinion about the work undertaken by the president in the previous paragraph. A sum of Rs. 23,000 was allotted to TVN and the remaining Rs. 17,000 were given to the TC members of SVP to do the work in their tank. The TVN-TC member used the earthmover for about 13 hours. Some money was given to PWD officials and Rs. 10,000 divided among the five members of WUA including the president and four TC members of TVN. This TC member claims that he added Rs. 2,000 to his Rs. 2,000 share and cleaned the distribution canal from the third sluice of the tank. When questioned about why they did not use these Rs. 10,000 to carry out work in the canals or in the tanks, he declared that the informal rule in such activities is that some money is for personal use. "As all of us had spent money during the elections on campaigning and buying votes, some money has to be taken by individuals". Buying votes is thus clearly paid back from the amount the government allocates for projects[9].

In SVP, TC members intend to recover funds to manage the work themselves (not through a contractor) as well as to keep some money as a contribution for the common needs of the village. This could be then added to the money generated through the auction of common resources, earmarked for the building of a school. Once the WUA had been set up, the president (from TVN) insisted that half the money generated by auctions should be given to the association. But the SVP people and TC members who are also the village *natamai* objected to this since they need the money (for the land bought for the school).

Owing to this difference which is augmented by the factional affiliation within the caste between the WUA president and the SVP TC members of the WUA (who are also village *natamais*), in 2006 the grass in SVP tank was auctioned by the informal village institution (*natamai*) as well as the WUA president. Both of them organised the auction of the same grass and both received money from their respective bidder. The one who gave money to the WUA president was not allowed by SVP people to cut the grass.

Similarly in 2007, when the people of SVP wanted to auction the thorn trees in the tank, the WUA president did not give his consent. The SVP-WUA members appealed to the Forest Department, they applied for an estimate to cut the trees, paid the amount to PWD and bribed the officials that gave permission to cut the trees. The auction brought in Rs. 75,000. The WUA president insisted that Rs. 50,000 should go to the WUA, though SVP village could have the remaining Rs. 25,000. The SVP *natamais* negotiated with the WUA president and in the end it was decided that Rs. 10,000 should go to the WUA.

Owing to these issues about the auction of resources, the WUA president wrote to the collector's office to know which of the tank resources the TA could auction. According to the President, grass and trees alone could be auctioned by the WUA, in the presence of an official from the PWD. The former then explained the need to open another bank account for the WUA exclusively in the name of the WUA president and another person from the managing committee, to deposit the money generated from such auctions. We can see here how the WUA would like to be financially independent as the informal institutions were, and how this collective resource management is a source of dissension between two villages and especially between this new formal institution (WUA) and the informal village institution.

The WUA president's action is not only criticized by SVP *natamai* (also members of the WUA). Even within the TC members of TVN, some ˉfrom the opposite faction to that of the WUA president—criticize the latter's decisions, reproach him for not heeding other members' propositions and accuse him of embezzling the money. The issue, based on the misappropriation of money, also reflects a relation of power, with the president being accused of being guided by his will to flaunt his authority. Another example related to the auction of grass in TVN is also a source of dissension within members of the WUA: some members of TVN revealed the incoherent decision taken by the WUA president (to allow the auction of grass even though he knew that the water would be soon released from the tank by PWD, and that this water release would interfere with cutting the grass). This example shows that the WUA president has to face various challenges, and it seems that he often favours his personal (in economic matters, but also in local sphere of power due to political and social factions) rather than collective interests.

Association Autonomy and Initiatives

According to the TVN-WUA president, it is not possible for the association to become financially as well as socially independent. He gives an example involving an irrigation canal, damaged because it ran through the market place, and about which farmers do not wish to solve the problem. But presenting the issue in such a way is not acceptable since many stakeholders are involved, with various concerns and at different levels of power. This short example illustrates that the arguments given by various actors to criticize or to validate the functioning of an institution have always to be contextualized, and that this question of "good" or "inefficient" functioning is a complex topic, as many social dynamics can be played through the issues the WUA has to face.

However, some interesting activities—that the command area farmers consider an achievement—did indeed take place in Koratur: with the amount (Rs. 40,000) made from a tree auction, the WUA could evict encroachers from the tank by using an earthmover. This was possible since they could obtain the support of the MLA, PWD and the police. In practical terms, that means the MLA sent a letter to the WUA president to support the activity, and PWD officials and policemen came to the village the day the earthmover was used to destroy the illegally cultivated crops. This initiative is a precedent in the district and has encouraged the neighbouring WUAs to initiate eviction of encroachers in their tanks.

Another interesting activity launched by the KOR-WUA is to fix a timetable for distributing irrigation water in the tank command area with the help of the *kambakaran*. The *kambakaran* will be given the key to the sluices and the responsibility to close them at 6 pm (so that the water will no longer be wasted during the night, since nobody uses it then). These men seize the opportunity provided by the fact that the WUA formalizes their service to ask for a permanent remuneration instead of the traditional practice of receiving a bunch of paddy during harvest, since this last type of remuneration is on the decline with an increasing number of farmers moving over to sugarcane cultivation.

These two examples of WUA show some similarities such as the misappropriation of project funds either for collective purpose or to reimburse the cost of vote-buying; conflicts with the informal village institution due to the revenue generated from common resources management. The dynamic that is underlying such institution functioning seems then linked to the financial benefit that could be generated from the various activities. However, whatever the differences in caste, factions, political party affiliation are, there is some loyalty to share the misappropriated monetary benefits.

However, some major difference between these two WUAs can be highlighted: in one of the village (TVN) of one WUA, the traditional irrigation institution of TVN is no longer involved in the tank management, WUA members being linked by political interest and objectives whereas in the other village of the same WUA (SVP), the persons involved in the tii are those who have been elected for the WUA. In the first case, the WUA is an opportunity for some village people to access to power, while the second case shows a continuity in power. As far as the other WUA is concerned, its case is interesting as members seized the opportunity of the WUA creation to reinstate the role of *kambakaran* (while the previous mentioned village tries to evict them), and thus to manage water in a better way. In this regard, it had the initiative of evicting encroachers, for which it got the support of the administrations. And when compared to the Puducherry cases, where eviction was the first objective of TRPP project, this case involved less social complications—may be also to the fact that Koratur tank is still used for irrigation.

Summary

Through these six village case studies, we have seen how traditional tank institutions were far from being only focused on tank irrigation and have to be understood as a village institution. As such, the power and decision making were in the hands of a local elite, generally big landholders or linage elders of the dominant village caste(s)—this has even evolved within the last 50 years.

However, women, marginal farmers, landless people and Scheduled Castes have been always deprived of the decision making functions. In order to mitigate such inequalities, the Tank rehabilitation project in Pondicherry imposed tank associations to have representatives from all categories in the village. In spite of such guidelines, the existing social system is so strong that its mechanism of dominance by few persons from upper castes is still running, the vulnerable persons having very little chance to voice their opinion or to be listened. In some way, the situation remains as an "old wine in a new bottle". One of the new characteristics of formal institutions (either Tank Associations of Puducherry or Water User Associations of Tamil Nadu), is the intrusion of the political dimension in the membership process, especially in the case of Tamil Nadu Water User Associations. Our few examples have nevertheless shown a diversity of situations: generalizing the process of the social dynamic is thus impossible. However, conflicts with other village institutions about the management of the revenue generated from the village collective resources of the tank are common. These formal associations, framed by outsiders and linked to government offices, give way to a new power position, particularly interesting as funding has to be managed. Thus some people grab the opportunity to evict the persons who had traditionally the power and thus contest the elections; but some influential people seize also the opportunity to reinforce their position. Whatever the situation is, very few formal associations are really involved in managing water in a better way.

NOTES

1. On the land of big landlords there were one or more families, either from the same caste or from another caste or from a lower caste with lower

economic status, who worked hereditarily. These people were called *padi azhu*.

2. In Puducherry, as a French colony up to 1954, a system of formal associations existed since 1839 (*See* Box 8). Irrigation was managed at that time by *Caisses Communes* and *Syndicats agricoles*. This system was in practice until 1963 when the Public Works Department was authorized to take over the full management of the operation and maintenance of the tanks (Loes Sank, 2003).
3. The spread of the grass that was inside the tank is said to have been confined to a small area unlike the present spread that is vast.
4. http://www.tn.gov.in/budget/archives/interim_budsph2006-07-1.htm
5. The TA published an announcement about the auction in local new papers.
6. They also propose to transform the tank into a bird sanctuary; an idea which is circulating among NGOs and various associations.
7. According the list prepared by the village administrative officer in November 2003, TVN had 629 voters and SVP had 400 voters.
8. Not all from the same political party.
9. Regarding the man who lost the election, he is waiting for another opportunity, either at *panchayat* elections or another election of the same kind, where he will compete, try to win and then hope to regain the money he spent on previous elections.

REFERENCES

Bouma, G. D. 1998. Distinguishing Institutions and Organizations in Social Change. *Journal of Sociology* 34 (3): 232-245.

Hooja, Rakesh, G. Pangare and K.V. Raju (eds). 2002. *Users in Water Management—The Andhra Model and Its Replicability in India*. Jaipur: Rawat Publications.

Hooja, R. 2004. Below the Third Tier: Water Users Associations and Participatory Irrigation Management in India. *Indian Journal of Federal Studies 9th issue* (1/2004), http://www.cfsindia.org.in/jour4-1_4.htm

Janakarajan, S. 1993. In search of Tanks: Some hidden facts. *Economic and Political Weekly* 28 (June 1993): A53-A60 (review of agriculture).

Janakarajan, S., Rajagopal, A. 2002. *State in perplexity: The politics of water rights and system turnover in Tamil Nadu*, Working Paper No. 171, Madras Institute of Development Studies, Chennai, pp. 1-8.

Ludden, David, 1989. *Peasant History in South India.* Oxford: Oxford University Press.

Messer, Norman and Philip Townsley. 2003. *Local institutions and livelihoods: guidelines for analysis.* Rome: Rural Development Division, Food and Agriculture Organization of the United Nations, 131 p.

Mollinga, P. 2001. Water and politics. Levels, rational choice and South Indian canal irrigation. *Futures* 33 (8): 733-752.

Mollinga, P., R. Doraiswamy and K. Engbersen. 2001. Participatory irrigation management in Andhra Pradesh, India. *International Journal of Water* 1 (3-4): 71-90.

Mosse, David, 1999. Colonial and contemporary ideologies of community management: the case of tank irrigation development in South India. *Modern Asian Studies* 33 (2): 303-338.

Ostrom, Elinor. 1992. *Crafting Institutions for Self-Governing Irrigation Systems.* San Francisco: Institute for Contemporary Studies.

Pain, Adam. 2004. "Understanding Village Institutions: Case Studies On Water Management From Faryab And Saripul". Series/Report no.: Afghanistan Research and Evaluation Unit Case Study, http://ageconsearch.umn.edu/bitstream/14639/1/cs04pa03.pdf.

Palanisami, K. 2000. *Tank Irrigation-Revival for Prosperity.* New Delhi: Asian Publication Services.

Palanisami, K., Easter, K.W. 2000. *Tank Irrigation in the 21st Century—What next?* New Delhi: Discovery Publishing House.

Prabhakar, P.I. 2008. "The Illegal Practice of Cultivation in Irrigation Tanks: An Approach to Legal Pluralism through Actors' Narratives", pp. 465-496 In: Eberhard, C. (ed), *Law, Land Use and the Environment: Afro-Indian Dialogue, Natural Resources Management Comparative Perspectives.* Puducherry (India): French Institute of Pondicherry (IFP Collection Sciences Sociales No. 13).

Reddy, V. Ratna and P. Prudhvikar Reddy. 2005. "How Participatory is Participatory Irrigation Management (PIM)? A Study of Water User Associations (WUAs) in Andhra Pradesh", Working Paper of Centre for Economic and Social Studies (CESS), Hyderabad.

Schenk-Sandbergen, Loes and Niren Choudhury. 2003. *From heroines to beneficiaries: from beneficiaries to heroines? The impact of a small-scale irrigation project on gender in West Bengal, Terai, India.* IDPAD. New Delhi: Manohar.

TN-WRCP Implementation Completion Report, 2005: http://www-wds.worldbank.org/external/default/main?pagePK=64193027&piPK=64187937&theSitePK=523679&menuPK=64187510&searchMenuPK=64187283&siteName=WDS&entityID=000160016_20050607115943

Uphoff, Norman T., Ruth Meinzen-Dick and Nancy St. Julien. 1985. "Improving Policies and Programmes for Farmer Organization and Participation in Irrigation Management", Water Management Synthesis Project Paper No. 1, Irrigation Studies Group, Cornell University 7.

Uphoff, Norman T. 1986. *Improving International Irrigation Management with Farmer Participation: Getting the Process Right.* Boulder, Co: Westview Press.

Vaidyanathan, A. (ed). 2001. *Tanks of South India.* New Delhi: Centre for Science and Environment.

CHAPTER

10

Gender and the Community Participation Process in the Tank Rehabilitation Project, Pondicherry: Mainstreaming or Male-streaming?

Loes Schenk-Sandbergen

'The tank is like a mother, giving food and money to me for almost 50 years'.
—Quotation of an old landless men in Vadhanur Tank

10.1. Introduction: Reforms and Mainstreaming

Farmers' direct involvement in irrigation, under the label of Participatory Irrigation Management (PIM) is the most common type of reform in the Indian irrigation sector (Gulati *et al.*, 2005). "Managing irrigation together" became a powerful message in common property irrigation interventions (Maloney and Raju, 1994; Sengupta, 1991; Shivakoti *et al.*, 2005). PIM and all resulting reforms were prompted by what Mollinga (2000) terms the "three-fold crisis": financial crisis, technical crisis and public image crisis of the irrigation sector under public management. 'Community participation' became the rediscovered, magic solution to support, with a prominent role of NGOs and Participatory Rural Assessment (PRA) methodologies, the formation of Water Users' Associations (WUAs), and the 'privatization' of common goods.

Parallel with these trends, gender development policies formulated by donors focus worldwide, in line with Gender and Development (GAD) strategies, on 'mainstreaming' women in the core activities of development projects (Kelkar, 2005; Rao and Kelleher, 2005). 'Gender mainstreaming' has become the global gender rhetoric with the theoretical goal to achieve gender equality.[1] What are the results of 'mainstreaming' women, in particular poor rural women in the new 'managing-irrigation-together' type of irrigation organizations and projects?

It is no surprise that it is widely documented that women's participation in Water Users' Associations all over the world is very low, despite high involvement of women in irrigated agriculture and agricultural decision making (Meinzen-Dick and Zwarteveen, 1998, 2001; Zwarteveen, 2006; Schenk-Sandbergen and Choudhury, 2003). An overview in Meinzen-Dick and Zwarteveen (2001: 73) shows that it is zero per cent in Nepal and Pakistan, six per cent in India and 15 per cent in Sri Lanka. WUAs have much less official and legal gender backing than in the institutions of the *Panchayati Raj* where state intervention has ensured women's formal political entitlement by the reservation of seats (one-third with existing reservation provision for SCs/STs) (Anandhi, 2002).

The community participation approach of the Tank Rehabilitation Project Pondicherry pursued from the start of the project to mainstream gender, in particular women and men of poor landless households, in the formation and functioning of the Tank Associations and all related activities. It will be shown that groups of landless Dalit women were initially enthusiastic members and participating stakeholders in the Tank Associations, but over time the costs were too high and their benefits almost zero, and in the process they withdraw. The reasons for the withdrawal were that many lost their customary use-right of a small plot of land in the tank, their access to tank resources became even limited, and few manual labour and income was generated during implementation of the technical work and from tank resource development. Despite the sincere and intense pro-gender and pro-poor community participation approach, conducted by NGOs and Community Organizers (COs), the mainstreaming achievements of the beginning years of the project faded away.

The main focus in this contribution is to show the social dynamics of the community participation approach in the TRPP in a time perspective following the subsequent phases of the tank rehabilitation process. The community participation aimed to match the rehabilitation of the tanks with the mainstreaming of women and landless people in order to ensure that: their customary rights regarding the use over the tank resources would not be endangered; to improve the livelihood of landless and poor women and men to initiate employment generation through the physical work to be implemented; to promote the development of tank resources as income for operation and maintenance of the tank, and simultaneously as income generation for members of poor households; to prevent that weaker sections might be disadvantaged by the project, and if so, to initiate mitigating (compensation) and corrective measures, and to empower women, in particular poor landless women in their practical needs of employment and income.

In short the community approach tried desperately to prevent that the tank rehabilitation became another example of an out-side intervention in which the favourite gender mainstreaming policy and approach is not working and in practice burns down to bare 'male-streaming', resulting in the instrumental effort to get some participation and representation of land owning women in the Tank Associations.

The composition of this contribution is as follows. In paragraph two some gender theory and concepts are sketched relevant for our focus on irrigation reforms and gender, poverty, community and community participation. In paragraph three the dire reality of the 'approaching tragedy of the commons', as foreseen by Vandana Shiva (1988), is sketched. The caste/class and gender character of tank rehabilitation and social context of the project area are described in paragraph four. In paragraph five the set-up and achievements of the community participation process in the first phase of the formation of Tank Associations is the central theme. Paragraph six shows gender dimensions of the membership and representation of women in the Tank Associations (TAs) and in particular in the Executive Committees (ECs). The role of women community organizers is highlighted in paragraph seven. The negative impact of the tank rehabilitation after the planning phase by evicting poor encroachers and creating no employment opportunities for the poor and landless women during the implementation of the physical work and the phase of development of tank resources is presented in paragraph eight. Efforts to empower poor women in an additional 'gender and equity' programme, which was added after the mid-term review, and shortly after it the collapse of the community participation and the gender programme is briefly sketched in paragraph nine, followed by some concluding remarks in paragraph ten.

10.2. Theory and Concepts: 'Irrigation Truly is a Men's World'

To position gender in irrigation studies I want to refer to the recent impressive Ph.D. thesis of Margreet Zwarteveen, Wedlock or deadlock? Feminists' attempts to engage irrigation engineers (2006). Her research is, according to me, a milestone of dedicated and concerned irrigation feminist research. She states:

"Irrigation truly is a men's world. Rights to water and irrigated land are mostly vested in men, and decision-making about irrigation is likewise dominated by men. Ways of speaking and thinking about irrigation can also be characterized as masculine in that they render the work and activities of women invisible, and denote the analysis of social and gender relations as irrelevant. The irrigation professional is also masculine: most irrigation professionals are men, and professionalism is associated with masculinity" (291).

No Land and Water Rights for Women

Rules for membership in Water Users' Groups are problematic as patriarchy and male dominance are well-known causes for the absence of land and water rights for women. The concept 'water user' is deeply ingrained in Indian culture as the 'one who owns the land'. There is no concept of 'water user' as the 'one who does the work', and is actually the user of the water. Women, in particular landless women, who do the transplanting, weeding and harvesting work, are not perceived as 'water' users and therefore not perceived as eligible members of Water Users' Associations.

The law and practice is that the holder of the land, who therefore is considered as the 'water user', legally controls water rights. Men inherit land, and in case of acquiring land by buying, 'the head of the household', the male, becomes automatically the owner of the land. Formal land titles are adjudicated in his name. Although women have legal rights to inherit, and own land, the practice is different (Sharma, 1980; Agarwal, 1994; Rao, 2005; Rahman and Van Schendel, 1997). The existing patrilineal descent system, inheritance practice and the patrilocal post-marriage residence pattern cut women off from house, land and water rights.

Imported Western gender ideologies and perceptions on irrigation as a male domain are also very decisive. Even in matrilocal Laos where the majority of women inherit the house and land, I found that mainly men were members of the newly formed Water Users' Organizations (Schenk-Sandbergen and Choulamany-Kamphoui, 1995b; Schenk-Sandbergen, 1998). It were however the women and girls providing the financial means for the strengthening of the irrigation schemes and maintenance by weaving till deep in the night beautiful skirts which were sold in the market. In the traditional Lao water management groups the participation of women was equal with men. The new public domain, created by donor-driven institutional reforms in water management transfer, has decreased the formal participation and decision making role of Lao women in irrigation, although their informal role to influence the decision making process was substantial (1995: 85).

The cost-benefit factor to participate in water management organizations is different for women compared to men. Women are much more under time pressure because of their heavy, time consuming household tasks and economic workload. Meetings are often convened at times and places impossible to attend for women without loss of status, food and income. Moreover, poor women are socialised not to sit or speak in front of men and in case they speak, their needs will hardly be listened too. The dynamics and changes in perceptions of cost-benefits of women of the land-owning class and landless women in various project phases of the TRPP will be shown in this chapter.

Class and Gender Differences in 'Communities'

In particular in Tamil Nadu, Andhra Pradesh (Sharma, 2003) and Karnataka policy reforms are developed since the 1990s to rehabilitate tanks[2], with the financial support of international donors (EC, World Bank), and to hand them over to local communities as common property resource. In 1990 Palanisami wrote a small study on "Tank Irrigation in South India", and asked "what next?"

In 2000 he published a book with co-author Easter under the title *"Tank Irrigation in the 21st Century: what next?"* The conclusion is that studies have shown that investment in tank management will yield high rates of return, farmers are capable, are eagre to participate but concerned about a.o. legal status (2000: 182). On the 'what next' question I will say that the sociological analyses of winners and losers in the transfer of management of tanks in a class/caste and gender perspective is missing and should be taken up 'next'. In the following a modest effort will be made.

The strategy of the TRPP[3] is that the rehabilitation of the tanks will be carried out together with the local 'community', and by handing over the irrigation tanks and its assets to Tank Associations, new water users' institution formed by the project, which will operate and maintain these tanks after rehabilitation. In order to ensure that the TAs are able and willing to operate and maintain the tanks, they will have to be carefully established after which they are involved in the planning and implementation of the rehabilitation works. After the TAs have been formed, the PWD engineers are supposed to plan and to design the technical works with the participation of the 'community' through the TAs. The TA is free to select a contractor to implement the work under supervision of working committees formed by the TA. To make the future operation and maintenance of the tank system financially sustainable, the TA has to develop and to manage tank resources (tree planting and fish rearing) and thus generate its own funds. The message for the farmers is: "stop using deep tube well water during one season to restore groundwater". The Chief Engineer of the Public Works Department, who is the responsible implementing authority of the project, will implement the above process[4]. To manage the project a Project Management Unit (PMU) was formed, headed by the project director (Chief/Superintending Engineer, Public Works Department Circle II) and advised and guided by the Dutch expatriate project co-ordinator, better known as team-leader, who was residing permanently in Pondicherry. Short-term consultants from the counterparts were advising the project. In that capacity I visited the project eleven times as a sociologist/community organizer/gender specialist.[5]

The role of 'communities' in natural resources management is a popular concept in the new reform 'transfer of management' to the communities policies, that, according to FAO research, is used in more than fifty countries to protect forests in a better way (Agrawal and Gibson, 2001:1). The assumption that 'communities' are better managers of tanks than the state bureaucracies, or the Public Works Department, is a strong conviction. According to Shah (2003) these perceptions are not based on a careful inquiry about internal power dynamics in the community resulting in inequitable distribution of the resource. It is reported that despite the fact that tank WUAs are promoted as non-political institutions, "elite capture" and political involvement dominate their functioning (Reddy and Reddy, 2005).

The 'community' means in male irrigation thinking, broadly, the land-owning male farmers having land in the command area. As pointed out in many studies, 'communities' are not romantic homogeneous groups with common interests and objectives of women and men (Guijt and Shah, 1999). Women, in particular women of poor landless households, are often excluded as part of the 'community', or 'stakeholders' in the process of transfer of management of water and land resources (Sarin, 1998; Viegas, 1993). There are only a few very interesting documented cases linking poverty and gender with water management reforms, and of higher women's involvement in Water Users' Associations and *Pani Panchayats*, but they originate mainly from women's only organizations (Koppen, van, and Mahmud, 1995; Koppen, van, 1998; SEWA, 1997; Bhowmik and Jhabvala, 1996; Schenk-Sandbergen, 1995). Reema Nanavaty (2000) rightly emphasizes that the priority should not be to rehabilitate the water resource but 'we need to protect and restore the lives and work of women as water users, mangers and owners. If we do this, water as a resource will be protected and restored.' Gender differentiated access and control to resources analysis are widely used in Gender Assessment Studies, aiming to support empowerment policies and strategies (Rao *et al.*, 1991; Young, 1993; Moser, 1989).

10.3. The Tank as Common Property: The Approaching Tragedy of the Tank?

After being a while in the Puducherry region one easily learns to recognise the location of a tank. It is a beautiful landscape with a lot of green colours, and shows an earthen dyke (bund) with rows of Palmyra palm trees planted on it.

Flora and fauna are abundant around the tank. Many beautiful black-white, big orange, small yellow butterflies, dragonflies, and other insects, rice-birds and eagles are flying around. Frogs and fish make glittering bubbles in the water. In Keelparikalpet tank the white and red water lilies in the very pure water offer a more splendid view than the paintings of Claude Monet. People use the bund as threshing ground. Palm trees and other trees are marked with blue numbers to show they are government property and auctioned to users. Sheep are sleeping in the shade of the trees. Black pigs are roaming in muddy places with stagnant water. There are sometimes ruins of old granite stones, which have been placed as a water metre. The old farmers know that when the water reaches the upper table stone, the tank is half filled up but that rarely happened in the past decades.

Tanks are under state control, managed by the Public Works Department. The tank and the surrounding land belong to the category of 'common property'. The importance of the tank as a 'common property' that has to be protected and looked after is often symbolized by the temples between the trees at the bund with the statues of big red horses and goddesses and god: Iyyannarappan and his two wives: Porkilai and Poorani (twins). They have to protect the tank. It is believed that in the night they ride on horses with big swords and protect the common property. The author Mosse writes:

> *"Tanks are not simply sources of physical inputs (irrigation water). They are also repositories of symbolic resources. Tanks, like village temples, are public institutions expressive of social relations, status, prestige and honour. They are part of the symbolic 'production of locality'."* (Mosse 1997: 472)

In the project the main focus of the engineers of the Public Works Department is on the future use of the irrigation water in the tank by the farmers (*ayacutdars*) having land in the command area of the tank (*ayacut*). However, our observations[6] of the tank use, and a study of one of my Dutch students (Van Vuuren, 2002) showed that there are many more users of the tank land and water as: grazing of cows, goats and sheep by poor women, shepherds and children; rearing ducks, (duck nomads[7]) and pigs; collecting grass (for fodder and thatching roofs) and firewood by poor women and girls; making mats and baskets from reeds and special species of trees; making pots from the tank soil; making bricks (brick kiln); catching fish; collecting crabs and snails as food by poor women; catching rats and snakes by rat hunters; wage labour by landless people working in the fields in the tank; cultivating encroached land; washing cloth (*dhobis*); drinking water for the cattle; open toilet (large majority of the population with segregated places for women and men); bathing; collecting medicinal herbs, fruits, Palmyra fruits; toddy tapping; collecting seeds from the Palmyra palms for own use.

We[8] learned that for the poor the tank has strong personal and emotional connotations. That is why I came across several poor people who associate the tank with a 'mother', who cares for them with love. My following field note testify this perception:

> *"One old man carries firewood he fetched from the tank on his head. He sells it for Rs. 5.00: his daily income. He states that ' the tank is like a mother, giving food and money'. He survives already 50 years, just because of the tank. And with him the poor people of 18 other villages located around the tank. He also does sometimes wage labour in the tank (if they call him as he is old, thin and weak): weeding and harvesting paddy during two months a year. He is afraid for any change in the tank and its management as it might have a negative impact on his very vulnerable livelihood system."*

The above list of uses shows, that the access and control over the tank area and the tank resources is crucial for the survival of many poor people, mostly landless *Dalits* living around the tank. Some old landless women, shepherds, and in particular children spend their whole day in and around the tank to let their goats, cows and sheep graze and to collect grass and cooking fuel. The meaning of the tank in their life is more important than for the *ayacutdars* who are the favourite stakeholder group in the perception of the implementing authorities and project designers. We became very concerned about the social impact of the 'transfer-of-management' to the TAs as it might endanger the 'common property' as survival base for the poor and landless women.

Jodha (1986) has documented the importance of 'common property' resources to the village poor in India. His research shows that the benefits of using the commons were greater for the village poor than were the benefits of government programmes targeted for their welfare. It is also in this perspective that the debate on the 'approaching tragedy of the commons' (Shiva, 1988: 83) seems to be relevant for tank rehabilitation programmes. At the moment there is so to say a 'free' access for landless women in using and managing the resources of the tank as a common property. However, the 'free' access counts also for the rich and powerful who encroached the tank land, or, it is even sanctioned by paying a fee. *Commune Panchayats*[9] try to prevent encroachments on the bund by individual persons by planting seedlings of palm trees on it. The brick kiln owner, needing soil from the tank, has a strong interest to keep encroachers out of the tank. The Public Works Department has tolerated this process for many years. The rich use the tank as a 'goldmine' and just appropriate the resources, the poor encroachers use the tank as a 'social safety net', a last resort to survive. A bit of imagination is necessary to predict what changes will happen because of the transfer of management of the tanks to the mainly 'goldminers'. Therefore, we tried to include also the poor and women in the Tank Associations to be able to voice their views and to defend their interests at the formal level.

10.4. The Class/Caste and Gender Character and Social Context of Tank Rehabilitation and Tank Management

It is significant to realise that the surface water in tanks and channels from rainfall and rivers, which is so important to the recharge of the groundwater, is a common resource. It is only after this water goes underground that it emerges as a private good (Meinzen-Dick, 1984: 69). This fact is crucial for the analysis of access and control over tank water. A common good as rainwater is privatized as soon as it becomes groundwater. It is appropriated by those who can afford the investment of deep bore wells to uplift the ground water. In social terms it means that, when the objective of the TRP Project, to increase the recharging capacity of the tank will be successful, the richest and most powerful landowning farmers having borewells in the tank command area will be the main beneficiaries as they will get double benefits: they can use more groundwater in their own fields and sell more water to poorer farmers.[10] This is a reality in the Puducherry context, as the village communities are not at all a homogeneous groups of farmers.

The caste/class situation is very visible in the villages around the tanks. The land-owning class, often from the same caste, lives in the main streets in good houses; the landless *Dalits* live in clusters apart from the upper-classes/caste wards, outside the village in colonies or pets in small houses and huts with thatched roofs. Even in Puducherry with a reputation of modern, rich and advanced, the spatial caste/class segregation is broadly similar as in other parts of India.

Generated data on landholding size in the project tank *Ayacuts* of all the 84 tanks are shown in Table 10.1.

Table 10.1: Land Holding Size in Project Tank Ayacuts (ha)

Holding Size	*Holdings in %*	*Area in %*
< 1.0	60	22
1.0 – 2.0	24	26
> 2.0	16	51
Total	100	100

Source: Mid-Term Review Mission Draft Report, May 2002.

The polarised class context is also reflected in the findings of our household survey which I could only process in the beginning years of the project.[11]

Community organisers interviewed 4299 households around 16 tanks that had been taken for rehabilitation in 2002. The data show that the number of landless households is alarmingly high in the project area. With the exception of Mannadipet where the percentage is low because of urbanisation (28%), at least half, but up to 80 and even 91 per cent of the villagers are landless. Seasonal migration to the sugar cane fields in Andhra Pradesh of mostly men, to earn an additional income, is common[12] and some have found employment in nearby industries where only a few could find a permanent job. Alankuppam tank is an exception, as the majority of the villagers found jobs in light industries in the Auroville Ashram enclave. In Keelkarikalpet we found a percentage of 78 per cent of male agricultural wage labourers. Focus group discussions made clear that the same percentage, or even higher, counts for women agricultural labourers. Many of these women expressed their desperate need for wage labour and credit facilities from the project.

The percentage of land owning farmers, who will be the main beneficiaries in the tank rehabilitation project ranges from only six percent to 31 per cent of the population in the villages. Landless groups in the village are dependent on the land-owning classes, who are often also prominent political leaders, to get employment, loans and access to water and land.

Caste Context in Villages Around the Tank

Caste is an important indicator for land and deep tubewell ownership. Main land-owning castes in the project area are: Reddiars, Vanniyars or Gounders,[13] and to a certain extent Nadars. The following table shows that the higher, more powerful, the caste the larger the average landholding per farmer in each caste category in one tank.

Table 10.2: Average Landholding per farmer in each Caste Category (Keezhparikkalpet)

Caste	*Number of farmers*	*Land extent in Acres*	*% landholding by caste*	*Average landholding per farmer (acres)*
Vanniyar	31	54.9	51.9	1.77
Reddiar	5	30.5	28.9	6.10
Nadar	5	14.3	13.6	2.86
Scheduled Caste	6	5.9	5.6	1.00
All Castes	47	105.6	100	2.25

Source: Data of Raghunath, NGO CERD Field Manual 2003.

In another tank village where the majority of the villagers are landless and poor, the following caste/land ratio can be given. The majority of the population (68%) in the village is of Scheduled Caste Community. They own only about 1.2 per cent of the total land, whereas Vanniyars (19 per cent of the total households) own 34.7 per cent and the Reddiars (5 per cent of the total households) own 59.4 per cent of the total land. The ownership of deep tubewells shows the same trend. In one tank the eight Reddiar families own eight deep tubewells (50 per cent of the total), 30 Vanniyar families own 7 tubewells (44%), 13 Nadar families own one tubewell (6%) and 109 Scheduled Caste families own zero deep tubewells.

Gender Context in Villages Around Tanks

Many Reddiar women do not do agricultural work to ensure the 'prestige' of the family. They felt that, 'men will look down on them' if they have to attend meetings of the TA with a large majority of other caste men. The Vanniyar women are the farmwomen of small landholdings, who do the weeding, harvesting and transplanting of rice in their own land. Mostly they have to buy water from the Reddiars. Vanniyar women have a direct interest to become a member of the Tank Association, as it will benefit them to use tank water in one season. A few became a member of the Executive Committee and as we will see in paragraph eight, they participated very well in the planning and implementation of the technical works. We quickly discovered that women leaders were scarce and the few identified were already involved in Self-Help Groups.[14]

Dalit women are for the majority the agricultural labourers in the project area.[15] They are dependent on labour opportunities from the land owning caste in, and outside, the Ayacut and of farmers in nearby villages. Women told that they are starving for around six months, as there is no agricultural wage labour. The extremely low daily wage rate of Rs. 17[16] for *Dalit* women showed the need for employment against a fair wage for these women. As shown in the previous they use the tank almost daily for grazing of goats, fetching firewood and as their open toilet. With the neglect of the tank water and tank land they have full access to these tank resources. At the start of the project we discussed the possible consequences of the tank rehabilitation with them. The argument that there might be more agricultural work with better irrigation in future did not appeal to them. They say that the landowners will bring migrant labourers from other places, which are even cheaper. Some very poor *Dalit* women became very emotional when we talked on the tank rehabilitation. The crucial point was that in several tanks many landless women (and men) have occupied a small plot of land in the tank (30, 40, 50 cents…etc.) for already many years to cultivate some rice for three to four months. This saves them lots of money in the months they have no work and face serious survival problems. The women 'encroachers' were very worried for a negative impact of the tank rehabilitation. They were afraid that they will be evicted and loose access over their small plot of land in the tank area. Being landless, the access to some handkerchief piece of land is the most important in their life. It is a question of a food-safety-net, status, of dignity and identity. They felt threatened that their small plot will be taken away, when the earth moving work will start in the tank. This is exactly what happened, as we will see in the next paragraph. They were the allies of the rich farmers who encroached the tank and installed even deep tubewells in the tank, allowing them to cultivate rice in the tank and to buy the water from them. In case a new institution as the Tank Association will regulate the use of the tank, they feared to lose their 'customary right' to have a small plot of land to grow food. What they expressed is clearly the 'approaching tragedy of the commons': the fear that they will be evicted from the tank with bulldozers and police. And that is exactly what happened, although in many tanks the people vacated 'voluntarily'. In many focus group discussions in villages, poor women expressed that they will vacate the encroachments and co-operate enthusiastically with the project but as compensation they expect employment and income from the project in the implementation of the technical works. They wanted wage labour to feed their children. Secondly, they want fishponds and latrines. Thirdly they expect from the project to support them in forming and giving seed-money to Self-Help credit groups in order to develop some income generating activity. In the following we will see whether the project could provide any of these urgent requirements.

10.5. The Community Participation Approach: Including Women and the Poor

Main Concept of the Community Participation Approach

The main concept of the community participation approach in the Pondy project was that all stakeholder[17] groups in the villages around the tank should be mobilized and involved in the project including women and the poor. They all should become a member of the TA, with the intention that its members participate in the planning, design making, implementation and development of tank resources, operation and maintenance and management of the tanks. This approach emerged during our experience in the pilot tanks at the start of the project. The meaning and use of the tanks as a common good for the landless people and the caste/class and gender context were analysed and showed the necessity to add a strong commitment in the project showing our concern for vulnerable groups in the villages around the tank. We envisaged that the main beneficiaries in the TRPP will be the land-owning farmers, the *ayacutdars*, with land in the command areas: approximately 13,500 farming households. They form a small minority of the population in the villages around the tanks. The large majority of the village households is landless, and lives in poverty. In particular the women of landless households are in urgent need of food for their children and earning money. We feared that in the end the large farmers will dominate the Tank Associations and will be the winners. They will have double benefits, while the poor and landless *Dalit* women, will hardly gain anything, or even worse, were robbed from their small share in the public 'common good' which provided some livelihood security before the project was initiated. It was also encouraged that the TAs will act as a catalyst to initiate general development programmes in the villages.

In previous tank rehabilitation projects this "all-stakeholder" approach was so far not followed[18]. Under British colonial rule in Tamil Nadu, the French in Pondicherry, and the tank improvement projects of the past 20 years, the focus was exclusively on the mobilisation and organisation of the farmers having land in the *ayacut* of the tank (Maloney, 1995: 4; personal communication with resource persons). As the *ayacutdars* are supposed to be the only tank users, the organisation of this kind of WUA's is not so complicated, compared with our Pondy approach to mobilise and organise all the existing tank users groups in the villages around the tank. This requires much more time, knowledge, dedicated NGO staff, high quality and honest Tank Association members and leaders with a feeling for social justice and women's issues. Key point of the followed 'all-stakeholder' approach in Pondy was that it challenged caste/class, land ownership, political, religious and gender relations as it pursued to give a voice and rights to the poor and landless women for whom the tank is often for their survival. It was hoped that representation created a communication channel in which they could protest against activities of the project, which definitely might have a negative impact on their group.

As the Executive Committees could not be too large (11 to 20 members) we had to reduce the categories of stakeholder groups, and as the rich powerful farmers were the main contributors in the costs of the rehabilitation (approx. 15%) and operation and maintenance, the following proportionally representation in the EC was decided: *ayacutdars* (60%), non-*ayacutdar* farmers (10%), other sections including the landless women (30%). Women as a category should have a representation of at least three members in the ECs (*Ayacut*, non-*ayacut* and landless category).

The Organisation of the Community Participation

The Community Participation started in the beginning of 1999 and more than five years experience has been build up in the social wing of the PMU till the closing of the EU in 2004. The leading Non-Governmental Organisation had posted the residential sociologist as the main adviser in the Project Management Unit for community organisation matters. He worked in close cooperation with the Government of Pondicherry appointed sociologist, the nodal officers of the local NGOs and me. The NGOs recruited and appointed the community organisers (COs): men and women. Three pilot-tanks were selected.[19] Three field offices were established with the start of the second batch of tanks by the NGOs. NGOs were engaged by the PMU (3 NGOs in the pilot phase, 5 NGOs in batch 1, 2 and 3, and 4 NGOs in batch 4) to carry out the community participation work. Tasks, roles and responsibilities were substantial, but unfortunately cannot be described in the limited pages available. The following overview shows the increase in tanks and community organizers and consequently in the community participation work over the years. Over the years there was a clear inflation in the Ratio Tank/CO. The impact of this inflation rate was an increasing workload for Community Organisers.

Table 10.3: Overview: Number of Tanks and Number of COs (1999-2004)

Years/batch	*Number of tanks*	*Number of COs (gender specific)*
1st year 1999, pilot tanks	3	6 (2 women, 4 men)
2nd year 2000, batch 1	15 (3+12)	14 (6 woman and 8 men)
3rd and 4th year * 2001, 2002, batch 2	43 (3+12+28)	29 (6 women and 23 men)
5th year **, 2003, batch 3	72 (3+12+28+29)	29 (6 women and 23 men)
5th and 6th year, batch 4	82***	37 (13 women and 24 men)

** includes 1 sluice and 1 drainage Association*
*** includes 6 sluices and 1 drainage Association*
**** including 10 sluices and 1 drainage Association*

Methodology of the Community Participation

In the first years of the project a 12-step methodology of community organisation have been developed, which is documented in the excellent field manual on Community Organisation (T.P. Ragunath *et al.*, 2003). Many training sessions were organised to discuss and to analyse the 12 steps. The 12 steps include: introduction in the tank village, preparation of social map and a time line of the village, making a tank profile through transect walks and writing the history of the tank, making study tours to other already rehabilitated tank, organising village meeting, performing street-plays[20], door-to-door household visits, formation of stakeholder groups, selection of representatives, making an overview of stakeholder groups, Tank Association, Executive Committee and Office Bearers elections, study tour for Executive Committee members, formalising the Tank Association. Of each step the community organisers have to give an output to the staff in the office of the Project Management Unit.

Community organisers got substantial training in the methodology of community organisation, Participatory Rural Appraisal and my counterpart and I organised numerous gender training workshops for community organisers on gender awareness raising, gender theory and international gender policies, gender activities in the TRPP, improving the participation of women in planning and implementation, etc.[21] Moreover we organised regular meetings with the women representatives in the Executive Committees on their role, achievements, constraints, etc.

The community organisers have showed patience, wisdom, courage and a good sense for village life. Some organisers lived in modest thatched hut in the village for a few months without electricity. Moreover, they were able to use Participatory Rural Appraisal methods and tools in a flexible and creative way and to produce information and invented new exercises and teaching material themselves. Nodal officers and community organisers were the pillars of the community organisation and participation. Nodal officers were the contact persons of the involved NGOs and guided and supervised the work of the community organisers. Some community organisers felt that they were first the friend of the people, then became the enemy (or trouble maker and/or victim) and became the friend again in the 'tree-and-fish' stage. The role of the community organisers depends on various aspects as, for example, the power relations around the tank, the opposition capacity of encroachers, attitude and quality of the members in the Tank Associations, personality of engineers, etc.

10.6. Achievements of the Pro-Women and Pro-Poor Community Participation Approach

Achievements of community organisation and social aspects are difficult to assess in figures and percentages. In general only a small percentage of social reality can be measured in quantitative terms. This counts also for the results of the community participation in the TRPP. Many achievements can only be observed, or described in a qualitative way. Nevertheless I am of the opinion that the all-stakeholder approach has been very successful in the beginning years of the project in almost all of the villages. In particular landless women groups (Self-Help Groups) came forward and groups of the Youth clubs to facilitate the process of member ship recruitment, and they joined the Governing Body of the Tank Associations with enthusiasm.

The formation of the Tank Associations and Executive Committees were small 'revolutions' in itself. The fact that during the formation of the ECs the various caste, and class groups, women and men of all status groups, meet and sit, talk and discuss together in the otherwise highly polarised, and political faction ridden villages of Puducherry, was a major achievement of the NGOs, and in particular of the Community Organisers[22]. This is unique and has never happened before, as the villagers are not familiar with the forming of a democratic body in the Puducherry context. In contradiction to other states in India, the elections of village *Panchayats* are unknown in Pondy for about three decades, which make it very difficult to create an acceptable leadership. This achievement should be valued the more in view of the many hindrances that had to be overcome as: Political factions, hostility between villages, hierarchical behaviour, dominance of rich farmers, gender blindness, etc.

It is difficult to measure the changes in the mental attitude and mindset of the people and to know how far the community participation has boosted them to become self-confident and self-reliant in using and maintaining the tank in future. Transparency, honesty, common responsibility and democracy should be basic social values of Tank Association members, which should be rooted by the motivation work of the COs. It requires a lot of effort to let for instance Office bearers of the TAs understand that they should be advised and controlled by the members of the General Body and not by the Government staff.

The data of 44 tanks in 2003 showed that of the members of the Governing Body, 58 per cent were men and 42 per cent women, which is a great achievement from a gender 'mainstreaming' point of view. The highest membership rates in the General Bodies in 2003 were provided by the landless (40%), followed by the *ayacutdars* (24%), non-*ayacutdar* farmers (12%) and other professional categories (24%).[23]

In the villages small Tank Association offices have been inaugurated, opened and equipped to start the work. Moreover, in some TAs the president belonged to the small farmer category without borewells. However, it cannot be denied that the majority of the Office Bearers of the TAs belong to the rich and land owning classes and castes and are all men, with one exception of a woman Office Bearer in Keezhur tank.

10.7. Women's Participation and Representation in the Executive Committees of the Tank Association

The first step to involve the women of poor and landless household in the Tank Associations was to encourage husband and wife, both, of each household, to become a member of the Global Body of the TAs and to pay the membership fee of minimum Rs. 5. With gender awareness training and division of labour games we explained to Community Organisers, engineers, villagers, the important role women play in agriculture and their use and interest in the tank, and tried to convince them that women also should become a member of the TA and representatives in the Executive Committee and Office Bearer.

The representation of women in 62 Executive Committees of the TAs ranged from 13 to 33 per cent. On average it was one/sixth of the total. The only exception was a tank, which was under supervision of a woman nodal officer, where out of 11 executive members five women (four landless) were in charge. In the poorest tank, four landless women were elected (no farm women) on the total 17 members. However, the total number of women office bearers was only 5.4 per cent, almost all in the position of vice-president or joint-secretary. No woman got the treasurer position. It was strongly advised to the male decision makers in the PMU to pursue women as treasurer as it can be expected that women do the job seriously and honestly. Moreover it is socially accepted that women are accountants. And women in Self Help Groups (SHG) have the experience in accountancy.

In view of the overall India target to pursue representation of women in Government and decision-making bodies of 33 per cent, the representation of women members in the Executive Committees of the TAs was only half of this target. This is, however, a logical outcome of the proportional allocation of representatives of stakeholder groups in the EC. However, fewer male farmers and more farmwomen would have been possible.

Role and Perceptions of Women Executive Committee Members

Many workshops and empowerment trainings were organised for women EC members to exchange experience. Many women faced the problem that they were scared in the beginning to sit in the meeting with all the men. They say, "as a woman we cannot freely speak up and have freedom of movement." Now it is better. The role of the Women Community Organisers has been vital in overcoming this cultural gender constraint. They were a role model, moving around in the villages and conducting meetings with farmers[24]. It was interesting that several women EC members were widow. Some women got the full support of their husbands to become an EC member. One EC member told that a farmer had said to resign from the EC if any woman will be appointed. Not only the men constrained the EC members in coming forward but also women of 'opposite' parties undermined their and their families reputation by spreading bad gossip that they are 'loose' women interested in other men. During our training poor women complained: "Today, again we lose one day wage labour, sitting here and talking, but we don't get compensation of the Project."[25] Women landless EC members have struggled to get manual labour in the tank planning and implementation for the poorest landless women in their tank villages. In two tanks the toddy tapping on the bund has been abolished and the alcohol shop has been removed to protect poor women from violence and economic suffering.

Many women EC representatives are also leaders of Self-Help Groups which are basically saving and credit groups. Women's SHGs have been very important for the TRPP for the following reasons:[26]

1. The Community Organisers used SHGs meetings as a platform to inform about and to discuss with the women the aims and activities of the project.

2. Candidates from the SHGs were nominated and elected as women representatives for the Executive Committees[27].
3. SHGs have assisted as volunteer of the COs to collect membership contribution for the TAs.
4. SHGs women helped with organisation of meetings.
5. SHG women provided meals and food during the implementation of the work.
6. They were the platforms for the initiation of tree planting in the tanks.
7. The establishing, or reviving of SHGs by the three NGOs showed to the village women that the Project cares for their problems, and that might enhance the social profile of the project.

Substantial support (training, seed-money) for Self-Help Groups should be pursued as an integrated part of the project; to support and to facilitate the development of the village in a broad sense by taking initiatives as provision of drinking water, health improvement for children, education, removal of alcohol shops, etc.

However, Self-Help Groups were considered by the PMU authorities as women's organisations 'outside' the project. Reviving and initiation of groups was done by the NGOs in their 'own' time with their own small staff on a very meagre budget.

10.8. Women Community Organisers

Another gender achievement was that roughly one-third of the community organisers were women. The gender dimensions of the community organisers' work were assessed in several workshops. Advantages and disadvantages of being a male or a female community organiser were discussed. In some wards of villages, women community organisers cannot organise meetings in the evening, as alcoholism is rampant. On the other hand, the help of women community organisers is required in villages where a male community organiser is posted to call meetings with women or self-help groups. The advantage of women community organisers was brought out very clearly by a male community organiser. He observed that, 'when a woman community organiser calls a meeting with farmers they will come all, while that is not sure in case a male community organiser does the same.'

Real atrocities and cultural and social constraints might hamper that in particular young unmarried women community organisers can live alone in the villages themselves. The disadvantage of not living in the village is that evening meetings are difficult to conduct. As most farmers and landless labourers (women and men) are working during daytime, the early evening meeting seems the most convenient for the villagers. However, two women community organisers found their own solution acceptable for their parents and the community.

First, one female young unmarried community organiser was not allowed by her parents to live alone in the village. However, when a Dutch female student doing her M.A. thesis in anthropology on the 'use of the tank' took the initiative to reside in the tank village, the parents of the community organiser agreed that the two girls could stay together in the village. Later one more female community organiser joined them as her tank was near to the village. The grandmother, an old lady of 75 years, of one of the community organisers came after some time to their residence and lived with them. She did the household chores and did the cooking for the girls, which was very nice as they hardly had time to take time to do the household work and to prepare meals. Men community organisers can take food in small restaurants but for women, certainly young and unmarried, this is considered as not proper and indecent. In the mean time the two girls have resigned as their health came down[28] and it was time for marriage proposals.

Deliberately, we did not place women community organisers in a position as assistants of men community organisers. In that case we would just reproduce, and even re-enforce existing gender ideologies, which we pursue to change in our community participation work. For the public imagination it should be clear that we consider man/woman community organisers as equal, capable to do both the mainstream work. The ideal situation from a gender equality perspective is that teams of two community organisers, a man and a woman, work in the villages around one tank to reach out in an optimal way to the whole population. It was regrettable that in batch two, three and four there was hardly one community organiser per tank. The criteria used by the project managers, to estimate the number of COs required is based on the number of hectares in the *ayacut* to be covered. It makes more sense to base the number of community organisers required on the size of the population, and anticipated political/power relations in the village, large farmer/tubewell dominance, number of encroachers, etc.

The appointment of a woman nodal officer has enhanced the gender component in the project. In the tanks allotted to her, the highest number of women are elected as members in the executive committees, and even one woman is an office bearers in a Tank Association. Her NGO has juvenated dormant groups and initiated numerous new women Self-Help Groups in the villages of the tank rehabilitation project. But, this also happened in the tanks of the male nodal officer of CERD.

10.9. Negative Impact of Tank Rehabilitation on Gender and Poverty

Let us now have a look at the project phases after the Tank Associations were formed and the 'community planning' was over. It was in the phase of implementation of the physical works, and when the tank resources should be developed, that poor women started to realise that they will be very negatively impacted by the project and that their high expectations to get work and income will not be rewarded. The following three points were disastrous:

1. Eviction of poor encroachers without any compensation.
2. Little manual labour could be generated for the landless, particularly women during physical work.
3. Little labour could be generated in tank resource development: tree planting and fish raising.

Eviction of Poor Encroachers without Compensation

To start the rehabilitation work the mandate was, to evict all the encroachers in the waterspread, *Ayacut*, on the bunds, in the feeder canals, in short everywhere and to prevent at all costs that they will return after the works were over. So, the tank project had already very serious negative consequences for the poor and landless women as they were evicted even before any bulldozer could start the real rehabilitation (e.g. desilting and bund strengthening) work. Encroachment was considered illegal and eviction was carried out with bulldozers and police, but as said encroachers also left 'voluntarily'. Compensation did not exist as, according to the implementing authorities, it will reward people for illegal acts and encourage future encroachment.

A clear eviction of encroachers' integrated policy of PMU and PWD did not exist. This created many problems for the NGOs and Community Organisers as they were forced to play the role of the police. From 'friends' of the people they became an 'enemy' of poor people which was detrimental for their community participation role to involve also the poor in the 'managing-irrigation-together' concept.

The following case in one of the pilot tanks illustrates the problems faced regarding eviction of a few rich and many poor households.

In Vadhanur tank there are a few very powerful farmers having batter land and borewells installed in the tank. They earn a fortune by selling irrigation water to 254 poor families who have encroached small plots of land in the tank for home consumption of rice. When the borewell owners came to know of the TRPP, they quickly established their own Water Users Society (7 members) to show their power. They don't want to co-operate, as the Project is a threat for them, as they know that in order to rehabilitate the tank they have to be evicted out of the tank.

Many landless people depend on these borewell owners for buying water, and for wage labour in the tank. Therefore, the majority did not come on meetings and did not want to co-operate with the TRPP. After the eviction of the 254 poor encroachers and borewell owners, this last group convinced their 'clients' to take revenge. The sabotage and opposition of these borewell owners resulted in serious threatening of, and assault on the CO, the burning of the sugarcane fields of one of the office bearers of the Tank Association, and a three days looting of fish, with sticks and bare hands by the poor landless men, women and children. The fish was raised by the TA as income generation for the operation and maintenance of the tank. During the implementation of the technical works permanent police patrols were posted in and around the tank.

The total number of poor encroachers in the tanks is not known. In some villages evicted landless women were very angry and during our visits we were surrounded by large crowds of desperate poor women, begging for some employment in the project. 60 poor women were evicted who encroached a tank for already 30 years. They lost 150 kg of rice (two bags) having a value of Rs. 1000, which is a fortune for them.

Two widows, one with small children and other whose husband died recently as he tried to rescue two children in the river, told that paddy cultivation on a small plot of land in the tank is their survival. It was felt that there is something as a 'compassion' encroaching that is tacitly allowed by the community in case of a real family calamity. A policy, to mitigate the damage of the category 'compassion encroachers', was even lacking in the project. In one tank there were 13 poor encroachers. The authorities have given them the land years ago as a reward for the Family Planning Operation. They are not even aware that their cultivation in the tank is illegal.

The rehabilitation of the tanks created immense survival problems for hundreds of poor households and brought immense tension and unpopularity for the Community Organisers after the TAs were formed. Many poor were disadvantaged by the project. What happened with the promise that some income and employment should be provided in a later stage?

Little Employment for Poor Women During Implementation of the Technical Work

Some work that has to be done for tank rehabilitation as water hyacinth (Eichornia) removal, bush clearance, desilting of supply channels, sluice and surplus weir repair and strengthening and sectioning of the dykes can be done by manual labour. Meetings with the chief engineer and sub-divisional officers in all Circles, COs and with office bearers of the TAs were organised to discuss the plans, ideas and scope for manual labour. However, very little labour was created for landless women and somewhat more for men.

One of the main reasons was that cost estimates leave no scope to deploy wage labourers from landless families in the village but only for use of machines. In fact machines were cheaper than manual work. It is a shortcoming that in the project design no quota are fixed by the donor of the total amount of the costs of the tank rehabilitation that should be spent on giving manual labour to the members of poor landless households (men and women) in the villages of the tank. This should be made compulsorily in the cost estimates of the engineers, which should be controlled by the Tank Association and Project Management Unit.[29]

Added to this, the land owning communities in the village are against giving good wages to manual labourers fearing the rates for agriculture works would go up. They rather opt for use of machines and this is encouraged by the fact that the rates given to the Tank Association are adequate only for machine operations. Though in some villages there was an inadequate labour force due to seasonal migration, in other villages, poor *Dalit* and Vanniyar women fought for getting labour to desilt the village pond.

Nevertheless, there were exceptions depending on the social awareness of the Office Bearers, the efforts of the NGOs to insist on manual labour, and the self-confidence of the landless women representatives in the ECs to demand and to defend their interests. Mrs. T., landless EC woman member, had insisted on manual work for bund strengthening. 80 per cent of the households are landless in her village. 220 men worked for 9 days. Why only men? The answer is that the work was too heavy for women as thorny bushes had to be removed. The NGO was very proud on the work as the bund shape and quality of the work was much better by manual work compared with machines. The removal of water hyacinth was carried out by men and women in one tank and was an enormous success. The planting of trees was also mainly done by men as the soil of the bund was slippery and the work heavy. Poor women have contributed a lot to the formation of TAs, but the benefits for the women in terms of employment, income or other practical activities were far below expectation.

Development of Tank Resources: Trees and Fish

The Tank Association should pursue financial sustainability, as after the 'transfer of management' they have to finance the annual maintenance of the tank and its structures, once the rehabilitation is over. The TA has also to mobilise its own resources for paying its water distributor. The TA has theoretically the following resources to become sustainable: (1) Interest from the Corpus fund, (2) Income from tank usufructs like fish culture and tree auctions but also boating and tourism are considered, (3) Sale of tank silt, (4) Water cess collected from command farmers, and (5) Other contributions.

It is expected that the main source of income has to come from the tank usufructs in developing tree plantations on the tank land and to raise fish in the water of the tanks. But in practice this expectation was very problematic. Concerning usufruct rights of tank resources (trees and fishing) the existing situation is that the *Commune Panchayat* auctions fish and trees for a nominal amount to village leaders[30]. They can re-auction the resources for a higher amount. The proceeds accrued are mainly used for the operation and maintenance of temples in the villages. The people consider this as a gift from the tank to the whole community. A major constraint in tanks was the fear of the traditional religious elite (temple trust) that they will lose their customary usufruct rights to maintain temples and to finance religious celebrations as the Tank Associations claimed also part of the usufruct rights for the maintenance of the tanks. So, the temple maintenance has to share with the tank, and endless meetings were conducted to avoid conflicts.

A hot gender issue in the resource development of the tank is the auctioning of the Palmyra palm trees on the tank bund. These trees belong to the *Commune Panchayat* as they grow on Government land. The re-auctioning of these trees is a very profitable business as the trees are used for toddy tapping which is often sold in small toddy shop located at the edge of the tank bund. We met many drunken men. Alcoholism is rampant and causes increased poverty, heavier workloads for poor women to earn income and domestic violence. In many meetings it was expressed by poor women that it is unacceptable that the Tank Associations are supposed to become sustainable at the cost of the suffering of poor women. This is even more urgent because it appeared that in tanks men had planted new Palmyra saplings as they are interested in toddy, while the women wanted to plant Tamarind and Neem trees which are beneficial for food and medicinal use.

To our delight we found that in Bahoor Sitheri tank the auctioning of toddy tapping was stopped at the tank bund because of the gender awareness of the president of the TA and a woman representative in the EC. The palm trees were usually auctioned for Rs. 30,000 to a political leader and moneylender who made much profit with the toddy business. The TA blocked the auctioning. The toddy mafia beat up the president and the police took no action. But the women of the village came to protect their president. After the abolishment of the toddy trade the palm trees are directly auctioned to the poor toddy tappers for Rs. 12,000. They can fetch the fruits and palm leaves to sell. The abolishment of the toddy business means a considerable loss of money for the TA but they and the *Dalit* women are happy. It was observed that the number of drunken men has come down. When the toddy is not sold at their doorstep they 'automatically' drink less.

The Project Director of the TRPP agreed that a notification should be sent to all presidents of Tank Associations to follow the Bahoor model and to abolish the toddy business. But economic profit was more important than empowerment of poor landless women and the notification was never sent.

Building up of new fish and tree-resources around the tank was a major activity in the project although the budget available for subsidy was very meagre. I participated in tree planting ceremonies, and conditions for survival of the saplings looked fine, but in my next visit I was informed that the saplings did not survive for many reasons: no water, no care, men appointed, etc. In a few tanks tree plantations were more successful due to the dedication of landless women caretakers looking after the saplings for a very meagre salary and the motivation they got from the supervising NGO.

In another tank the CO had taken much effort to form a tree-planting committee of landless women. Numerous meetings were conducted to determine the species and locations suitable for tree planting. A map was drawn with a tree-planting plan. In the middle of this process the PMU authorities gave the tree planting to the Forestry Department, which gave employment to its own male labourers. Digging the pits and even the watching of the trees is done by the Forest Department staff. The community participation was undermined and as we will see it restricted also the use of the tank for poor women and children. Moreover, this happened in the tank in which more than 250 poor families were evicted and as revenge looted the fish-ponds of the TA. My field notes reveal the following:

> *We re-visit the tank and see a huge horrible fence of barbed wire along a substantial area in the water spread of the tank. We speak with three girls (16, 17, 15 years) and a boy (9). The girls dropped out of school after 8th standard. They now have to fetch the grass for the cow from the tank and to carry the heavy head load to the house. Earlier, before the tree plantation by the Forest Department, there cows could graze in this part of the tank. Now this is forbidden. They tell that they are now more afraid of snakes, as the trees and bushes take away the open clear sight they had before. The Forest Department promised that the trees would grow in two years, and that after that time the cattle can graze free again in the tank. The girls foresee that the control on them will be less as the cows can hide behind the bushes. For the same reasons they will also feel more insecure as 'somebody' can hide behind the trees.*

The Forest Department is also involved in other tanks and is forming so-called Village Forest Committees. A notification is issued in which guidelines for the implementation of Joint Forest Management in Puducherry are stipulated. This all shows that with the resource development of the TA, the use of the poor women of the 'common good' became more restricted and the participatory intention of the project came down. Worse, employment was even given to male outsiders while local poor women's groups were trained by the NGOs for making nurseries and tree planting.

In 2004 the Government of Pondicherry has transferred the usufruct rights for fishing to some TAs, but only for one year, which is of course too short to develop a sustainable fish resource. In several tanks I have attended small ceremonies to release fingerlings and the yields were diverse.

It is hoped that other resources of the tanks should not be taken away or commercialized at the cost of the poor and weaker sections. It is stipulated in the bylaws of the TAs, that 'usufruct' rights of poor people should be maintained according to 'customary' rights when they do not endanger the functioning of the tank. Up to now it is totally uncertain whether the poor will have any profit in future of the development of the resources. It should be kept in mind that benefits of development of tank resources will go in the first place to the purse of the TAs to be able to finance operation and maintenance of the tank in the future. Will they be able to yield a surplus, and will they be willing to invest that in productive activities of the weaker sections in the village?

10.10. Alternatives: The Gender and Equity Programme and Gender Specialist

In many tanks it appeared after the implementation phase of the physical works that poor *Dalit* women and Self-Help Groups were no longer interested to be a member of the Tank Associations, and did not want to renew their membership for the next year and to pay the fee. Many of them had lost their small plot of land in the tanks and others did not get the employment they expected. They turned their back to the Tank Associations and the project, and it was clear that much of the build up enthusiasm and their involvement in the community participation seeped away.

To create space for empowerment we pursued to appoint a full-time gender specialist and to give her a budget to start small income generating activities and empowerment training at all fronts. This was successful. The Mid-Term Review in May 2002 agreed with these suggestions. It was recommended to appoint a Gender and Equity specialist under the Technical Assistance to promote the full participation of women and to improve the quality of life of the Below Poverty Level (Rs. 2000 monthly per household) population in the tank villages. From June 2003 onwards the additional gender programme was initiated. The appointed gender specialist had an excellent NGO experience and knew the TRPP and the Pondicherry Government set-up very well.

With a small budget all kinds of gender programmes were taken up in the field of training and gender entrepreneurial experiments. One of the most successful components was the leasing of land for *Dalit* women and to use the System of Rice Intensification Cultivation Methodology to get higher yields. For six Self-Help Groups 50 per cent subsidy was given to buy power tillers, which can be rented out to farmers and in this way the group, can earn money.

Also commercial vermin-compost production was taken up to achieve supplementary income for rural households and to encourage eco-friendly farming and avoid soil degradation. Other activities were fodder cultivation, goat rearing, and vegetable/flowers/medicinal plants cultivation, soap, cleaning and washing powder production. The concept of 'tank shop' was developed and women got an opportunity to sell their products in the tank shop while in the mean time information on the aims of the tank rehabilitation project was given. Fish cultivation in ponds by women Self-Help Groups in which they got the usufruct rights from the *Commune Panchayat*, was successful. Other activities were closely related with the TRPP as numerous meetings for gender sensitizing of engineers, exposure trips, showing rainwater harvesting models, etc.

In July 2004 the donor European Commission 'suddenly' closed the Technical Assistance of the project, and EC assistance ended in December 2004. Main reason was that the Government of Pondicherry had not taken proper action to increase the electricity rate step-by-step to limit unrestricted deep tubewell pumping and to deplete the groundwater. This was a precondition mentioned in the official agreement of the EC and GoP at the start of the project. For political reasons the GoP had made little effort to increase the tariff. One effort led to immediate hunger strikes, protest demonstrations and road blockages by angry farmers.

The gender specialist was dismissed in July 2005 and the Gender and Equity programme was transferred to the male government sociologist. No gender and poverty activities and achievements are known from that time onwards. Community participation is reduced to a minimum and the NGOs and COs concentrate mainly on the technical engineering work. The Pondicherry Government extended the project for another three years and has decided to finance the project during these years. Out of 84 tanks planned for rehabilitation, 61 were completed in July 2004, and work in the remaining 21 must be in progress. It is reported by the NGOs that farmers use the water in some tanks for one season. In other tanks the water is not used for irrigation but only used to recharge the groundwater. Does this mean that the tubewell-owners are dominant and do not allow the poorer farmers to use the water as that means less income for them from water selling?

10.11. Some Concluding Remarks

In the previous paragraphs it is shown that the poor and landless *Dalit* women were initially enthusiastic members and participating stakeholders in the Tank Associations, but the costs were too high and their benefits almost zero and in the process they withdraw. Many lost their customary use-right of a small plot of land in the tank, the access to tank resources became even limited, and few manual labour and income was generated during implementation of the technical work and from tank resource development.

Tank rehabilitation and 'transfer of management' of tanks might have some positive effect on the recharge of the groundwater and the ecological environment, although that still has to be proven, but has no effect on reduction of poverty and gender inequality. Even worse, the 'transfer of management' of 'common property' excludes in the rehabilitation process in particular the poor and landless women and as a consequence re-enforces unequal caste/class and gender relations. As documented in deep tubewell and canal irrigation the large farmers are the winners and the landless women (and men) are the losers. The large farmers have double benefits of the tank project, while poor and landless women have to pay a high price for getting facilities from the 'managing-irrigation-together' policy even before the 'real' water management can start.

Lessons to be learned are that there should be a clear awareness, understanding and commitment on the part of the donor and implementing agency to reach out in 'managing-irrigation-together' projects to the most disadvantaged. The focus needs to shift to how participatory irrigation can provide the poor and landless women with sustainable alternatives for meeting their tank use and wider livelihood security by recognizing them as equally capable and responsible members of communities.

A policy commitment of gender 'mainstreaming' by donor gender advocates in embassies and ministries in the formulation of PIM land-and-water projects is by far not enough to ensure participation and empowerment of the poor and landless women. Gender 'main-streaming' in practice is not realistic, impossible and undesirable, in particular in an outspoken caste-class polarised and patriarchal context (Schenk-Sandbergen, 2002). Specific income generating and empowerment activities, in particular for poor landless women, should be identified, planned and designed in 'common property' projects with budget and staff. The worldwide call in water forums and international water conferences, to involve women in water management by 'main-streaming' gender, is easily made and sounds fine, but is broadly an illusion in the current male dominated practice. A solid only-women's project will be much more useful to improve the conditions of poor women.

Moreover, low women's participation in irrigation development and institutional reforms relates to their lack of resources in terms of land-and-water ownership. 'Transfer of management' does not address this crucial point.

As long as knowing and understanding the detailed caste/class and gender context is not a priority in irrigation policies and strategies, 'transfer of management' and Participatory Irrigation Management will never become a real participatory process including the whole community. And 'mainstreaming' women will remain 'male-streaming' without empowerment or real improvement of the living conditions of poor landless women.

NOTES

1. The strategy of mainstreaming is according to the UN report of 2002 (Gender Mainstreaming), "the process of assessing the implications for women and men of any planned action, including legislation, policies or programmes, in all areas and at all levels. It is a strategy for making women's as well as men's concerns and experiences an integral dimension of the design, implementation, monitoring and evaluations of policies and programmes in all political, economic and societal spheres so that women and men benefit equally and inequality is not perpetuated. The ultimate goal is to achieve gender equality."
2. For an overview of tanks of South India see the excellent publication of Vaidyanathan, 2001.
3. The Tank Rehabilitation Project Pondicherry started in 1999 and is still running although the Technical Assistance was closed by the donor (European Commission) in July 2004. The European Commission, the funding agency, has committed itself to an amount of Rs. 28.13 crores and the Government of Pondicherry to Rs. 4.40 crores. Rs. 2.20 crores have to be paid by 'community stake holders'.
4. There exists no Department of Water Resources in Puducherry as in neighbouring states.
5. I wrote 11 Technical Reports with detailed data on field visits.
6. In particular the observations of Dhanagopal, one of the Community Organisers working in the pilot tank Vadanur, were very useful.
7. Duck nomads are people having a fixed tank route every year. They release their flock in a tank for some weeks. After all the food resources are finished they move to another tank and release the flock there. In this way they survive. Any earthwork in the tank will impact them very much. They are a 'mobile' stakeholder group.
8. 'We' means the NGOs, the nodal officers, community organisers and partly the consultants of the social wing in the Project Management Unit.
9. The *Commune Panchayat* is the main administrative unit in Puducherry.
10. In general farmers were motivated to use tank water for irrigation as it is perceived as more fertile than the 'dead' tubewell water.
11. There were already Agronomical and Hydrological Surveys in the TRPP, so the Household Survey was not considered important as a database to analyse the social transformation process. The Household Survey was conducted and used by the COs as a support to know the people and to search for good representatives of the various stakeholder groups.
12. Women told that they join the men when there are no small children in the household. As soon as the children grow up the women stay behind in the villages.

13. See for the origin and status position of the castes in Tamil Nadu the excellent work of Mencher, 1978.
14. There are hundreds of Self-Help Groups in Puducherry (see paragraph 8).
15. One woman told that the tacit custom of weeding implies that they can take the weeds with them as fodder for the cattle. She has no cattle but gives it to neighbours. In return she can take the cowdung of their animals for cooking fuel.
16. Rs. 17 was the rate in 1999 but in November 2001, Rs. 20 is consistently mentioned as the rate for wage labour from Rs. 7.00 in the morning till Rs. 13.00 hrs afternoon. In 2004 the rate was still Rs. 20.
17. Stakeholders are people, groups, or institutions, which are likely to be affected by the proposed tank rehabilitation (either negatively or positively) or those which can affect the outcome of the intervention.
18. Under British Colonial Rule in the Madras Presidency the Tank Restoration Scheme began in 1883. Direct cause was the famines from 1866 onwards (Meinzen-Dick, 1984: 17). Tank improvement projects in Tamil Nadu were sponsored by the State Government and World Bank and the European Economic Community since the 1980s (*ibid.*, 5).
19. The Community Organisers working in the Project in the beginning were: Mr. Dhanagopal. Mrs. A. Rany, Mr. Paramanathan (he left in April 2001), Ms. Shamila, Mr. Nagalingam, Mr. Krishnamoorthy, Ms. Ilamathi, Mr. Bhuvaneswaran, Mr. Guansekaran, Mr. Manimaram, Mrs. Sreeprabha, Ms. Suchuda, Mr. Gopalakrishnan, Mr. Gopi, Mr. Manimaran.
21. NGOs have developed wonderful street-plays, which are performed in all tanks taken up.
22. In pilot tanks more than 80 meetings with the community were necessary to form the TA.
23. I wrote 11 reports as the sociologist/gender specialist on 11 missions. These data come out of Technical Report No. 31, 9th mission.
24. It was difficult to appoint women Community Organisers. At the peak of the project there were 37 COs, 13 women and 24 men.
25. In the Field School participants get Rs. 30 attendance fee per day.
26. See for a detailed background and history of Self-Help Groups, Schenk-Sandbergen, Technical Report No. 19, Report on the Sociologist/Gender Specialist's fifth mission, December 2000.
27. In some tanks where no Self-Help Groups existed it seemed difficult to find a woman leader ready to come forward as EC member.
28. In October 2001 the two community organizers got typhoid as a result of the bad drinking water quality in the village.
29. A quota of minimum 30 per cent is considered as a reasonable percentage.

30. It is not clear to me whether all tank resources are auctioned and what rewards are sluiced back to the *Commune Panchayat* and the PWD as the custodian of the tank.

REFERENCES

Agarwal, Bina. 1994. *A Field of One's Own, Gender and Land Rights in South Asia.* Cambridge South Asian Studies 58. Cambridge: Cambridge University Press.

Anandhi, S. 2002. "Interlocking Patriarchies and Women in Governance: A Case Study of Panchayati Raj Institutions in Tamil Nadu" in Kapadia Karin (ed) *The Violence of Development, The Politics of Identity, Gender and Social Inequalities in India.* New Delhi: Zed Books.

Gender Mainstreaming, An Overview. 2002. New York: United Nations.

Gupta, Deepak and L.P. Srivastava. 1999. *Status of Participatory Irrigation Management in India: A Documentary Critique.* ISS Manuscript Report 13. New Delhi: Institute of Social Sciences.

Jodha, N.S. 1986. "Common Property Resources and Rural Poor in Dry Regions of India", *Economic and Political Weekly* 21 (27), July 1986: 1169-1181.

Kelkar, Govind. 2005. "Development Effectiveness through Gender Mainstreaming, Gender Equality and Poverty Reduction in South Asia", *Economic and Political Weekly* 40 (44&45), October 29, 2005: 4690-4699.

Koppen (van) Barbara and Simeen Mahmud. 1995. *Women and Water-pumps in Bangladesh: the impact of participation in irrigation groups on women's status.* Department of Irrigation and Soil and Water Conservation, Wageningen Agricultural University and Dhaka: Bangladesh Institute of Development Studies.

Koppen (van) Barbara. 1998. *More jobs per drop: Targeting irrigation to poor women and men.* Ph.D. thesis. Amsterdam: The Netherlands Royal Tropical Institute.

Maloney, Clarence and K.V. Raju. 1994. *Managing Irrigation Together, Practice and Policy in India.* New Delhi: Sage Publications.

Meinzen-Dick Ruth Suseela. 1984. *Local Management of Tank Irrigation in South India: Organization and Operation.* Studies in Irrigation, Number 3. Ithaca: Cornell University.

Meinzen-Dick, Ruth and Margreet Zwarteveen. 1998. "Gender Participation in Water Management: Issues and Illustrations from Water User Associations in South Asia" in Douglas Merrey and Shirish Baviskar (eds), *Gender Analysis and Reform of Irrigation Management: Concepts, Cases and Gaps in Knowledge,* Proceedings of the Workshop on Gender and Water; 15-19 September 1997, Habarana, Sri Lanka. Colombo: International Water Management Institute.

Meinzen-Dick, Ruth and Margreet Zwarteveen. 2001. "Gender dimensions of Community Resource Management. The Case of Water Users' Associations in South Asia" in Agrawal Arun and Clark C. Gibson (eds), *Communities and the Environment. Ethnicity, Gender and the State in Community-Based Conservation,* pp. 63-88. New Brunswick, New Jersey, USA and London, UK: Rutgers University Press.

Mencher, P. Joan. 1978. *Agriculture and Social Structure in Tamil Nadu, Past Origins, Present Transformations and Future Prospects.* New Delhi: Allied Publishers Private Limited.

Mid Term Review Mission Draft Report. May 2002. Tank Rehabilitation Project, Pondicherry. Commission of the European Communities, WS Atkins Consultants.

Moser, Caroline. 1989. "Gender Planning in the Thirld World: Meeting Practical and Strategic Gender Needs", *World Development* 17 (1), December 1989: 1799-1825.

Mosse David, 1997. "The Symbolic Making of a Common Property Resource: History, Ecology and Locality in a Tank-irrigated Landscape in South India", *Development and Change* 28: 467-504.

Nanavaty, Reema. 2002. *Protecting and restoring: not the resources but resource users, managers and owners. Women Professionals in Water Sector. Water and Nature: Women's Work is Central.* SEWA Papers presented at the Second World Water Forum, March 17-22, The Hague, The Netherlands.

Palanisami, K. 1990. *Tank Irrigation in South India: What next?* ODI-IIMI, Irrigation Management Network Paper 90/2e. Colombo: IIMI.

Raghunath, T.P., Pushpalatha, S. and T. Srikant. 2003. *Field Manual on Community Organisation for Tank Rehabilitation Project Pondicherry.* Pondicherry: Centre for Ecology and Rural Development.

Rahman, Mahbubar and Willem van Schendel. 1997. "Gender and the Inheritance of Land, Living Law in Bangladesh" in Breman Jan, Kloos Peter and Ashwami Saith (eds.), *The Village in Asia Revisited,* pp. 237-276. Delhi: Oxford University Press.

Rao Aruna, Anderson Mary B. and Catherine A. Overholt (eds). 1991. *Gender Analysis in Development Planning, A Case Book,* pp. 1-31. West Hartford: Kumarian Press.

Rao, Aruna and David Kelleher. 2005. "Is there life after gender mainstreaming?", *Gender and Development* 13 (2), July 2005: 57-69.

Rao, Nitya. 2005. "Women's Rights to Land and Assets, Experience of Mainstreaming Gender in Development Projects", *Economic and Political Weekly* 40 (44 and 45), October 29: 4701-4708.

Technical Proposal. 1997. Pondicherry Tank Rehabilitation Project, India, European Commission, Antea, Arcadis, Pradan.

Sarin, Madhu. 1998. "Community Forest Management: Whose participation...?" in Guijt Irene and Meera Kaul Shah, *The Myth of Community, Gender Issues in Participatory Development,* pp. 121-130. London: Intermediate Technology Publications.

Schenk-Sandbergen Loes. 1986. *People, Trees and Forest in India,* Annex 2. Report of the Mission of the Government of the Netherlands on the Identification of the Scope for Forestry Development Corporation in India, Den Haag.

Schenk-Sandbergen Loes. (ed). 1995. *Women and Seasonal Labour Migration,* Indo-Dutch Studies on Development Alternatives. New Delhi: Sage.

Schenk-Sandbergen Loes and Outhaki Chalamany-Khampoui. 1995b. *Women in Rice Fields and Offices: Irrigation in Laos, Gender specific case studies in four villages.* Heiloo: Empowerment.

Schenk-Sandbergen Loes. 1998. Gender, Culture and Land Rights in Rural Laos, Gender and Development Programme, School of Environment Resources and Development, Bangkok: Asian Institute of Technology, Gender Studies, *Monograph* 7, 81 p.

Schenk-Sandbergen, Loes. 2002. Why gender 'main-streaming' means 'men-streaming' and informal 'women-streaming' in practice: a case study of the Tungabhadra Irrigation Pilot Project. IDPAD (research report).

Schenk-Sandbergen Loes and Niren Choudhury. 2003. From heroines to beneficiaries: from beneficiaries to heroines? The impact of a small-scale irrigation project on gender in West Bengal, Terai, India. IDPAD. New Delhi: Manohar.

Schenk-Sandbergen Loes. 2004. "Some Gender Dimensions in managing Water Scarcity" in Vaidyanathan, A. and H.M. Oudshoorn (eds) *Managing Water Scarcity, Experiences and Prospects.* IDPAD. New Delhi: Manohar.

Sengupta, Nirmal. 1991. *Managing Common Property, Irrigation in India and the Philippines.* Indo-Dutch Studies on Development Alternatives-6. New Delhi: Sage Publications.

SEWA (Self Employed Women's Association). 1997. Managing Water for the People by the People. A participatory Assessment on SEWA's Experiences with Rural Water Resource Management in Gujarat, Ahmedabad.

Sharma, Ursala. 1980. *Women, Work and Property in Northwest India.* London: Tavistock Publishers.

Sharma, Abhishek. 2003. *Rethinking Tanks: Opportunities for Revitalizing Irrigation Tanks. Empirical Findings from Ananthapur District, Andhra Pradesh, India.* Working Paper 62. Colombo: International Water Management Institute.

Shiva, Vandana. 1988. *Staying alive. Women, Ecology and Survival in India.* New Delhi: Kali for Women.

Tank Rehabilitation Project Pondicherry (n.d.), Terms of Reference.

Vuuren (van) Sonja. 2002. 'Bron van Leven', een onderzoek naar het gebruik van land en water rondom een collectieve watertank in de context van overlevingsstrategieen van landloze vrouwen in Pondicherry, Zuid India, Source of Life, research on the use of water and land around a collective watertank in the context of survival strategies of landless women in

Pondicherry, South India. Master Thesis, Anthropology, University of Amsterdam.

Viegas, Philip and Geeta Menon. 1993. "Bringing Government and People Together: Forest Protection Committees of West Bengal – Role and Participation of Women" in Singh Andrea M. and Neera Burra (eds) *Women and Wasteland Development in India*, pp. 171-210. ILO. Delhi: Sage Publications.

Young, Kate. 1994 (repr.) *Planning Development with Women, Making a World of Difference*. Development Studies. London: Macmillan.

Zwarteveen, Margreet Z. 2006. Wedlock or deadlock? Feminists' attempts to engage irrigation engineers. Ph.D. thesis. Wageningen: Wageningen Agricultural University.

CHAPTER

11

Water Tanks as Ecosystems: Local Ecosystemic Perception for Integral Management of Water Tanks in Tamil Nadu, South India

Pere Ariza-Montobbio, Elena Galan-del-Costillo, Tarik Serrano-Tovar, Victoria Reyes-García[1]

Introduction

Water tanks are complex human-made ecosystems involving many natural resources and providing a wide variety of functions. Furthermore, different stakeholders actively use water tank's resources and functions in different ways (Reyes-Garcia *et al.* 2011). The different ecological, social, and economic elements involved in the ecosystem are closely related and dependent on the existence of the water tank. The central presence of the tank gives the essential structure to the territory.

The geography of South India is characterized by the presence of a large number of tanks, making water tank ecosystems one of the most extended ecosystems in this region. Historically, the main function of tanks was to store and distribute water for irrigation purposes, but water tanks provide many resources other than water for irrigation. Water tanks also entail uses and functions other than those related to agriculture. Uses refer to activities that involve human's active utilization and benefit. Function refers to services generated by water tanks independently of the human direct benefit, such as the ecosystem services. Palanisami (2000, 2001), have shown that water tank ecosystems provide multiple uses (agri-uses, non-agri-uses as domestic uses). Here, we classify water tanks uses and functions in economic (i.e., agriculture, livestock, fishing,), ecological (i.e., groundwater recharge, prevention of soil erosion and floods), and socio-cultural (i.e., domestic, leisure, festivals). Water tank functions and uses are not independent one from each other for one element can share socio-cultural, economic and ecological aspects. For example, rice cultivation has both ecological and economic uses. The ecosystemic resilience that water tanks provide in case of disturbances (like droughts or floods) involves the recuperation of the society, the economy and the natural habitats and species. Thus, this ecological function has also an impact in the economy and the society.

In this chapter, we built on the assumption that water tanks are complex ecosystems that require an integral management. The section has two related goals. First, we assess people's perceptions of water tanks. Specifically we analyze whether users conceive tanks as a water storage system, or whether they consider them as complex ecosystems with many closely dependent elements. Second, we examine whether or not the current management of water tank ecosystems in our study sites considers all the resources, uses, functions, and stakeholders related to the water tank, as well as the relations among all them. For the empirical study, we use a primary body of data from two villages in a rocky zone area of the Villupuram District (Tamil Nadu).

Examining the social perception of water tanks is important to improve water tank management. If water tanks are perceived and mainly used as an irrigation infrastructure, then political authorities might argue that other forms of irrigation (i.e., groundwater, dams) might substitute water tanks without further consequences for the local population. In contrast, if—as it happens to be the case—people perceive and use the many functions and resources of water tank ecosystems, then the substitution of water tanks by other forms of irrigation might have severe collateral effects on the local population.

In the first part of this chapter, we use primary and secondary data to describe water tanks as ecosystems with different resources, functions, uses and the main stakeholders using water tanks. In the second part, we assess how the studied population perceived water tanks. In the last part, we describe and analyze the current management of water tanks in our study cases.

11.1. Description of Water Tanks as Ecosystems

Water tanks provide different habitats creating a heterogeneous net of interconnected territories. As we can see in Fig. 11.0, water tank ecosystems include the physic area occupied by the tank, the surrounding spaces, the resources associated to the tank, the human population living in the area, and the connected water bodies. The physical area occupied by tank changes from rainy to dry season.

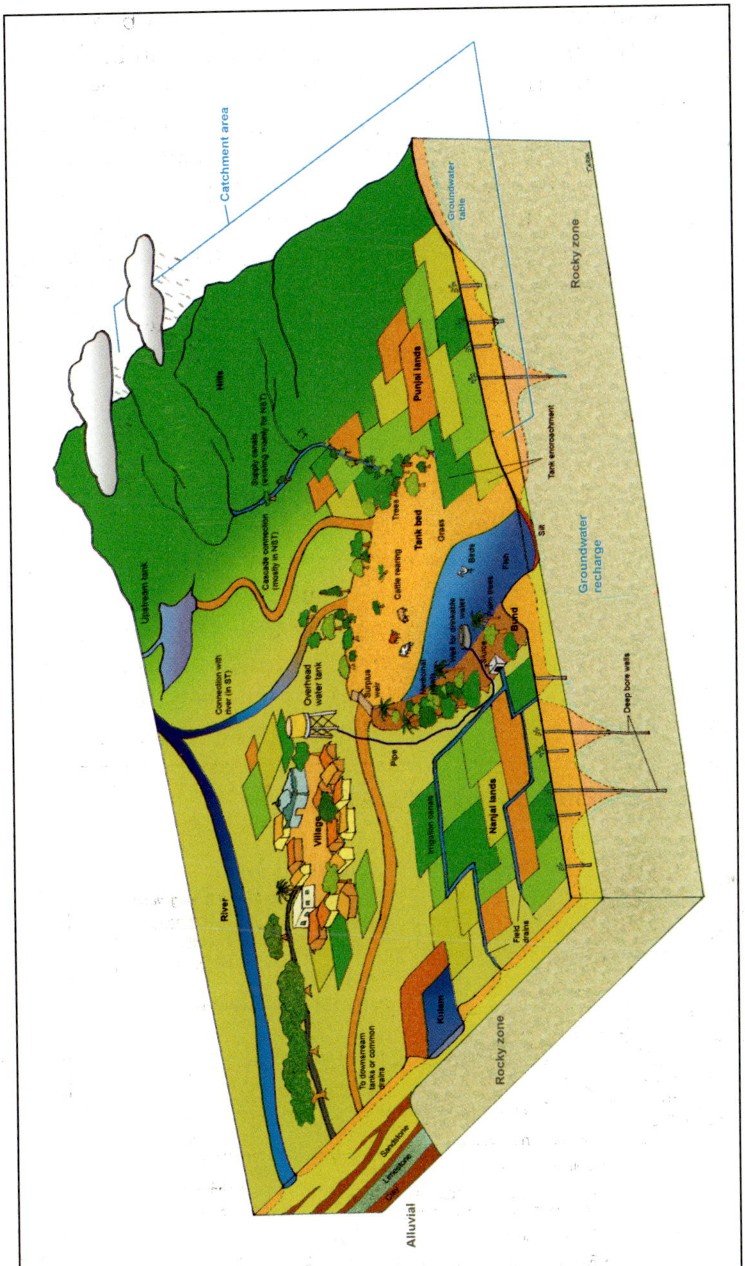

Fig. 11.0: The tank ecosystem

During the dry season, when the tank is at its lowest, the tank bed becomes a common open land (*porombokku*) where several natural resources can be found. Occasionally, the tank bed is also encroached by some farmers who grow seasonal crops on it. The surrounding spaces of the tank include the agricultural fields, the natural spaces like forests, and the human settlements. Both the natural resources and the human population living around the water tank are key elements of the ecosystem, because they depend on it and continuously modify it. Last, water tanks are linked to an extensive net of water bodies, including small ponds (*kulams*), wells, other water tanks, and, in some cases, rivers. The net of depending territories and related water bodies amplifies the ecosystemic relations of a water tank to a large scale. Thus, the boundaries of the water tank ecosystem are not limited to the area surrounding the tank, but are diffuse.

(a) Uses and Functions of Natural Resources in Water Tanks

Water stored in tanks is the main natural resource associated to them. Historically, water in water tanks was mainly used for agriculture, but also for cattle and for drinking. Because water tanks ecosystems are formed by several landscapes, many other natural resources—other than water—are associated to water tank ecosystems, increasing the value of the ecosystem from a social and ecological point of view. Here we distinguish between the economic, socio-cultural, and ecological functions of water tank ecosystems, although we acknowledge that some uses and functions might overlap (Fig. 11.1).

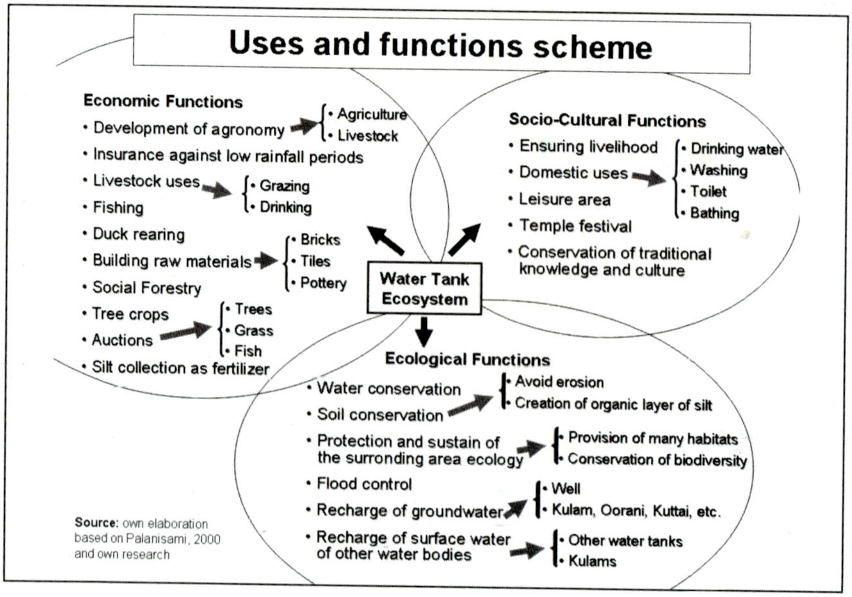

Fig. 11.1: Diagram of Tank Functions and Uses

Water tank have a clear economic use for villagers, using or not irrigation. Most natural resources associated to water tanks are considered village's common property. Villagers exploit natural resources in two different ways so the benefits are shared among the entire village. First, villagers celebrate public auctions to award the exploitation of natural resources of economic importance (i.e., trees, grass) to a single user. The money paid on the auction is commonly used as a fund to the village and is invested for the common benefit (mostly to fund temple festivals), or shared between the members of the village. Second, the exploitation of minor natural resources (i.e., silt for manure or medicinal plants) is open access. Any person from the village can access those resources. For example, the collection of silt from the tank bed to be used as manure for the agricultural lands is open to any villager. In sum, the wide variety of natural resources found in tank ecosystems have an economic use for many people, other than farmers.

Besides its economic uses, water tank ecosystems also provide many social and cultural uses and functions. Water tanks show the footprint of the relation of the Tamil culture with the environment through its use as a sacred place where traditional rituals take place[2]. For example, during some festivals the tank-bed is used as a stage for dramas, and sometimes cattle is sacrificed in the tank-bed. Tanks are also the stage of Hindu weddings and burial ceremonies. Tanks are also related with the preservation of traditional religious rituals, the potter community makes some sacred goddess statues with the soil of the tank. Tanks also have an active role in the preservation of ethnobotanical knowledge, especially medicinal knowledge, because many wild medicinal plants are found in the tank's surroundings. Last, other important social use, sometimes not mentioned by the local population but very common, is as a toilet place. Bushes around the water tank provide a hidden and private area. Women appreciate more than men the privacy of bushes in the water tanks to use them as toilets.

Water tanks are also central in the ecology of the area, providing the necessary *ecological functions* for a healthy environment. Water tank ecosystems include many types of different habitats, such as lacustrine, cereal cultivated fields, forests, tree plantations, shrub lands, riparian, or urban. Biodiversity of tank ecosystems can potentially be very high. Biodiversity guarantees the necessary functional units to provide and to ensure the supply of environmental services. Conservation of biodiversity is also an ecological insurance for the future that allows the ecosystem to regulate itself to the proper stable environmental conditions. The provision and creation of different habitats and the conservation of biodiversity increments tank ecosystems strength against ecological disturbances, generating high socio-ecosystemic resilience. The conservation of animal and plant species is a consequence of the ecological functions that tanks provide. Tanks are bird's sanctuaries and they help to preserve the avifaunal biodiversity. Tanks are very important for wetland and field birds, but also for migratory birds, which have adapted to use the tanks as a necessary place to rest and eat during their migrations. The dead storage of the tank has also the function to keep alive fish and other aquatic animals in the tank during the dry season.

In sum, water tanks are fundamental to conserve the dynamics of the ecology of the area. As explained before, water tanks are a central part of the territory, supporting and connecting many of its different ecological compartments (i.e., water, biota and soil), and creating a balance between the human actions exploiting the territory and the reposition and rehabilitation of the natural resources.

Below we describe the natural resources associated to the tank and their main economic, social, and ecological functions. Previous authors have provided descriptions of the agricultural use of water from water tanks (Mukundan, 1988; Janakarajan, 1991; Palanisami and Easter, 2001; Vaidyanathan, 2001; Mosse, 2003), so here we only provide a brief description of the resource water and we focus on other six natural resources dependant on the existence of water tank ecosystems: trees, grass, other non-timber plants, fish, other wild animals, and soil.

Water: Historically, the development of tanks in South India is linked to the development of agriculture (Mosse, 2003). South India climate is determined by intense monsoons and severe long dry periods. Agriculture may not have been developed with the current crop production in the area without water tanks. Thus the most salient economic function of water tanks is the storage and provision of water for the irrigation of crops. The existence of tanks allows the growth of crops along the year providing a supply of water for irrigation during part of the dry season and depending on ground characteristics recharging the groundwater (Mosse, 2003).

The most evident ecological function of water tanks is to preserve rainwater for periods of drought. Water tanks preserve water in the surface but also recharge groundwater through percolation. Furthermore, water from tanks might recharge other water bodies like other tanks or *kulams* through channels or groundwater flows.

Another important ecological function of tanks is to prevent floods. The climate characteristics of the area, with very heavy rains during very short periods make all South India a flood-risk area. Historically, water tanks controlled and protected from floods keeping closed the run-off waters and distributing the surplus water during periods of heavy rains. In case of poor maintenance of water tanks and canals, risk of floods becomes a major problem to the entire population and territory.

Water from tanks has also had a key socio-cultural function: to provide the basic supply of drinking water for the population. Through the history, villagers have used water from the tank and dependent *kulams* to drink. Water bodies were not polluted with fertilizers and pesticides, thus the water was apt for human consumption. Nowadays most of the drinking water come from piping networks, generally supplied in rocky zone with water from a big well constructed inside the tank beds. Piping networks use the groundwater available from the tanks' recharge. Thus, to this day, drinking water in rural villages of rocky areas is associated to water tanks.

Washing vessels and clothes are other very extended domestic uses of the water in the tank. Women wash vessels directly on the tank, on irrigation canals, wells, or *kulams*. Villagers also use tanks to wash or refresh cattle, and even to wash vehicles.

Last, bathing in the water tanks is most common when tanks are full of water, at the end of the rainy season. Bathing in the tank does not only have hygienic reasons, but is also done for leisure. When the tank bed is empty of water, there is a flat open space that is used in many cases as a playground for children or to practise sports like cricket.

Trees: A variety of trees are key elements of the water tank ecosystem due to their important uses and functions. Trees may be located almost everywhere in the tank ecosystem, but they mainly grow in the tank's bund, the tank bed, at the edges of agricultural fields, on the streets of the villages, and besides the main roads. The most important uses of trees are the human consumption of fruits and the use of logs for firewood and timber. For example, in the study area, enormous tamarind trees are commonly found besides the roads and their fruits are consumed locally. Wild trees such as *Prosopis juliflora* and *Acacia nilotica* are mainly used as timber and firewood. Fruits like coconuts and palm nuts are also consumed or used to make liquor. The big leaves of the palm trees are used to construct light roofs and walls of the traditional houses in the villages. During the 1960s the Forest Department of Tamil Nadu implemented programmes to encourage the growth of fast-growing tree species inside the tank bed. Following this programme, species such as *Prosopis juliflora* or *Acacia nilotica* are auctioned and harvested every 20 years (Palanisami, 2001).

Trees also provide important ecological functions such as creating a microclimate which refresh the temperature of the area. They purify the air, and they host many animal species.

Grass: Grass is specially abundant in most water tanks during the dry season, when they occupy most of the tank bed. One of the main species of grass in the water tanks of Tamil Nadu is a perennial grass called Vetiver (*Vetiveria zizanioides*). Vetiver grows when the surface water on the tank bed empties after the rainy season, leaving an open space in the remaining area.

This long and hard grass has important economic uses since it is used as construction material to build the roofs of many houses. It is also used as fodder for the cattle. Weed infestations and encroachment of the tank bed are now reducing the quantity of grass available. Another types of soft grass also grows on the tank bed. Soft grass is important for the grazing activities of the villagers owning livestock, because it provides fodder for cattle even in the season when no other food is available.

Other non-timber plant species: Bushes and aquatic plants found in water tanks also have different uses and functions. For example, many plants have medicinal properties. Some are used for cooking as spices or flavours, to make oils and lubricants or liquid fuels, or even to prepare poisons. Villagers also use bushes as domestic fuel.

Fish: Fish growing in the various water bodies associated to the water tank is an important source of food for villagers. Fishes grow naturally in the different water bodies, although new development programmes incite villagers to often introduce them to increase fish production. During the dry season, most of the tanks are emptying of the remaining water. In some cases the villagers empty completely the water of the tank, in order to catch the fishes with ease. Right to access fish differs from village to village (free access to all, customary rights for community fishermen who sell them in local markets, right given to the person(s) who took the auction and sell them). Whatever the case be, villagers consume fishes.

Other wild animals: In addition to fish, water tank ecosystems are the natural habitat for many other wild animals. The diversity and quantity of fauna vary a lot depending on the area. The tank provides an aquatic ecosystem in a dry and hot area where it creates a proper habitat for many different lacustrine and aquatic animals. Birds like ducks, herons, partridges, and pigeons are important groups of wild animals inhabiting the tank ecosystem. Some villagers may hunt those birds for meat or feathers. Villagers also take the eggs of some species, which are considered as delicatessen. Crabs, snails, and shellfishes are also commonly collected from the tanks and other water bodies like *kulams* and consumed or used to prepare medicines and manures. Many other types of animals like little mammals, snakes, frogs, and lizards live in the tank ecosystems, and they can be exploited or not by humans depending on the village.

Soil: The soil of the tank ecosystem is also a valuable natural resource for villagers. The natural processes in the tank ecosystem generate several types of soils, being the silt the most important for human uses. Silt has a high value as fertilizer for cultivated lands. Silt is mostly found on the tank bed, in the central part where water accumulates forming a layer of black soil very rich in organic matter and nutrients. The run-off water collects many different materials, such as leaves, branches, and humus, in their way from the catchment area to the tank. Those materials accumulate and decompose on the tank bed, creating a new layer on the land. Villagers collect the silt from the tank, transport it, and spread it in their cultivation fields as fertilizer. Villagers also exploit the red clay from water tank ecosystems to make bricks and tiles. The red clay is available in many tank beds due to its accumulation in stagnant water areas, collected from catchment areas rich in this material.

By controlling surface water flows, tanks also provide the ecological function of avoiding soil erosion, thus preserving the necessary layers of soil to keep the lands fertile. The creation of a layer of silt on the tank bed and subsequent distribution along the fields by farmers also contributes to create and maintain the quality of soil in the area.

(b) Stakeholders in Water Tank Ecosystems

The relations between society and environment are visible in the economic and social activities, the tank management and the relationship of the different stakeholders within the ecosystem, being active elements in the tank ecosystem dynamics.

Because water tanks involve many resources, uses, and functions, they also affect a variety of groups of users and beneficiaries. Each group has a different interest on the uses, functions and resources of the tank ecosystem, and therefore, interacts in a different way with it. Some of the groups have more power than others to manage and make decision that will affect the transformation of the ecosystem. Some groups influence only the ecosystem transformation without being directly involved in its management. Below we describe the different groups of interest found in the area.

Farmers are important stakeholders in the tank ecosystem and its management. The interest of the landholder on the water tank ecosystem depends on the type and surface of land owned. Agricultural lands around tanks are usually classified as irrigated (*nanjai*) or dry land (*punjai*). For example, according to the Tamil Nadu Farmers Management of Irrigation Systems Act (TNFMIS) only farmers having nanjai land are legal members of Water Users Associations (WUA)[3]. In practice, not all the nanjai farmers are involved in the management of the tank through this institution. Some rich farmers, well's owners, and encroachers, may not formally participate in the WUA, but they influence water tank management through their influence in village's social life, given that they have a high rank in the local social hierarchies. Farmers having *punjai* land are not legally involved in the WUA, but they are also important stakeholders in water tank ecosystem, because they are groundwater users and potential participants in the auctions of natural resources.

Differences of interest among farmers can also be due to the total surface owned. Big, small, and marginal farmers[4] occupy different social positions and roles within the society. The social position established by socio-economic differences (social class) affects the uses and interests of different groups on the tank ecosystem. The big farmers have a greater opportunity to be the highest bidder in the auctions. The marginal farmers and landless people are more dependent on the open access natural resources offered by the tank ecosystem. Last, the hierarchical structure of the caste system also affects the role and the access to resources by different caste groups as it will be explained later.

Landless people constitute a diverse group which have interests on the tank uses, functions, and resources that depart from those of farmers. The agricultural labourers and shepherds are direct users of tanks although they do not own land. As persons working in the fields, agricultural labourers depend on the tank for their daily work, like farmers. Shepherds use the water and the grass for rearing cattle. Some landless people use the grass in the tanks for roofing their house and their stall. Fishermen and duck owners might not use the tank daily, but they do in some seasons. Some nomads often hunt wild animals or fish if there is no auction procedure in the tanks.

There are also some *gender* differences referring to the uses of the tank ecosystem. Women make more domestic uses, washing clothes and vessels and collecting firewood, apart from their agricultural tasks, as we will see in the results in our case study.

Finally, the rest of villagers with non-agricultural occupation (shop owners, taxi and lorry drivers, students, road construction workers, tailors, building constructors, teachers, innkeepers, etc.) are also indirect beneficiaries of water tanks. All the residents in a village with a water tank benefit from diverse functions of the water tank such as recharge of drinking water wells, uses of wells, revenues of the auctions spent in the temple festivals. Villagers also benefit from the many ecological services of water tanks (i.e., prevention of floods). Hence, the entire village benefit from the water tanks in a direct or indirect way.

The complex village society is then an important element of the tank ecosystem that transforms it through its involvement with the economic and socio-cultural uses, within the net of ecological relations of the tank ecosystem elements. The high number of stakeholders involved in the water tank ecosystem makes necessary to understand multiple evaluation languages and perspectives, making important the post-normal science (Funtowicz and Ravetz, 2000), a new paradigm of science based on the people's participation of the social and environmental problems' management.

11.2. Assessing the Ecosystemic Perception of the Society

To compare the local perception of water tanks with their current management, we first examined whether people perceive water tanks as ecosystems or whether they mainly perceive them as irrigation infrastructures, without understanding the relations of the diverse elements. To do so, we conducted field research in the upper and western part of the Kaluvelli watershed, in the Villupuram district of Tamil Nadu.

(a) Methods

We collected data during a four-month internship with the IFP (February-May 2007). Research was conducted in two villages, Endiyur and Attur, both at about five kilometres to the east of the nearest city, Tindivanam. We selected the villages according to six criteria. First, both villages have non-system tanks that belong to the Public Works Department (PWD) and Panchayat Union (PU). Non-system tanks are more dependant on rain water than tanks connected to rivers. Second, both villages are located in a pediplain rocky zone. Wells in rocky zones are less deep than wells in alluvial area due to the hardness of the rocks and the lack of the technology to dig deeper wells. The first two criteria assure that villagers depend on the tank for irrigation, and thus increase the potential importance of the tank for the villages. Third, in both villages there were functioning Traditional Irrigation Institutions. Fourth, both villages celebrate auctions of natural resources from the water tank. Auctions might influence the importance of the tank for the livelihood and for the potential ecosystemic perception of the local population. Last, villages differ in the presence of modern irrigation institutions (fifth criterion). In Endiyur there were two Water User Associations implemented by one NGO (Palmyra) that promotes tank rehabilitation's activities. In Attur there were no modern irrigation institutions implemented. The presence of modern institutions should affect villagers ecosystemic perception of water tanks. The presence of the NGO is another important criterion.

The Setting

Endiyur is a suburban village with more than 3000 inhabitants in about 665 households (Census of India, 2002). About 95 per cent of Endiyur's population belongs to the vanniar caste (Most Backward Caste, MBC). The remaining 5 per cent belong to other MBC castes such as *assari*, *ambition*, and *dhobi*. The main economic activity in Endiyur is agriculture (80 per cent of the active population) (Suseela *et al.*, 2006). Most of the households in Endiyur (73%) are formed by small or marginal farmers and only 10 per cent by big farmers. About 17 per cent of households in Endiyur do not own any land.

Attur is a small village with about 1350 inhabitants in 290 households. The village is divided between the *Ur* and the Colony. In the *Ur* live people from the Backward Castes, BC (Naidu, 24%), the Most Backward Castes (Vanniar, 37%), and Other Castes, OC (Reddiyar and Muslim, 4%). In the Colony live people from the Scheduled Castes (SC) (most of them Adi Dravidar, 34%) and some Christians, BC (1%). As in Endiyur, the main economic activity in Attur is agriculture. Most of the households belong to small and marginal farmers (58%), and only 5 per cent to big farmers, who mostly live in the *Ur.* About 37 per cent of the households are landless and most of them live in the Colony.

Both villages have two non-system tanks. The two tanks in Endiyur are called Periya Eri and Chinna Eri, and the two tanks in Attur are called Attur Union Tank and Aranguti Eri. The PWD is responsible for the two big tanks (Periya Eri and Attur Union Tank) and the Panchayat Union for the two small ones (Chinna Eri and Aranguti Eri). At the local level, the institutions involved in the tank management are the *nattamais*, the village Panchayat, and the *nanjai* farmers themselves. In Endiyur there are also two WUA, one for each tank.

Data Collection

We collected qualitative and quantitative data. Qualitative data was collected through the research, but mostly during the first month of fieldwork, and contributed to provide a general description of (1) the water tank ecosystem and (2) the role of all the stakeholders and institutions involved in the use and management of the tank ecosystem. Methods of qualitative data collection used for a ecosystemic description include participant observation, participatory rural appraisal (PRA), open-ended interviews with key informants, and a qualitative environmental diagnosis.

To assess the local perception of water tanks, we collected quantitative data using free listings, pile sorts, and knowledge and use tests. We conducted free listings to generate a comprehensive list of local functions of water tanks (Weller, 1998). The free listing method consists in generating lists of elements that are related with a cultural domain. In our case the objective of the free listing was to get a list of all tank ecosystem resources, uses, and functions. We used free listings in both villages with a stratified sample of 54 people that included the various groups with expected variation in the perception about the water tank ecosystem, such as men and women or people with different types of occupation and land ownership. Every person was asked "Why do you think the water tank is important for the village?" We selected some items from the results and added some other elements from the literature and from the qualitative data collection, to elaborate cards with drawings representing village's elements. We prepared 28 items cards (see Fig. 11.3) to conduct the pile sort. Pile sorts are used to assess the relations that people make among the different elements of the water tank ecosystem. We asked a total of 44 people drawn from a stratified sample in each village to sort the pictures in piles according to their similarities (Bernard, 1995).

To better understand differences in people's knowledge of tank ecosystems, and to check if knowledge relates with daily activities and uses, we carried out a knowledge and use test. This method was useful to understand the differences between different stakeholders. We asked a stratified sample of informants (n = 109) to answer 11 theoretical questions about the resources and functions of the tank, seven questions about their use of water tank resources and functions, and two questions about water tank management. Theoretical questions included questions that aimed at assessing informant's perception of the ecosystemic nature of tanks, such as, ecological functions of the trees in the bund. Thus a higher score in this part indicates higher knowledge of the ecosystemic nature of the tank. The questionnaire about the use of the water tank included a set of questions on the many possible uses of tanks, such as washing clothes and using silt from the tank as manure. Thus, a higher score in the practical part indicates more diverse uses of the tank's resources and functions. The management part included opinion questions about who should be in charge of managing the water tanks. All the questions were in a multiple-choice format. People's answers were classified according to their high or low knowledge or use about water tanks having a numeric mark in each question. To analyze responses we generated two different scores adding the correct responses to each part of the questionnaire, one for the knowledge questions and one for the use questions.

To analyze the data collected through freelisting, we calculated: (1) The frequency of each item; (2) The average rank of how each item was on average ordered in the list of respondents; and (3) a Saliency Index that takes into account at the same time the frequency and the average rank of the element in the respondents' list. The Saliency Index evaluates the importance of an item in the all the lists (*see* Table 11.1). To analyze data from pile sort, we use non-metric multidimensional scaling to create a chart (*see* Figure 11.2) with clusters of elements. To analyze responses to knowledge and use test, we did two analyses. One was to analyze each single question to explore specific knowledge about one ecosystemic relation. The other one was to sum for each individual the marks of the questions of each part of the questionnaire to get a mean to compare knowledge across different groups. To analyze the knowledge and use scores, we used t-test and chi-square test.

Bias, Errors and Limitations of the Study

This study can have some potential biases in the data collection, mainly systematic errors that push the result to some direction unintentionally. The most important biases are the translation and interviewer bias. Bias in translation might be due to misinterpretation of questions and answers due to lack of linguistic abilities. Interviewer bias can be due to the evaluation of the answers, and the way to explain the questions and to help in case of trouble in the understanding among the informant and the interviewer. Results might also be affected by an unintentional confirmation (or non-neutrality) bias.

Interviewers might have remembered better information of things that confirms their hypothesis and the expectant results in the way of interpretation.

Table 11.1: General Results of the First 15 Elements from the Free Listing Sorted by Saliency (N = 51)

	Item	*Frequency*	*%*	*Avg. rank*	*Saliency*
1.	Crops	43	80	2.093	0.700
2.	Irrigation	25	46	2.840	0.377
3.	Drinking water for cattle	29	54	3.759	0.372
4.	Fish	33	61	5.515	0.285
5.	Drinking water	23	43	3.826	0.283
6.	Recharge of wells	20	37	3.700	0.260
7.	Water storage device	14	26	3.000	0.190
8.	To wash clothes	19	35	6.211	0.175
9.	Grass to make roofs	15	28	4.933	0.172
10.	Trees	21	39	6.000	0.168
11.	Grass	16	30	5.500	0.150
12.	Firewood	16	30	6.438	0.135
13.	Bathing	15	28	6.333	0.128
14.	Silt used as manure	14	26	6.929	0.111
15.	Auction of fish	14	26	7.143	0.102

(b) Results of the Study

We divide the results of our study in two main sections. The first substantive finding relates to the tank ecosystem approach and the local perception of the water tank ecosystems. The second substantive finding relates with the relations between the ecosystem perspective and the management of water tanks.

Local Perception of Tank Ecosystem

In the freelisting exercise, informants listed a total of 51 different reasons why water tanks are important for the village. On average each informant listed eight items in free listing (SD = 3). The shortest list had two items and the longest seventeen. None of the respondents mentioned only the purpose of storing water, which can be interpreted as that nobody would think that the tank is a device with the only purpose to store water. The length and diversity of items in the lists are the first indicators that people do recognize many of the different uses and functions of water tanks.

The maintenance of crops was single out as the most important and salient item in our lists. 80 per cent of the people interviewed mentioned crops, and on average crops were mentioned as the second item in free listing (Table 11.1). But although crops was the most important item in free listing, it is worth mentioning that as much as 20 per cent of the people interviewed did not mention crops when being asked to list the reasons why water tanks are important for the village. If we put together "crops" and "irrigation", we still find that as much as 7 per cent of the people interviewed did not mention any of these two uses.

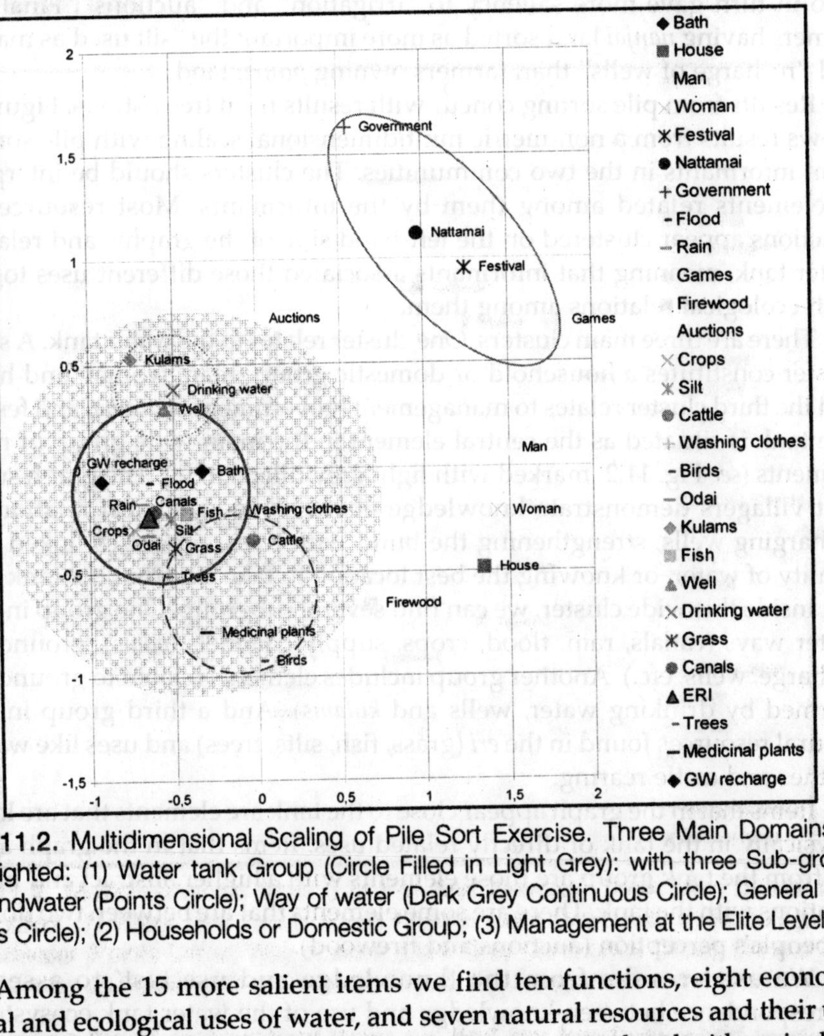

Fig. 11.2. Multidimensional Scaling of Pile Sort Exercise. Three Main Domains are Highlighted: (1) Water tank Group (Circle Filled in Light Grey): with three Sub-groups: Groundwater (Points Circle); Way of water (Dark Grey Continuous Circle); General Uses (Lines Circle); (2) Households or Domestic Group; (3) Management at the Elite Level

Among the 15 more salient items we find ten functions, eight economic, social and ecological uses of water, and seven natural resources and their uses. We highlight here the high saliency of two elements not related with agriculture: fish and drinking water (Table 11.1).

We run the same data using only parts of the sample. We found that using trees as firewood and washing clothes were more salient items for women than for men in the sample. The people with occupations related with agriculture mentioned more "recharge of wells" and less "drinking water" and "fish" than people working in other occupations. For landless people "drinking water for the cattle" and "grass for the roofs" were more salient than for landholders, who in turn gave more saliency to "irrigation" and "auctions". Finally, the farmers having *nanjai* land sorted as more important the "silt used as manure" and "recharge of wells" than farmers owning *punjai* land.

Results from pile sorting concur with results from free listings. Figure 11.2 shows results from a non-metric multidimensional scaling with pile sort data from informants in the two communities. The clusters should be interpreted as elements related among them by the informants. Most resources and functions appear clustered on the left-hand side of the graphic and related to water tank, meaning that informants associated those different uses together with ecological relations among them.

There are three main clusters. One cluster relates to the water tank. A second cluster constitutes a household or domestic group (man, woman and house), and the third cluster relates to management (government, *nattamai* and festival). The tank is situated as the central element in the main wide group of related elements (*see* Fig. 11.2, marked with light grey filled circle). The graph suggest that villagers demonstrate knowledge about some ecological relations like recharging wells, strengthening the bund with trees, processes linked to the quality of water, or knowing the best location to take silt from the tank bed.

Inside this wide cluster, we can find several subgroup. One group includes water ways (canals, rain, flood, crops, supply channels (*odais*), groundwater recharge, wells, etc.). Another group includes elements related to groundwater (formed by drinking water, wells and *kulams*). And a third group includes natural resources found in the *eri* (grass, fish, silts, trees) and uses like washing clothes and cattle rearing.

Items that in the graph appear close to the tank are elements that are located physically in the tank or directly related uses. Items that in the graph appear far from the tank group are those elements with a higher abstract and indirect relations with the tank. There are some elements that are between two domains in people's perception (auctions and firewood).

We used results from the 'knowledge and use test' to assess the correspondence between knowledge and use of the water tank ecosystem by the villagers. We found a positive and statistically significant correlation between individual knowledge and use scores. Results from a Pearson correlation between individual knowledge and use scores had a coefficient of 0.34 ($p = 0.0003$; $n = 109$). The low correlation coefficient can be due to recent changes in some practices, such as the use of artificial fertilizers. For example, we found that the 80 per cent of the people related with agriculture know where is the better place to take silt to use as manure (near the sluice, located in the bund), but only the 48 per cent of the farmers really use this resource from the tank. This means that the low values of correspondence between knowledge and use shown by the villagers are in general related to external factors, outside the villagers' control. These external factors are mainly related to economic decisions, when there is not competition for the external alternatives of use and resources away from the tank ecosystem, like the artificial fertilizers instead of using silt as manure.

We also analyzed differences in knowledge and use scores according to individual attributes (Table 11.2). We found that farmers and agricultural labourers had a higher knowledge score (mean = 18.2) than people with other occupations (mean = 14.6). The difference of means between the two occupational groups was statistically significant ($p < 0.1$). We also found that landowners, *nanjai* farmers, people who had received training or participated in some schemes like NREGA, and people who had undertaken management duties at whatever level had significantly higher knowledge scores than people without those characteristics. When dividing the population according to their sex, we found that men had significantly higher scores than women. Analyzing the data according the caste of the respondents we did not find significant differences in the knowledge among the Scheduled Caste people and the other caste groups.

Table 11.2: Comparison of Mean of Knowledge Scores in Multiple Choice Tests in Different Groups

		Mean	*SD*	*Obs*		*Mean*	*SD*	*Obs*	*T-Test of comparisons of means (p)*
Occupation	Farmers and agricultural labourers	18.2	3.9	74	Other occupations	14.6	3.9	35	0.000
Land-wnership	Land owners	18.1	3.9	75	Landless	14.7	4.0	34	0.0001
Type of landholding	*Nanjai* farmers*	18.8	3.9	46	*Punjai* Farmers	15.8	4.0	63	0.0001
Training	Have received training	20.0	4.3	17	No Training	16.5	4.0	92	0.0011
Participation*	Yes	18.7	4.5	59	No	16.4	3.9	37	0.007
Gender	Male	17.8	4.1	48	Female	16.5	4.3	61	0.05
Caste	Scheduled caste	18.2	4.4	19	Other castes	17.2	4.3	77	0.33

Note: *Participation refers to people who's report having been involved in tank management duties at whatever level (Panchayat, Grama Sabha, SHG, NGOs, and WUA) and in some schemes (NREGA). *Nanjai* includes the farmers having *nanjai* land and farmers having *nanjai* and *punjai* lands. The *punjai* farmers category includes the people who only have *punjai* land.

We conducted a similar analysis for use scores. Recall that the scores of use do not relate to a higher intensity of uses, but to a higher diversity of uses. We did not find significant differences in use scores associated to individual socio-economic characteristics (see Table 11.3). Most of the groups in the sample have similar scores in the questionnaire about the use of the water tank, regardless of their occupation, land ownership, type of landholding, training in water tanks, and gender with one exception. When dividing the population according to caste, we found that people who belong to a Scheduled Caste had significantly higher diversity of uses of the water tanks than the people who belong to other casts (p = 0.03).

Table 11.3: Comparison of Mean of Use Scores in Multiple Choice Tests in Different Groups

		Mean	*SD*	*Obs*		*Mean*	*SD*	*Obs*	*T-Test of comparisons of means (p)*
Occupation	Farmers and agricultural labourers	3.9	2.3	74	Other occupations	3.4	2.4	35	0.12
Land ownership	Land owners	3.5	2.2	34	Landless	3.8	2.4	75	0.28
Type of landholding	*Nanjai* farmers	4	2.4	46	*Punjai* farmers	3.6	2.2	63	0.20
Training	Have received training	3.3	2.7	17	No training	3.8	2.2	92	0.20
Participation	Yes	2.3	1.4	59	No	2.3	1.3	37	0.36
Gender	Male	4.1	2.1	48	Female	3.5	2.4	61	0.10
Caste	Scheduled Caste	2.7	1.3	19	Other castes	2.1	1.3	77	0.03

Discussion

In the following section we discuss the general results for the entire population and the results by categories about the ecosystemic perception of water tanks.

The evaluation of such an abstract concept as the perception (or awareness) of the ecosystemic nature of water tanks is a difficult task. Results from our three surveys suggest that people do have an integrated perception about the ecosystem, although we set the factors that determine the differences between groups of people.

We found that respondents' occupation is related to people's perception of the ecosystemic nature of water tanks. Villagers working with agriculture (farmers and agricultural labourers) have a higher ecosystemic perception of the water tank than villagers who do not work in agriculture. The finding can be explained due to the experience in the field, which puts people in contact with elements of tank ecosystem. For example, people with agricultural occupations know more and have a better understanding of the ecological relations between the environmental compartments, as the concept of "water quality" asked in the first question of the knowledge test shows. Thus, 50 per cent of people working in agriculture could explain that nutrients dissolved in water are different depending on the source of water, while only the 25 per cent of people with other occupation knew it. Similarly, more farmers and agricultural labourers could also explain the ecological functions of the dead storage 34 per cent of people working in agriculture mentioned that the dead storage was useful to recharge wells *versus* 11 per cent of people with other occupations. We also found differences related to the occupation of the informants on the knowledge about the best place to take silt from the water tank, and the knowledge and participation in the auctions of natural resources. 89 per cent of farmers and agricultural labourers knew the location of silt *versus* 58 per cent of the people with other occupations. Farmers and agriculture labourers also knew the procedure of the auctions, having a more active participation (offering a higher bid) than people with other occupations. Performing cross tabs and Chi-square test we found a significant difference between these two different groups of occupation.

Results from Tables 11.2 and 11.3 show that, among the people who are dedicated to agriculture, landowners have a more integrated perception of the water tank ecosystem than landless agricultural labourers. The finding can be explained due to landowners' social position and role in management activities. Furthermore, among landowners, *nanjai* farmers show a higher holistic vision than *punjai* farmers. The difference can be due to the closer relation of *nanjai* farmers with the tank through the use of its surface water. For example, when examining the question on the silt location, we found that 100 per cent of *nanjai* farmers knew the best place to take the silt. We found a significant difference performing cross tabs and Chi-square test.

We also found that participation and training influence the social perception of water tanks as ecosystems. People who have been trained by NGOs have a more ecosystemic perception of water tanks than people who have not received such training. People who had carried out management duties at any level (Panchayat, Self Help Group managing, WUA, financial duties or agricultural managing, for example), were able to make more complex relations of different elements. This eases their perception of the elements of the ecosystem and their relations. For example, 73 per cent of the people experienced in management activities or trained by NGOs knew the ecological functions of trees located on the bund, but only 35 per cent of the non-experienced people knew it. There is a significant difference between these two groups when performing cross tabs and Chi-square test. We can see also the same result in the whole analysis of the knowledge scores (*see* Table 11.2), where it is found a significant difference in knowledge among the people who participate in managing activities or have been trained by some NGOs and other institutions.

We also found gendered differences in tank ecosystem perception. Women use different resources and services of the water tank than men due to the gendered division of domestic labour. Services used by women include washing clothes and collection of medicinal plants and firewood. Women's different perception is shown in the free-listing exercise where those items are more salient in women's than in men's freelisting. Men are typically in charge of the management tasks, and this is the reason why we did not find significant differences when analyzing the quantity of uses in the uses test score by gender. The management task done by men creates a more holistic vision of the ecosystem in general, and this is shown in the significant differences in the knowledge scores.

Despite important differences in people's knowledge according to socio-economic characteristics, we did not find differences in the diversity of people's uses of the tank. The finding can be interpreted as that although uses change from one group to another, everybody in the village use the tank for some reason (irrigation, auctions participation, collection of silt and medicinal plants, consumption of natural resources, washing clothes, bathing, drinking cattle or roofing with grass). This important finding contributes to consider the relevance of the water tank functioning as ecosystem and the need of an integrated management considering all the resources, uses, functions and stakeholders.

11.3. Water Tanks Ecosystems' Management

In this last section, we analyse the current management in both case study and the opinion of the people about the management. The goal of the analysis is to evaluate whether or not the current management is considering the resources, uses, functions, stakeholders and their relations.

We divide the activities related to tank management into three main groups:

- *Maintenance of the infrastructure*: Water tank's infrastructure includes the supply channels (channels to lead the water from the catchment area to the tank), the tank itself (tank-bed, bund, sluice, surplus weir), and the irrigation channels (channels to distribute water to fields). Tanks whose irrigated area is above 40 ha are under the control of PWD, whereas the rest of tanks are under the control of Union Panchayat. These institutions are in-charge of the maintenance of the tank and supply channels but not of the irrigation canals. Farmers usually undertake the irrigation canals' maintenance; however, new schemes such as NREGA (National Rural Employment Guarantee Act)) and tank rehabilitation programmes involve other institutions in the tank's maintenance. For example, the purpose of the NREGA is to give work to rural population in sectors that have some collective impact such as road construction, water resource conservation, soil conservation, drought proofing (including forestation) and flood control. This scheme is implemented through the Village Panchayat whose members decided to desilt and strengthen the bund of the tanks without any involvement of the WUA. NGOs have also worked in the area, conducting tank rehabilitation programmes in Endiyur, where two WUAs were created as the tank was rehabilitated. Attur refused the NGO's conditions, especially due to the economic reason, so there was no water tank rehabilitation in Attur.

- *Use of water for irrigation*: Management activities related to the use of water for irrigation consist in the release of the water from the sluice and its allocation and delivery to the fields. Traditionally, the person in charge of the allocation of water in both villages was the *thotti*, although his role on this duty has been lost. Now a days *nanjai* farmers are in charge of this task.
- *Management of the other resources*: Other resources in the water tank also require management. Some of them (grass, fish and trees) are auctioned every year. The auctions are managed by the *nattamai* (a caste leader) and the Village Panchayat, and the benefits are kept by the *nattamai* for communal uses.
- *The other natural resources and uses of the water do not have restriction*: They are in open access. The lack of regulation over those natural resources may contribute to the damage of the tank ecosystem, because it can generate negative impacts on water (for example washing clothes or using the tank as a toilet) or it may lead to the depletion of those resources (like the soil for making bricks).

During our research, we identified several conflicts about natural resources of the water tanks. Many of those conflicts related to historical caste conflicts in Attur village, among the Colony (where the Panchayat President belongs since 2006) and the *Ur* (where the *natamais* belong). Traditionally, benefits from the auctions in Attur were divided in three parts, two parts going to the *Ur* and one part going to the Colony. However, nowadays the Panchayat President belongs to the Colony, and Colony residents claim for an equal share of the benefits of the auction, and the *Ur* (represented by the *nattamais*) do not agree with such request. As the *nattamais* and the Village Panchayat have traditionally belonged to the *Ur*, the inhabitants of the Colony could never participate in the management of any part of the water tank ecosystem, and now, with the SC reservation for the Panchayat President, the SC people of Attur Colony have been empowered for the first time in their history. Because of those conflicts, the auction of the natural resources from the tank was stopped. It is important to consider that the conflicts may hinder a potential integral management of water tank ecosystems.

We conducted interviews with 109 people to assess their opinion about who should have the power over the tank and the rights and duties to manage it and why (*see* Fig. 11.3). We found that the biggest group of respondents (29 per cent of the sample) thinks that the entire village should be involved in managing the tank. This is a very participatory way to manage village's own resources but it does not exist nowadays. The second biggest group of answers corresponds to the 21 per cent of people who think that the government should manage the tank. Respondents in this group argued that the government has resources to maintain the tank. Respondents might also have been confused with the current implementation of the NREGA programme from the government to repair the water tanks. The programme might have made people think that the government is making an active management of tanks.

Seventeen per cent of the sample thinks that the *nattamais* should manage the tank. In the case of Endiyur and Attur Ur, the *nattamais* are local managers of the village, and they do not represent any political party nor receive any wage for their duty. Moreover, in Endiyur the *nattamai* is elected every year, changing among the five Vanniar caste groups of contributors to the temple (most of the inhabitants of Endiyur belong to vanniar caste, the other castes and the Muslim people belong to another sixth group, that does not participate in the *nattamais* election). Another group of villagers (15%) thinks that the Village Panchayat should manage the tank. Their reasons are that the Village Panchayat is a local institution recognised officially by the rest of external institutions like the government, and they are elected democratically.

Some people also mentioned the PWD as the institution who should manage the tank (5%). Villagers think that the PWD is a department formed by technicians and specialized in these themes, being part of the government, and giving money to repair the tanks.

Three persons said that all the categories proposed to manage the tank should participate together, without leaving some of them outside of the management. We are not sure about the reliability of this answer because it sounded like an "easy" answer. It is also possible that respondents were thinking in a coordinative management where all the stakeholders at the different levels are important and they should take part in the management.

Finally, two persons choose the *nanjai* farmers as the best managers of the tank (they were *nanjai* farmers). The low number of people who pointed at *nanjai* as the most appropriated managers contrast with the current policy of leaving the management rights of the tank to the WUA formed only by the *nanjai* farmers, leaving all the other local stakeholders outside the management. When people were asked why they did not consider that farmers should manage the tank, some of them said that farmers are too selfish to carry on this task alone.

To finish we highlight that the nature of WUA obstructs the integral management of water tanks ecosystems. WUAs are only formed by *nanjai* farmers, excluding all the other stakeholders of the management of the tank. These exclusions originate management with limited visions of the ecosystem, giving preference only to the irrigations requirements of the nanjai farmers. In this sense, even the name of the institution supposed to manage the tank reflects the exclusion. The name of "Water Users Association" might suggest that the tank is just a water deposit, and the only people in charge of managing it should be the irrigation users. Among the *nanjai* farmers of the WUAs observed, there are also situations of "participatory exclusions" (Agarwal, 2001), which affects women and the smallest (and poorer) *nanjai* landowners, due to local conflicts of interest between different groups of users, like their membership to different political parties, and the different income based conflicts. These exclusions affect the behaviour of the group and give preference to some sectors of farmers, discouraging other sectors to take part in the menagement.

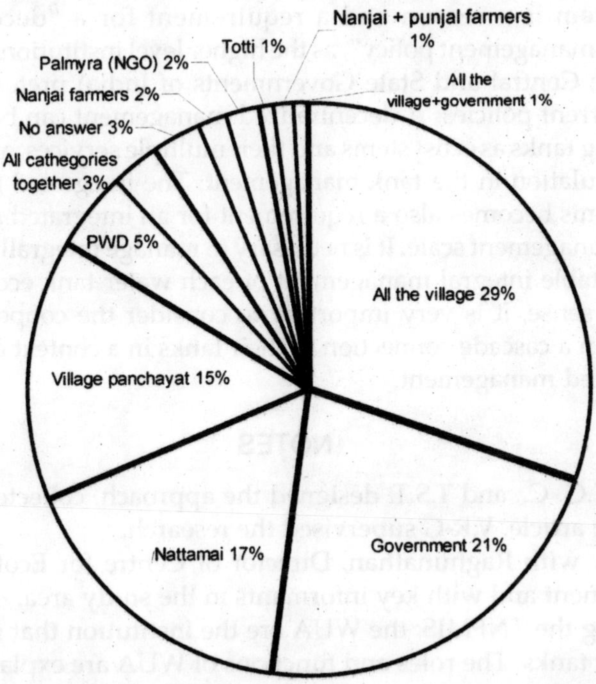

Fig. 11.3: Villagers' Opinion about who should manage water Tanks (N = 109).

Conclusions

The ecosystemic approach to the water tank system allows the consideration of all the resources, uses, functions, and stakeholders related to water tanks. The proposed approach enhances the value of water tanks serving the population and giving the essential structure to the territory. This vision suggest that water tanks are irreplaceable by other irrigation sources. As we have shown, the beneficiaries of water tanks uses, resources, and functions are more than WUA's members.

Unfortunately, the current system of management does not consider all those beneficiaries. We have analyzed how different groups of stakeholders perceive and use water tanks, and the socio-economic attributes that are associated to those perceptions. We found that knowledge of water tanks varies across socio-economic groups, however, we found that all the respondents use the tank differently but with similar intensity. The finding has implications in the water tank's management. A possible explanation for the serious deficiencies of the current management of water tanks relates to the exclusion in the participation in the management of a large number of stakeholders, a problem which will not be solved with the concept of the WUA implemented in Tamil Nadu.

Finally, it is pertinent to note that an integral management of the entire water tank ecosystem is a reason and a requirement for a "decentralized and participatory management policy", as the higher level institutions (i.e., the World Bank and the Central and State Governments of India) pretend theoretically with their current policies. A decentralized management can be only achieved by considering tanks as ecosystems and their multiple services, and by including the local population in the tank management. The integrated management of tank ecosystems becomes also a requirement for an integrated approach in the watershed management scale. It is necessary to manage integrally the watershed to have a suitable integral management of each water tank ecosystem at local level. In this sense, it is very important to consider the cooperation between villages within a cascade connection of their tanks in a context of decentralized but coordinated management.

NOTES

1. P.A.-M, E.G.-C., and T.S.T. designed the approach, collected the data and wrote the article. V.R-G supervised the research.
2. Interview with Raghunathan, Director of Centre for Ecology and Rural Development and with key informants in the study area.
3. According the TNFMIS, the WUA are the institution that should manage the water tanks. The roles and functions of WUA are explained in chapter 6 of this book.
4. The common classification of landholding within administration and economists consider the following criteria: > 5 ha: big farmers, 2.5-5ha: small farmers, <2.5 ha: marginal farmers.

REFERENCES

Agarwal, B. 2001. "Participatory Exclusions, Community Forestry, and Gender: An Analysis for zSouth Asia and a Conceptual Framework", *World Development*, Vol. 29, No. 10: 1623-1648.

Bernard, H. Russell. 1995. *Research Methods in Anthropology. Qualitative and Quantitative Approaches*. Walnut Creek: Altamira Press. Census of India. 1995.

Census of India 1991, Tamil Nadu. Delhi: Office of the Registrar General.

Census of India. 2002. *Census Info India 2001, Vol.1* (CD format). Delhi: Registrar General and Census Commissioner.

FERAL (Foundation for Ecological Research, Advocacy and Learning). 2005. *Resource handbook. Planning and monitoring collective local resource management in the Kalivelli watershed*. Pondicherry: Foundation for Ecological Research, Advocacy and Learning.

Funtowicz, S. and Ravetz, J. 2000 (1st ed.). *La Ciencia Posnormal, Ciencia con la gente*. Barcelona (Spain): Icaria.

Government of India. 2005. "National Rural Employment Guarantee Act, 2005", *The Gazzette of India*, New Delhi, September 7 of 2005, Number 42.

Government of India. "Tamil Nadu Farmers' Management of Irrigation Systems Act, 2000, Act 7 of 2001", *Tamil Nadu Government Gazette*, March 5th of 2001, Number 178, p. 22.

Janakarajan, S. 1991. In search of Tanks: Some Hidden Facts. Working Paper No. 97. Madras: Madras Institute of Development Studies.

Mosse, D. 2003. *The Rule of Water. Statecraft, Ecology and Collective Action in South India*. New Delhi: Oxford University Press.

Mukundan, T.M. 1988. "The very systems of south India". *PPST Bulletin* 16.

Palanisami, K. and K.W. Easter. 2000. *Tank Irrigation in the 21st Century—What next?* New Delhi: Discovery Publishing House.

Palanisami, K. *et al*. 2001. *Sustainability of tank irrigation systems in South India*. Coimbatore (India): Water Technology Centre, Tamil Nadu Agricultural University.

Reyes—Garcia V. *et al*. 2011. Local perception and uses of water tanks. A case study in two villages of Tamil Nadu, South India. *Society and Natural Resources 24(5) 485-499*.

Suseela, P., Sobhana, D. and K.D. Chandra. 2006. *Field work report (from 25.02.06 to 18.03.06) of Endiyur, Tamil Nadu*. M.A. Sustainable Development, First Year, Pondicherry University.

Vaidyanathan, A. (ed). 2001. *Tanks of South India*. New Delhi: Centre for Science and Environment.

Weller, Susan. 1998. "Structured interviewing and questionnaire construction" *in* Russell Bernard H. (ed.) *The Handbook of Methods in Cultural Anthropology*. Walnut Creek: Altamira Press.

INDEX